THE WORLD SUMMIT ON SUSTAINABLE DEVELOPMENT

The World Summit on Sustainable Development
The Johannesburg Conference

Edited by

LUC HENS
Vrije Universiteit, Brussels, Belgium

and

BHASKAR NATH
European Centre for Pollution Research,
London, UK

 Springer

A C.I.P. Catalogue record for this book is available from the Library of Congress.

ISBN 10 1-4020-3652-3 (HB)
ISBN 13 978-1-4020-3652-1 (HB)
ISBN 10 1-4020-3653-1 (e-book)
ISBN 13 978-1-4020-3653-1 (e-book)

Published by Springer,
P.O. Box 17, 3300 AA Dordrecht, The Netherlands.

www.springeronline.com

Chapters 1, 3, 4, 5, 8, 9, 10, 11, 12, 13, and 16 were previously published by Springer, under the
Kluwer Academic Publishers imprint, in *Environment, Development and Sustainability*, Volume 5,
Nos. 1-2, 2003.

Printed on acid-free paper

Printed in the Netherlands.

TABLE OF CONTENTS

PREFACE

The Johannesburg Earth Summit, which took place in the summer of 2002, confirmed the irreversible nature of the process that is founded upon the concept of Sustainable Development initially given form at Rio de Janeiro ten years earlier. This process is to be welcomed, while at the same time recognising the tremendous work that has taken place in converting this concept into a more concrete vision.

The Sustainable Development concept relates to every human activity, covering the social, economic and ecological dimensions, which are often in conflict.

Consequently, it is most important to include in research programmes some thought of the way people behave. In theory, the general elements of this inclusion are relatively easily defined. However, assessing the effects of one or another decision on all the interactions between the social, economic and ecological dimensions involves significant difficulties. All the more since we have to recognise, in all modesty, that humanity has not always excelled in the art of forward studies. In fact, the Precautionary Principle was introduced partly as a reaction to the sometimes blind confidence in technology and logic (even if it is sometimes invoked in an exaggerated manner).

Nevertheless, the duty to act for the sake of present and future generations is pressing. Throughout history mankind has had to adapt and to innovate. Now, at the beginning of the 21st century the urgent need for such adaptations is obvious. Indeed, we see that challenges and deadlines are increasing in the near term, at least compared with the duration of the history of mankind. Starting from today we must implement development management tools based on the universally recognized concept of Sustainable Development. It is up to each one of us, within the framework of our family or social obligations or within our working environment, to rebase our actions and our life style choices on a holistic approach. There is no unique or single solution but a group of solutions that work together to achieve the same aim.

Commitment to this individual holistic approach is the first step in trying to prevent Earth becoming uninhabitable for mankind.

However, alongside this commitment there is another essential element, namely the knowledge that arrives, as always, through study and reflection. This is all the more necessary, as I stressed previously, because of the necessity to take into consideration the extremely complex character of environmental issues. As UN Secretary General Kofi Annan stated last year: "In all these areas [of the Environment] there are things we can do now with the technologies already at our disposal, provided we give the right incentives. Science will bring us many more solutions if we make the right investment in research. Knowledge has always been the key to human development. It will also be the key to sustainability."

Therefore, it is neither correct nor appropriate to set in opposition Science and Sustainable Development. Science, with the proviso that it is ethically driven, can be one of the possible ways to take into account and to tackle the challenges that we face.

Moreover, I think that it is essential, or even vital, to develop solidarity, not only between North and South as advocated in Agenda 21, but also between generations. In a world which, all too often, leads to individualism and selfishness, we must remember that Man is essentially a social being and nothing that happens in the world should leave him indifferent.

Laurent of Belgium
President of the IRGT
Royal Institute for the Sustainable Management of Natural Resources and the Promotion of Clean Technology

LIST OF CONTRIBUTORS

K. BACHUS
Higher Institute for Labour Studies
Catholic University of Leuven
Kapucijnenvoer 33 Block H 4th floor
B-3000 Leuven
BELGIUM

J. BARBER
Integrative Strategies Forum
Rockville, MD
USA

A. BEGOSSI
Núcleo de Estudos e Pesquisas
Ambientais (NEPAM)
Universidade Estadual de Campinas
Campinas, SP, 13081-970
BRAZIL

C.J. CLEVELAND
Department of Geography
Center for Energy and
Environmental Studies
Fredrick S. Pardee Center for the
Study of the Longer-Range Future
152 Bay State Road
Boston University
Boston, MA 02215
USA

F. DIAS DE ÁVILA-PIRES
Instituto Oswaldo Cruz
Rio de Janeiro
BRAZIL

F. GHINA
International Coral Reef Action
Network (ICRAN)
c/o United Nations Environment
Programme-World Conservation
Monitoring Centre (UNEP-WCMC)
219 Huntingdon Road
Cambridge, CB3 0DL
UNITED KINGDOM

L. HENS
Human Ecology Department
Vrije Universiteit Brussel, Human
Laarbeeklaan 103
B-1090 Brussels
BELGIUM

C.V. HOWARD
Developmental Toxic-Pathology
Research Group
Department of Human Anatomy and
Cell Biology
University of Liverpool
Liverpool L69 3GE
UNITED KINGDOM

I. LÁNG
Hungarian Academy of Sciences
Roosevelt tér 9
Budapest
HUNGARY

CONTRIBUTORS

F. MESTRUM
Université Libre de Bruxelles
Av. F.D. Roosevelt 50
B-1050 Brussels
BELGIUM

D. MWANZA
Water Utility Partnership for
Capacity Building in Africa
Abidjan
IVORY COAST

A. NAJAM
Department of International
Relations
Center for Energy and
Environmental Studies
Fredrick S. Pardee Center for the
Study of the Longer-Range Future
152 Bay State Road
Boston University
Boston, MA 02215
USA

B. NATH
European Centre for Pollution
Research - Unit 2F
289 Cricklewood Broadway
London NW 2 6NX
UNITED KINGDOM

L.W. OLSON
Technology Center
Arizona State University
Mesa, 85212 Arizona
USA

M. PALLEMAERTS
Institute of European Studies
Pleinlaan 2
B-1050 Brussel
BELGIUM

H. PEETERS
Ethibel – Stock at Stake
Strategic Advice and Project
Development
Rue du Progrès 333/7
B-1030 Brussels
BELGIUM

A.G. STEWART
Cheshire and Merseyside Health
Protection Team
Microbiology Laboratory
Countess of Chester Health Park
Chester, CH2 1UL
UNITED KINGDOM

A.W. STRIGL
Austrian Institute for Sustainable
Development
University of Natural Resources and
Applied Life Sciences
Lindengasse 2/12
A-1070 Vienna
AUSTRIA

R. WHITFIELD
Envirostat
66, Weltje Road
London W6 9LT
UNITED KINGDOM

E. WILKINSON
Central Liverpool Primary Care
Trust
Liverpool L3 6AL
UNITED KINGDOM

LIST OF FIGURES

LIST OF TABLES

LIST OF BOXES

INTRODUCTION

By all accounts, the United Nations Conference on the Human Environment, held in Stockholm in 1972, focused the international community's attention on environmental concerns as never before and provided the impetus for the ascent of those concerns to the top of the international agenda. Until that conference, and even after it for some time, environmental concerns had been almost the exclusive preserve of environmentalists, ecologists and conservationists who, according to many, took an idealistic if not somewhat romantic view of nature conservation and environmental protection. They were often regarded with disdain by the then political establishment that viewed them as an irritant and potential or manifest obstacle to unfettered pursuit of economic development mainly through environment-degrading industrial activities.

The next and a very important milestone in the international environmental calendar was the publication in 1987 of the Brundtland Commission Report, entitled *Our Common Future*, under the auspices of the World Commission on Environment and Development. It was a key document, not least because it firmly established the paradigm of sustainable development at the top of the international agenda as the only means by which both intergenerational and intragenerational equity (the core requirements of sustainable development) could be secured.

Sealed with this international imprimatur, environmentalism and concerns over environmental damage inflicted by human activities began to shed its hitherto romantic image as perceived by many. Even hard-headed international structures as well as economists and others began to take much interest in these and related issues and problems, and this resulted in a veritable avalanche of assorted publications.

However, *Our Common Future* is first and foremost a political document, and the definition of sustainable development it gives is political too. And so it is proving to be extremely difficult to translate that definition into a unique operational definition for the practical implementation of sustainable development.

The lack of a unique operational definition of sustainable development ushered an open season for all to indulge in do-it-yourself definitions mainly to suit their own circumstances and to serve their own purposes. Unfortunately, this state of affairs has been serving to corrupt the intended meaning of sustainable development and to devalue its currency – the intended meaning being how the present generation ought to behave *vis-à-vis* the environment, and precisely what it ought to do in practice, and how, in order to secure both intergenerational and intragenerational equity. Interestingly (some would say perversely), and due probably to the hegemony over such matters which economists and politicians gained in the mid 1980s or thereabouts, sustainable development began to be defined in important documents not in terms of the sustainability of the natural environment or integrity of nature's life support systems, but in terms of economic sustainability. Typically, the following from Article 2 of the Treaty of the European Union (1992) illustrates this well: EC's environmental-policy objectives to include the goals of 'sustainable and non-inflationary (economic) growth respecting the environment'.

The UN Conference on Environment and Development, held in Rio de Janeiro in June 1992, was the next key milestone, and the document called Agenda 21 was probably the most important of its outputs. It drew up in considerable detail the blueprint for progressing towards global sustainable development. The Rio Conference was remarkable also for the hope, excitement and anticipation it generated world wide. Most concerned people everywhere actually believed that Agenda 21, ratified by most of the nation states, would usher a new developmental paradigm to protect and sustain the natural environment in the interests of both present and future generations. But it was not to be, as it turned out.

Indeed, most of the environmental problems have been exacerbating since Rio to make the global environment less sustainable today than it was ten years ago, and this constituted the background to the World Summit on Sustainable Development (WSSD), held in Johannesburg during 26 August and 4 September 2002. The reports of the Secretary General of the UN, presented to the first Prepcom of the WSSD in New York, painted a negative picture of the way the world had moved away from sustainable development since Rio.

At the global level today there is less concern over population growth than ten years ago, largely based on the scientifically uncertain prediction that world population would stabilise by 2080. The industrialised countries consider their stable economic growth and improved social conditions, which they have achieved while at the same time reducing environmental pressures in a number of areas, as their most important contribution to SD to date. During the last decade environmental policy and management regimes have been implemented in the developing countries. However, in many cases this institutional progress is yet to bear fruit in terms of improvements.

It is a matter of mounting concern that so far little, if any, progress has been made in addressing unsustainable patterns of production and consumption, and that poverty eradication is proving to be an apparently intractable challenge. Globalisation, and a credible and objective evaluation of its costs and benefits, is proving to be difficult too. While on the one hand the OECD countries continue to push forward ideas on how globalisation would benefit the global SD process by increasing world trade and investment and through trade liberalisation, on the other greater (often externalised) environmental costs of globalisation, and gross social inequity it is predicted to bring, points to an unacceptably high and growing sustainability deficit.

There is also mounting concern over resources of water, energy, land and biodiversity – all under threat today as never before, especially in sub-Saharan Africa and in the vulnerable ecosystems of small island states.

In view of the above, the objective of the Johannesburg Summit was to seek ways and means for reinvigorating the Agenda 21 process and to promote its practical implementation world wide, rather than to generate yet more voluminous documents. As the WSSD is described in detail in the next article, we will not dwell on it here, except to say that one of its principal outputs, the *Johannesburg Plan of Implementation*, is both comprehensive and goal-oriented for the realisation of Agenda 21 objectives, albeit perhaps a little too ambitious when judged against current and evolving geo-political realities.

This book contains a number of articles on selected key issues discussed and agreed at the WSSD. Written by specialists, the purpose of these articles is to discuss one or more of the following as appropriate: evolution of specific environmental issues or concerns, scientific background, deliberations at Johannesburg, and where do we go from here?

In particular, the authors were invited to compare the status of current knowledge of the key issues with how Johannesburg, and its Plan of Implementation, dealt with that knowledge.

This exercise is useful in at least two ways: first, it allows one to compare how the state-of-the-art relates to the information which diplomats and policy-makers used in their deliberations in Johannesburg. Clearly, such a comparison exposes the divide between scientific evidence and information on one hand and how they are used by policy-makers to make policy on the other, especially in areas in which hard scientific information is scarce, incomplete or tentative such as globalisation and production and consumption patterns. And second, this exercise identifies, or defines more precisely, the gaps in scientific information which must be filled in order for policy-makers to formulate more effective policies for SD. Indeed, filling such information gaps is a major part of the research agenda for SD. Interestingly, to this end a new contract between science and society was proposed at the WSSD. This new post-WSSD contract, it was suggested, should be based on and inspired by both social and environmental considerations, unlike the post Second World War contract that was driven almost exclusively by scientific, technological and economic considerations. The key issues to be included in the new contract emerged as these:

– addressing problems of poverty, population and health;
– sustainable use of water, energy and biodiversity;
– sustainable agriculture and food security;
– strategies and planning for SD;

- measuring SD;
- scientific support for local-to-global action;
- environmental management and sustainability;
- economic and social policy instruments for SD;
- fact-based education for SD.

This book is based on articles that have already been published in the journal "Environment, Development and Sustainability" (2003) that devoted a special issue to the WSSD. These papers have been complemented with a set of chapters on management of chemicals, health, Africa, governance, and partnerships that are essential in the Johannesburg discussion but previously unpublished.

The international community decided on a follow-up programme of the WSSD. In the ten years to come, the different main themes of the Johannesburg Plan of Implementation (JPoI) will be dealt with in more depth. Water and human settlements are the first themes the world will deal with. This book aims to provide essential background information for this decade to come.

The production of a book such as this is no mean task, involving as it does enormous scientific and administrative efforts. On the scientific side we are most grateful to the following colleagues who did such a splendid job of reviewing the papers published in this issue:

- Prof. Dr. Johan Albrecht, Universiteit Gent, Belgium
- Prof. Dr. Jan Otto Andersson, Abo Academy, Finland
- Prof. Dr. Yonathan Anson, Ben Gurion University of the Negev, Israel
- Prof. Dr. Giancarlo Barbiroli, Istituto di Merceologia, Università degli Studi di Bologna, Italy
- Dr. Lila Barrera-Hernandez
- Dr. Claudia R. Binder, Swiss Federal Institute of Technology, Zürich, Switzerland
- Prof. Dr. Karl Bruckmeier, Human Ecology Department, University of Göteborg, Sweden
- Prof. Dr. Emmanuel Boon, Human Ecology Department, Vrije Universiteit Brussel, Belgium
- Prof. Dr. Richard J. Borden, College of the Atlantic, Bar Harbor, Maine, USA
- Prof. Dr. Philippe Bourdeau, IGEAT, Université Libre de Bruxelles, Brussels, Belgium
- Dr. Mickel Christolis, Computational Fluid Dynamics Unit, National Technical University of Athens, Greece
- Dr. Donaat Cosaert, Flemish Institute for Scientific and Technological Aspects Research, Flemish Parliament, Brussels, Belgium
- Dr. Farid Dahdouh-Guebaz, Laboratory of General Botany and Nature Management, Vrije Universiteit Brussel, Belgium

- Dr. Morgan De Dapper, Geological Institute, Universiteit Gent, Belgium
- Prof. Dr. Fernando Dias de Avila Pires, Fundaçao Oswaldo Cruz, Rio de Janeiro, Brazil
- Prof. Dr. Theo K. Dijkstra, Department of Econometrics and Operations Research, University of Groningen, The Netherlands
- Prof. Dr. Jan W. Dobrowolski, Institute of Management and Protection of the Environment, Krakow, Poland
- Dr. Kwame A. Domfeh, School of Administration, University of Ghana
- Dr. Prabir Ganguly, Centre for European Studies, VSB, Technical University Ostrava, Ostrava-Poruba, Czech Republic
- Prof. Dr. Ir. Bernard Geeraert, Department of Mechanical Engineering, Catholic University of Leuven, Belgium
- Prof. Dr. Bernard Glaeser, Social Science Research Centre, Berlin, Germany
- Prof. Dr. Jackie Van Goethem, Royal Institute for Natural Sciences, Brussels, Belgium
- Dr. Leah Goldfarb, International Council for Science, Paris, France
- Mr. Lee R. Hatcher, AtKisson Inc., Seattle, USA
- Dr. Roberto Laserna, Centro de Estudios de la Realidad Economica y Social, Cochabamba, Bolivia
- Prof. Dr. Luc Lavrysen, Centrum voor Milieurecht, Universiteit Gent, Belgium
- Dr. Roderick J. Lawrence, Centre for Human Ecology and Environmental Sciences, University of Geneva, Switzerland
- Prof. Dr. Walter Leal Filho, Environmental Technology Department, Technical University of Hamburg-Harburg, Germany
- Prof. Dr. Gerry Marten, School of Policy Studies, Kwansei Gakuin University, Hyogo, Japan
- Prof. Dr. Patrick Meire, Antwerp University, Belgium
- Prof. Dr. Bedrich Moldan, Charles University, Prague, Czech Republic
- Prof. Dr. Armando Montanari, Italian Geographical Society, Rome, Italy
- Prof. Dr. Bart Muys, Faculty of Agricultural and Applied Biological Sciences, Catholic University of Leuven, Belgium
- Prof. Dr. Rudolfo Paz, Faculty of Mechanical Engineering and Production Sciences, Escuela Superior Politécnica del Litoral, Guayaquil, Ecuador
- Prof. Dr. Steven S. Penner, University of California at San Diego, USA
- Prof. Dr. Warren M.B. Pescod, Department of Civil Engineering, University of Newcastle, Newcastle-Upon-Tyne, UK
- Dr. P.K. Rao, Global Development Institute, Lawrenceville, NJ, USA
- Prof. Dr. Joe Ravetz, Centre for Urban and Regional Ecology, School of Planning and Landscape, Manchester University, UK
- Prof. Dr. Frank Rijsberman, International Water Management Institute, Colombo, Sri Lanka

- Dr. John P. Robinson, International Centre for Technical Research, London, UK
- Dr. S.O. Saaka, UNDP Capacity 21, Accra, Ghana
- Mr. Gordon Sillence, European Commission, Brussels, Belgium
- Dr. Michael Stauffacher, Swiss Federal Institute of Technology, Zürich, Switzerland
- Prof. Dr. Harro Stolpe, Faculty of Civil Engineering, Ruhr Universität Bochum, Germany
- Prof. Dr. Stoyan Stoyanov, Ecology Centre, University of Chemical Technology and Metallurgy, Sofia, Bulgaria
- Prof. Dr. Charles Susanne, Laboratory of Anthropogenitics, Vrije Universiteit Brussel, Belgium
- Dr. Emma L. Tompkins, School of Environmental Sciences, University of East Anglia, Norwich, UK
- Prof. Dr. Luc Vanliedekerke, Economics Department, Catholic University Leuven, Belgium
- Prof. Dr. René Van Grieken, Antwerp University, Belgium
- Dr. Olaf Weber, Natural and Social Science Interface, Swiss Federal Institute of Technology, Zurich, Switzerland
- Prof. Dr. Raoul Weiler, Club of Rome, Brussels, Belgium
- Dr. Willy Weyns, Flemish Institute for Scientific and Technological Aspects Research, Flemish Parliament, Brussels, Belgium
- Prof. Dr. Edwin Zaccai, IGEAT, Université Libre de Bruxelles, Belgium

On the administrative side sincere thanks are due to the secretarial staff of the Human Ecology Department of the Vrije Universiteit Brussel and, in particular, Mr. Serge Gillot and Mr. Glenn Ronsse.

Luc Hens and Bhaskar Nath

CHAPTER 1

THE JOHANNESBURG CONFERENCE

L. HENS[1]* and B. NATH[2]

[1]*Vrije Universiteit Brussel, Human Ecology Department, Laarbeeklaan 103,
B-1090 Brussels, Belgium;* [2]*European Centre for Pollution Research, Unit 2F,
289 Cricklewood Broadway, London NW 2 6NX, UK
(*author for correspondence, e-mail: human.ecology@vub.ac.be;
fax: +322 477 4964; tel.: +322 477 4281)*

Abstract. The World Summit on Sustainable Development (WSSD), held in Johannesburg during 26 August and 4 September 2002, was the biggest event of its kind organised by the United Nations to date. A major objective of the WSSD was to set out strategies for greater and more effective implementation of Agenda 21, negotiated in Rio ten years ago, than hitherto. An overview of the WSSD is presented in this chapter, including a scrutiny of its major outcomes.

Discussion begins with a detailed account of major UN environmental conferences and related events, such as Doha and Monterrey conferences, that led to the WSSD, followed by a brief discussion of the deliberations that took place at the preparatory meetings (PrepComs) of the WSSD. A detailed account and scrutiny of the following, that are the main outcomes of the WSSD, is then given.

- The "Johannesburg Declaration on Sustainable Development", which is a political declaration mirroring the will of the international community to move towards sustainable development.
- The "Johannesburg Plan of Implementation", which is the core document of the WSSD containing an impressive list of recommendations for accelerating the implementation of Agenda 21.
- "Type II partnerships", which are projects that allow civil society to contribute to the implementation of sustainable development.

The increasingly important post-Rio issue of globalisation, which has serious implications for a number of issues directly or indirectly impinging on global sustainability, was an important element in the contextual background to the WSSD. Reference is made to some of these implications.

Type II partnerships are an innovation of the WSSD. Although a good deal of confusion persists over their precise nature and *modus operandi*, they were nevertheless presented at the WSSD as powerful and more democratic instruments for the realisation of Agenda 21 objectives.

The analysis shows that the Summit contributed at defining sustainable development more precisely. The Plan of Implementation is most instrumental in showing how to make resource use and the generation of pollution less unsustainable. In this way implementing the recommendations of the Johannesburg Summit offers an important defeat, worldwide.

Key words: declaration, Johannesburg, plan of implementation, sustainable development, Type 2 partnerships, world summit.

Abbreviations: CITES – Convention on International Trade in Endangered Species; CSD – Commission for Sustainable Development; DESA – Department of Economic & Social Affairs (of the UN); EOLSS – Encyclopaedia of Life Support Systems (of UNESCO); EU – European Union, the; GDP – Gross Domestic Product; GEF – Global Environmental Facility; IPCC – Inter-governmental Panel on Climate Change; IUCN – International Union for Conservation of Nature (and Natural Resources); JPI – Johannesburg Plan of Implementation; MDG – Millennium Development Goals, the; NEPAD – New Partnership in Africa's Development; NGO – Non-governmental Organisation; ODA – Overseas Development Aid; OECD –

Readers should send their comments on this paper to: BhaskarNath@aol.com within 3 months of publication of this issue.

L. Hens and B. Nath (eds.), The World Summit on Sustainable Development, 1–33.
© 2005 *Springer. Printed in the Netherlands.*

Organisation for Economic Cooperation & Development; R&D – Research & Development; SD – Sustainable Development; SIDS – Small Island Developing States; UN – United Nations, the; UNCED – United Nations Conference on Environment & Development (Rio de Janeiro, 1992); UNCTAD – United Nations Conference on Trade & Development; UNDP – United Nations Development Programme; UNEP – United Nations Environment Programme; UNESCO – United Nations Educational, Scientific & Cultural Organisation; WCED – World Commission for Environment & Development; WEHAB – Water, Energy, Health, Agriculture & Biodiversity; WMO – World Meteorological Organisation; WSSD – World Summit on Sustainable Development (Johannesburg, 2002); WTO – World Trade Organisation, the; WWF – World Wild Fund for Nature (previously World Wildlife Fund).

1. Introduction

The World Summit on Sustainable Development (WSSD), held in Johannesburg during 26 August and 4 September 2002, was attended by 9101 delegates from 191 governments and 8227 representatives of major groups who deliberated on how to implement sustainability in more effective ways than during the last ten years, as well as by 4012 media representatives who reported on it.

Most of the indicators confirm that both environmental quality and sustainability have further deteriorated since the Rio Summit of 1992, and the WSSD was primarily concerned with why so little progress had been made towards achieving the Rio goals of sustainable development (SD). It was generally agreed that while Rio's Agenda 21 was a reliable and high-quality document giving guidance for implementing SD, its practical implementation fell far short of what was needed and agreed in Rio ten years ago.

This huge conference – biggest of its kind organised by the UN to date – was remarkable both for the scope and complexity of its organisation. The context of the WSSD deliberations encompassed the outputs of the UN environmental conferences to date, the UNCED documents, the Millennium Declaration, the Doha and Monterrey conferences and the WSSD "Prepcoms" among others. Box 1 lists the environment and development milestones building up to the WSSD during a period of thirty years.

Box 1. Environment and development milestones during 1972 and 2002.

1972
- United Nations Conference on the Human Environment, Stockholm.
- UNESCO Convention on the Protection of World Cultural and Natural Heritage.
- First report of the Club of Rome.

1973
- Convention on International Trade in Endangered Species and Flora and Fauna (CITES).

1976
- Convention on the Protection of the Mediterranean Sea Against Pollution.

1978
- The Governing Council of UNEP adopts principles of conduct in the field of the environment for the guidance of states for the conservation and harmonious utilisation of natural resources shared by two or more states.

1979
- Convention on the Conservation of Migratory Species of Wild Animals.
- The Geneva Convention on Long-Range Transboundary Air Pollution.
- First World Climate Conference, Geneva.

1980
- UNEP, in collaboration with IUCN and WWF, launches the World Conservation Strategy, considered the first comprehensive policy statement on the link between conservation and sustainable development.

1981
- Convention on Cooperation for the Protection and Development of the Marine and Coastal Environment of the West and Central African Region.
- Convention on the Protection of the Marine Environment and Coastal Area of the South-East Pacific.

1982
- Stockholm C 10 Conference organised by UNEP in Nairobi.
- Regional Convention on the Conservation of the Red Sea and Gulf of Aden Environment.

1983
- Convention on the Protection and Development of the Marine Environment of the Wider Caribbean Region.

1985
- Vienna Convention on the Protection of the Ozone Layer.
- Convention on the Protection, Management and Development of the Marine and Coastal Environment of the East African Region.
- Convention on the Protection of Natural Resources and Environment of the South Pacific Region.

1987
- Montreal Protocol on Substances that deplete the Ozone Layer.
- The Report, *Our Common Future*, published by the World Commission on Environment and Development.

1988
- The World Meteorological Organisation (WMO) and UNEP establish the Intergovernmental Panel on Climate Change (IPCC).

(Continued)

Box 1. *Continued*

1989
- Basel Convention on the Control of Transboundary Movements of Hazardous Wastes and their Disposal.

1991
- Establishment of the Global Environment Facility (GEF) with UNEP, UNDP and the World Bank as partners.

1992
- United Nations Conference on Environment and Development (UNCED) in Rio de Janeiro.
- Convention on the Protection of the Black Sea Against Pollution.
- Framework Convention on Climate Change.
- Convention on Biological Diversity.

1994
- United Nations Convention to Combat Desertification (in those countries experiencing serious drought and/or desertification, particularly Africa).

1997
- The Kyoto Protocol, adopted by 122 nations.

2000
- *We the Peoples: the Role of the United Nations in the 21st Century*: Millennium Report of the UN Secretary-General.
- The Cartagena Protocol on Biosafety implements a precautionary approach to trade in genetically altered crops and organisms.

2001
- The Stockholm Convention on Persistent Organic Pollutants requiring complete phase-out of nine persistent, toxic pesticides and limiting the usage of several other chemicals.
- Fourth Ministerial meeting of the WTO-Doha Declaration.

2002
- International Conference on Financing for Development: Monterrey Consensus.
- United Nations World Summit on Sustainable Development, held in Johannesburg.

The recent aspects of that context are summarised in what follows, along with a discussion of the main outputs of the WSSD, namely the Political Declaration, the Johannesburg Plan of Implementation (JPI), and the Type II Partnerships. They are evaluated with regard to both their intrinsic values and in comparison with the impacts of the Rio output.

2. The context and the antecedents

2.1. THE UN ENVIRONMENTAL CONFERENCES

The pioneering UN Conference on the Human Environment, held in Stockholm (Sweden) in 1972, focused on international cooperation for and on the environment. The conference theme was that environmental problems could be solved by science and technology, juxtaposed with Indira Gandhi's contribution that "poverty is the greatest polluter of the environment".

That Conference resulted in the establishment of environmental ministries and agencies in over 100 countries and marked the beginning of the explosive growth in the number of non-governmental organisations (NGOs) dedicated to environmental protection and germane issues. The UNEP was established in Nairobi to put the results of the conference in practice. In particular, the declaration of the conference, and the action plan it proposed with recommendations for international action, provided the impetus for the subsequent rapid development of international environmental law (Engfeldt, 2002).

Ten years later, the Stockholm + 10 Conference was held in Nairobi, Kenya. Whereas the approach of the Stockholm Conference of 1972 was technology driven, concerned mainly with local issues and problems and largely conditioned by Rachel Carson's *Silent Spring* (1962), the agenda of the Nairobi Conference reflected the practical and scientific concerns of the time. Indeed, it was at the Nairobi Conference that the social and economic drivers of environmental problems were recognised, leading to the establishment of the World Commission on Environment and Development (WCED). Chaired by Gro Harlem Brundtland, the WCED Report, published in 1987 (WCED, 1987), unravelled the relationship between environment and economy. It not only popularised the concept of SD, but also demonstrated most convincingly that anthropogenic environmental problems are fundamentally interdisciplinary and ought to be regarded as such.

Sustainable development was the focus of the UNCED, held in Rio de Janeiro (Brazil) in June 1992. Two issues prevailed:

- Link between environment and development, and
- Practical interpretation of the rather theoretical concept of SD, seeking to balance the modalities of environmental protection with social and economical concerns.

The Rio Conference generated these outputs (Johnson, 1992):

- *The Rio Declaration on Environment and Development:* It is a list of 27 principles on which SD policies are to be based. Most of these are still valid, notably the precautionary principle, the equity principles, and the principle of subsidiarity.
- Agenda 21 provides a remarkably sharp analysis of both the symptoms and the underlying causes of global unsustainability, as well as authoritative ideas on how to put SD into practice.

– Some of the most urgent issues discussed in Agenda 21 are those on the three
  Conventions that are related to Rio:

  • The Framework Convention on Climate Change, which addresses the issue
    of global warming.
  • The Convention on Biological Diversity, which urges action to be taken to
    prevent huge and continuing loss of biodiversity and forests.
  • The Convention to Combat Desertification (in those countries experiencing
    serious drought and/or desertification, particularly in Africa), which resulted
    from discussions at Rio but was concluded in March 1994.

In addition to these Conventions, Rio also resulted in the "Non-legally Binding
Authoritative Statement of Principles for a Global Consensus on the Management,
Conservation and Sustainable Development of all Types of Forests".

Ten years after Rio, it is clear that one positive outcome of UNCED has been the
recognition of the role of some major groups in implementing SD. This has created an irreversible bottom-up momentum which is having a considerable impact
on the democratic process itself. On the debit side, however, the practical implementation of SD continues to be thwarted by three main factors: vagueness of how
to measure SD; unrealistic expectations placed on the creation of the Commission
for Sustainable Development (CSD); and lack of funds needed to implement the
Rio "acquis" (the GEF provides funding only for the incremental costs of projects
related to these conventions, leaving over 90% of the Agenda 21 issues without
financial means for implementation (Upton, 2002)).

2.2. IMPLEMENTATION OF THE RIO AGREEMENTS

During the ten years between Rio and Johannesburg, raising wider social consciousness of the need for SD has proved to be a slow and time-consuming process.
Indeed, as indicated by most of the core indicators of SD, the situation today is
worse than it was ten years ago, especially with regard to the pollution of air, water
and soil, resource consumption, as well as poverty and north–south income disparity. However, some progress towards SD has been made in many parts of the world
with regard to some of the Agenda 21 issues, notably the following: slower population growth, reduced mortality rate, improved health, wider access to education,
and strengthened role of women.

The implementation of the Conventions has been a laborious process. For
example, nothing was done about the Convention on Climate Change until
Germany invited the partners to Berlin in 1995 for discussions, followed by a
meeting in Geneva in 1996. Subsequently 84 countries signed the Kyoto Protocol in 1997. And, so it was that for the first time emission reductions had been
quantitatively defined on the basis of the principle of shared but different responsibilities, and the signatory countries agreed on the instruments for reaching the
Kyoto targets (Orlando, 1998).

The road to Johannesburg took another four conferences (Buenos Aires, 1998; Bonn, 1999; The Hague, 2000 and Marrakech, 2001) to establish an action plan and mechanism to monitor and control the Kyoto agreements. Even when scientific evidence shows that Kyoto is only a beginning, and that prevention of climate change demands more drastic reductions in greenhouse gas emissions, the USA and Australia disputed these findings on the basis of their respective short-term, national economic interests. And so, when the WSSD began, it was still unclear whether a sufficient number of countries would be prepared to sign the Kyoto Protocol in order for it to be put into action. It is noted in passing that during 1990 and 2000 global carbon emissions grew by an average of 9.1%.

Attracting lesser media attention – and more effective probably for that reason – the development of the Convention on Biological Diversity followed a different path. Today there are some 180 parties to it who have committed themselves to undertaking national and international measures to achieve three objectives: conservation of biological diversity, sustainable use of its components, and equitable sharing of benefits accruing from the utilisation of genetic resources. Since the time of the adoption of the Convention, the parties to it have met six times in conference (Nassau, 1994; Jakarta, 1995; Buenos Aires, 1996; Bratislava, 1998; Nairobi, 2000; The Hague, 2002). The measures taken at these meetings include the following: adoption of programmes of work for a number of thematic areas and cross-sectoral issues; issuance of specific guidance for funding biodiversity projects; and establishment of *ad hoc* bodies to focus on the implementation of the provisions of the Convention (such as those relating to indigenous knowledge and safe biotechnology). These discussions have led to the adoption of the Cartagena Protocol on Biosafety (Cartagena, Columbia, 22–24 February, 1999). It is an international regulatory framework for reconciling the needs of free trade with those of environmental protection in a world of rapidly growing recombinant DNA industry (CBD et al., 2001).

However, in spite of this progress on the diplomatic front, the indicators of biodiversity continue to indicate accelerating species loss, and, continuing degradation of natural ecosystems. Earth's forests have been disappearing at a rate of 14.6 million hectares annually, while the proportion of coral reef loss due to human activities has increased from 10% in 1992 to 27% in 2000. This is why the WSSD was forced to come up with a more effective action plan to address the issue of biodiversity.

To ensure the follow-up of the tasks to be undertaken to achieve SD, agreed at Rio, it was recommended in Agenda 21 to establish a high-level CSD responsible for reporting on progress towards global SD to the Economic and SocialCouncil of the UN General Assembly. However, the CSD has not fulfilled the high expectations of the world, and so it was also a matter of debate at the WSSD. This is symptomatic of the fact that since Rio much greater progress has been made world-wide in environmental institution-building than in actually protecting the environment or pursuing effective policies for SD. Today, nearly all countries, both developed and developing, have government ministries and/or agencies in charge of, environmental affairs. Also, since Rio there has been a proliferation of institutions and organisations (including NGOs) of major groups such as women, indigenous communities, local authorities, business and industry, and scientists to support, promote and deal with environmental and SD issues at local, national, regional and global levels. The WSSD was expected to discuss a plan to harmonise these efforts for greater effectiveness.

2.3. THE MILLENNIUM DECLARATION

The Millennium Declaration of the UN, agreed at the Millennium Summit of September 2000, summarised the agreements and resolutions of the UN world conferences held during the last ten years to establish the Millennium development goals (MDGs). These goals are generally accepted as benchmarks for measuring actual development.

The eight MDGs and their targets are summarised in Box 2. The first seven goals are directed at poverty reduction. The eighth – global partnership for development – is about the means to achieve the first seven. The box also shows the specific quantified targets for each goal that need to be realised within a defined timeframe. For each target, these quantified indicators allow to monitor progress.

The environment is an essential component of the MDGs. Of particular importance is goal (7) for ensuring environmental sustainability. The targets of that goal refer to mainstreaming the environment in policy and programmes, reversing loss of environmental resources, and improving access to environmental services. The other goals, in particular (1), (4)–(6) for reducing poverty and improving health, are directly linked to SD. The goals and targets of poverty, environment and SD in the Millennium Declaration are recalled in the JPI.

2.4. THE DOHA DECLARATION

With increasingly intense activities for economic development, trade and globalisation have become central issues of SD, and many are convinced that the former offer opportunities for the latter. However, many countries, especially the least developed, have been bypassed and marginalised by globalisation. Although international trade has been increasing as a result of globalisation, it has so far benefited mainly the developed countries.

The Fourth Ministerial Meeting of the World Trade Organisation (WTO), held in Doha, Qatar, in November 2001, sought to address this inequity with a programme of work, called the Doha Agenda, which was agreed at that meeting.

This Agenda, which will govern international trade in the years to come, is based on the Doha Declaration which contains the five commitments listed in Box 3. In this declaration the environment is considered not only a topic relevant to trade, but one for negotiation too. This means that countries wishing to have stronger environmental safeguards have to be willing to make trade-offs with other countries

Box 2. The Eight Millennium development goals and targets.

(1) Eradicate extreme poverty and hunger
- Halve, between 1990 and 2015, the proportion of people whose income is less than $ 1 a day.
- Halve, between 1990 and 2015, the proportion of people who suffer from hunger.

(2) Achieve universal primary education
- Ensure that by 2015 children everywhere, boys and girls alike, will be able to complete a full course of primary schooling.

(3) Promote gender equality and empower women
- Eliminate gender disparity in primary and secondary education preferably by 2005 and in all levels of education no later than 2015.

(4) Reduce child mortality
- Between 1990 and 2015 reduce by two-thirds the under-five mortality rate.

(5) Improve maternal health
- Between 1990 and 2015 reduce by three-quarters the maternal mortality rate.

(6) Combat HIV/AIDS, malaria and other diseases
- By 2015 halt and begin to reverse the spread of HIV/AIDS.
- By 2015 halt and begin to reverse the incidence of malaria and other major diseases.

(7) Ensure environmental sustainability
- Integrate the principles of sustainable development into country policies and programmes and reverse the loss of environmental resources.
- Halve, by 2015, the proportion of people without sustainable access to safe drinking water.
- Achieve, by 2020, a significant improvement to the lives of at least 100 million slum dwellers.

(8) Develop a global partnership for development
- Develop further an open, rule-based, predictable, non-discriminatory trading and financial system (including a commitment to good governance, development and poverty reduction, both nationally and internationally).
- Address the special needs of the least developed countries (includes tariff-and quota-free access for exports, enhanced programme of debt relief for and cancellation of official bilateral debt and more generous ODA for countries committed to poverty reduction).
- Address the special needs of landlocked countries and small island developing states (through the Programme of Action for the Sustainable Development of Small Island Developing States and 22nd General Assembly provisions).

Continued

Box 2. *Continued*

- Deal comprehensively with the debt problems of developing countries through national and international measures in order to make debt sustainable in the long term.
- In cooperation with developing countries, develop and implement strategies for decent and productive work for youth.
- In cooperation with pharmaceutical companies, provide access to affordable essential drugs in developing countries.
- In cooperation with the private sector, make available the benefits of new technologies, especially information and communication technologies.

Box 3. The Doha Declaration.

In the Ministerial Conference of the World Trade Organisation (WTO), held in Doha, Qatar, in November 2001, the Ministers agreed to:

(1) The objective of sustainable development, with the aims of upholding and safeguarding an open and non-discriminatory multilateral trading system and acting for the protection of the environment and the promotion of sustainable development, can and must be mutually supportive.
(2) In agriculture to complete comprehensive negotiations aimed at substantial improvements in market access; reductions of, with a view to phasing out, all forms of subsidies; and substantial reductions in trade-distorting domestic support. Special and differential treatment for developing countries shall be an integral part of all elements of the negotiations and shall be embodied in the Schedules and concessions and commitments to be operationally effective and to enable developing countries to effectively take into account development needs, including food security and rural development.
(3) On market access for non-agricultural products to negotiate to reduce or as appropriate eliminate tariffs, including the reduction or elimination of tariff peaks, high tariffs, and tariff escalation, as well as non-tariff barriers, in particular on products of export interest to developing countries.
(4) On trade and environment, the mutual supportiveness of trade and environment shall be enhanced and with this view negotiations shall be conducted on the relationship between existing WTO rules and specific trade obligations set out in multilateral environmental agreements. Tariff and non-tariff barriers to environmental goods and services are to be reduced or eliminated.
(5) In all these arrangements special attentions are to be given to the least-developed countries.

that are not so convinced of the wisdom of placing environment at the centre of trade.

It is hoped that the fundamentally sound elements of the Doha Agenda, after they have been finally negotiated, will bring about a sustainable balance between global trade and the environment to promote both intergenerational and intragenerational equity. The WSSD was expected to take the Doha discussion to the next stage, and so the issue was taken up in the JPI.

2.5. THE MONTERREY CONSENSUS

Securing funds for implementing SD continues to be a difficult problem for the international community. Despite intense and protracted discussions in Rio, no satisfactory agreement was reached on how to finance the implementation of Agenda 21. The only agreement to emerge was that the GEF only finances the incremental costs of projects concerned with implementing the Rio Conventions. As a result, in the ten years since Rio there has been a serious shortage of resources to finance SD projects or policies. This is mainly because during this period the EU barely reached its target of 0.33% of its GDP for development cooperation; only the Netherlands, Luxemburg and Scandinavian countries met or exceeded the official target of 0.7% of GDP for development assistance. Although the USA has the largest budget of all donors in absolute terms, it only provides 0.12% of its GDP for development aid.

As a strategy, it was decided that the proceedings of the WSSD should not be allowed to be hampered by dispute or acrimony over discussions on finance. In March 2002, Heads of State and Government participated in the International Conference on Financing for Development in Monterrey, Mexico. The Monterrey Consensus recognised the common goal "to eradicate poverty, achieve sustained economic growth and promote sustainable development as we advance to a fully inclusive and equitable economic system". In this common pursuit of growth, poverty eradication and SD, the Monterrey Consensus proclaimed the following:

- Domestic financial resources to be mobilised for development.
- International resources, foreign direct investment and other private capital flows to be mobilised for development.
- International trade to be the engine of development.
- Greater international financial and technical cooperation for development.
- External debt relief.

In Monterrey the USA and the EU undertook to commit a total of 30 billion US dollars up to 2006, subject to good governance by the beneficiary countries. Monterrey also provided the impetus for the private sector to take its share of responsibility for implementing SD.

Because of these commitments, which had a profound impact on the preparatory work of the WSSD, only very limited discussions took place at Johannesburg on finance needed for implementing SD.

2.6. THE SUMMIT PREPARATORY COMMITTEES (PREPCOMS)

The goal of the WSSD, according to UN General Assembly (UNGA) Resolution 55/199, was to hold a ten-year review of the 1992 UNCED at the summit level to reinvigorate the global commitment to sustainable development.

Like the Rio conference of 1992, the WSSD was also preceded by four preparatory meetings. The first meeting of the Summit Preparatory Committee was held during 30 April and 2 May, 2001, at the UN headquarters in New York. With a strong organisational character, this meeting was concerned with the process of setting the WSSD agenda and determining its main themes. The latter emerged as the following: major groups, poverty, health, education, consumption patterns, human settlements, waste, finance and trade, protection of the atmosphere, energy and transport, oceans and seas, tourism and water.

PrepCom 2 was held during 28 January and 8 February 2002, again in New York. Impact of the major groups, and a number of statements by different countries, dominated the proceedings of this meeting. It became clear that the WSSD should also have a regional component with a sharp focus on Sub-Saharan Africa and Small Island States.

In PrepCom 3, which took place during 25 March and 5 April 2002, also in New York, different chapters of the JPI were drafted.

However, the most influential meeting was PrepCom 4. A ministerial-level meeting held in Bali, Indonesia, during 27 May and 7 June, 2002, it produced for the WSSD a draft plan of ten chapters focusing on the implementation of Agenda 21. It addressed the main challenges faced by the international community in implementing Agenda 21, as well as the opportunities (UN, 2002). As will be seen from Table I, the ministers and their negotiators fully agreed in Bali on 75% of all the paragraphs. The low level of agreement reached on globalisation and means of implementation is worthy of particular note. The latter, finance and trade in particular, was open for negotiation at the WSSD.

The Bali meeting also reached a "consensus" on the criteria for what are called Type II partnerships. These partnerships would eventually allow the major stakeholders of SD to realise the JPI, working in collaboration with relevant authorities. In fact, a call for proposals was launched world-wide shortly after the Bali PrepCom, thus making it possible to present a list of multistakeholder SD projects at the WSSD.

2.7. THE WSSD TARGETS

The events that led up to the WSSD, and shaped its contextual background, are summarised in Figure 1. It shows the different elements constituting the background, along with the vertical and horizontal relationships between the individual elements that contributed to the Johannesburg Summit. Although this figure only shows the main inputs to the WSSD, it is easy to demonstrate how the outputs of

TABLE I. Draft plan of implementation as it emerged from PrepCom 4 in Bali. Summary of elements (sub-paragraphs) on which agreement was reached.

Chapter		Fully agreed	
		Number	%
I	Introduction	3	60
II	Poverty	40	89
III	Consumption/Production	53	80
IV	Natural resources	136	88
V	Globalisation	1	7
VI	Health	28	97
VII	Small island developing states (SIDS)	20	87
VIII	Africa	38	80
VIII.bis	Other regions	8	89
IX	Means of implementation		
	Finance	2	11
	Trade	4	15
	Technology-transfer	9	75
	Science	13	72
	Education	15	83
	Capacity building	6	86
	Info/decision-making	13	72
X	Institutional framework	70	68
Total		459	75

other UN conferences on demography, children and youth, women, and HIV/AIDS have links with, and contributed to, the WSSD. Never before had the UN attempted to organise a conference with such a wide and complex context as that of Johannesburg.

The second most important input to the WSSD were the 22 reports of the UN Secretary-General, submitted to PrepCom 1, which assessed the implementation of Agenda 21. These reports identified the following serious deficiencies in implementation that needed to be addressed: fragmented approach to SD; lack of progress in addressing unsustainable patterns of production and consumption; inadequate attention to the core issues of water, energy, health, agriculture and biodiversity (collectively known by the acronym WEHAB); coherent policies on finance, trade, investment, technology, and SD; insufficient financial resources; and absence of a robust mechanism for technology-transfer.

However, as important as these documents to describe the content of the Summit is the general international spirit in 2002. This one is among others characterised by a phenomenal growth of economic globalisation and global security issues. This also explains the increasing role of an organisation as WTO in the international sustainability debate.

In consideration of the above, the WSSD was expected to reaffirm Agenda 21 as the main pathway to SD. Scarcity of ideas was not the problem, as acknowledged in the JPI, it was their implementation. The WSSD also stressed the importance of partnerships between countries as well as between governments and civil society. To this end what are called "Type II partnerships" have been proposed by

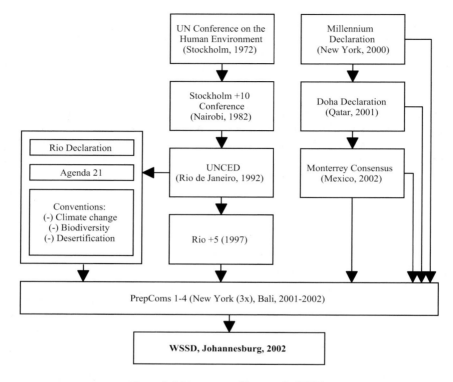

Figure 1. Main sources of input to the WSSD.

the WSSD. Johannesburg also provided a Political Declaration to overarch all the WSSD aspirations and to highlight its vision of global sustainability in an equitable world of peace and prosperity.

3. World Summit on Sustainable Development

3.1. ORGANISATION

The choice of Johannesburg as the venue for the WSSD is loaded with symbolism. It is indeed interesting to note that in the ten years since Rio, South Africa took gigantic steps from apartheid to secular, pluralistic democracy, and no doubt this fact had a significant influence on the choice of the venue.

The main conference centre was at *Sandton*, which is an ultramodern suburb of Johannesburg – a city most atypical of Sub-Saharan Africa and more like Los Angeles than nearby Kinshasa. Members of the UN, represented by their diplomats, experts, ministers, and, during the last three days of the conference, heads of state (or their representatives) met in conference at that centre. Apart from the official speeches, the real task of the conference was to reach consensus on the

JPI and the Political Declaration. Themes of the core events were mirrored and illustrated by side events, while a parallel conference was also organised in the main conference centre by the major groups and stakeholders in the SD process. The WSSD was covered by over 4000 UN accredited journalists and guarded by an army of policemen and security people.

In *Ubuntu* village, which is a thirty minutes' drive from the *Sandton* Centre, an exhibition was organised with workshops, special meetings and conferences by nearly 300 actors in SD. It covered a wide range of issues, including some of the main issues on the WSSD Agenda, such as those of water, health, rural and urban development, energy, science and technology, government (with special reference to local government), climate, social responsibility, and economics.

Water was at the top of the WSSD agenda. To reflect this, the *Water Dome* was set up. This impressive dome housed an exhibition of a wide range of water-related issues and activities – practices and policies from around the world; new technologies; projects and activities to promote the most efficient use of water; discussions; workshops; and cultural events. The dome hosted more than 100 exhibits, as well as over 50 meetings on different aspects of the water–environment-sustainability interface during the six thematic days (29 August–3 September).

Like the "Global Forum" in Rio, Johannesburg had its parallel conference too, organised by and for the civil society. In *Nasrec*, almost one hour's drive from *Sandton*, the "Global Peoples Forum" was organised under the leading theme of "A sustainable world is possible". An exhibition entitled "Ten years of broken promises" was also mounted. The same theme was discussed in workshops, and a call was made for action by civil society to improve matters. As it happened, this conference of the NGOs deliberated on the pressing issues of SD in greater depth, and with a sharper focus on the future, than the official UN conference at *Sandton*.

With these four conference sites and numerous exhibitions and meetings at still other locations, the Johannesburg conference was the most complex of all the environmental/SD conferences organised by the UN to date.

3.2. THE JOHANNESBURG DECLARATION ON SUSTAINABLE DEVELOPMENT

The Johannesburg Declaration on Sustainable Development is a political document that both overarched and concluded the conference. It was meant to clarify the Johannesburg vision of sustainable development and to pave the way for new negotiations.

On 1 September a draft text of the Declaration, containing 69 articles in 6 sections, was launched by the South African chair of the conference. The central part of the Declaration was the "Johannesburg Commitment on Sustainable Development", which is a pledge to implement the SD programme detailed in the JPI. After the text of the Declaration was launched, delegates were invited to comment on it, and the most pertinent comments were discussed during the short period remaining between 2 September and the close of the conference.

What finally emerged was a text of 37 articles – a watered-down version but more coherent of the original draft. The text outlines the path taken from UNCED to the WSSD, highlights present challenges, expresses a commitment to SD, underscores the importance of multilateralism, and emphasises the need for implementation. Box 4 lists the main sections and the keywords for each. Section 4, which still remains the core of the text, refers to a process of SD that is characterised by, among others, multilevel policy action, a long-term perspective, and broad participation. It recalls the main threats to SD, the main actors, and the core issues of water, energy, health, agriculture and biodiversity (WEHAB) among others, and puts emphasis on the small islands developing states (SIDS) and the least developed countries. Equal emphasis is also given to the most important instruments of SD policy, including capacity-building, technology-transfer, training and education, new partnerships, financial means, and good governance.

It was clear from the beginning that agreement on a negotiated Political Declaration was crucial to the success, and indeed credibility, of the WSSD. And so the agreement reached on it at Johannesburg was a great relief. However, although it refers to strategic approaches to the realisation of the JPI, it is not clear as to which doors, if any, it opens for new international negotiations. Therefore, the impact of the Declaration on future negotiations on SD would probably be limited.

3.3. PLAN OF IMPLEMENTATION OF THE WSSD

3.3.1. Introduction

The JPI, which is a negotiated document on which a consensus of all the UN members was reached, is the core document of the WSSD. It contains a list of actions, some with quantified targets, to be implemented to realise the Agenda 21 objectives set out in Rio.

3.3.2. Methodology for consensus-building

Consensus on the JPI was reached using the negotiating methodology of what is called the "Vienna Setting". Following this methodology, the draft of the Plan of Implementation, which emerged from the Bali PrepCom, was discussed in the main forum. As soon as a serious textual problem appeared, a group was formed to address it. For sensitive parts of the text, "packages" were often formed and discussed as entities. Although this approach allowed consensus-building over the text, often the negotiations became detached from the contents and intrinsic values of ideas. The Bali draft entailed, e.g., different references to "sustainability impact assessment" of projects, plans, programmes and policies. The idea was one of the outcomes of the Doha meeting. It was an element allowing to mainstream free trade in a way that it is at least not conflicting with SD. Next to this political commitment, it is hard to imagine that one can be opposed to this idea in a conference looking for instruments to implement SD. The contrary materialised. Sustainability impact assessment (together with strategic impact assessment) became part of a package deal and disappeared from the final text.

Box 4. The Johannesburg Declaration on Sustainable Development: structure and keywords.

(1) From our origins to the future
- Pillars: environment, social and economic development
- Levels: local, national, regional, global
- Pledge to implement a plan for poverty eradication and human development

(2) From Stockholm to Rio De Janeiro to Johannesburg
- Agenda 21, the Rio principles
- Major UN conferences
- Vision of SD

(3) The challenges we face
- Overarching objectives: poverty eradication, unsustainable patterns of production and consumption, natural resource base, and social and economic development
- The North–South divide
- Continuing environmental degradation
- Globalisation as a challenge
- Credibility of democratic representatives

(4) Our Commitment to Sustainable Development
- Characteristics: multilevel policy action, long-term perspective, broad participation, respect for human diversity
- Actors: multi-stakeholders, indigenous people, labour organisations, private sector, local governments, women, regional groupings, and alliances
- Threats to SD: hunger, malnutrition, foreign occupation, armed conflicts, illicit drug problems, organised crime, corruption, natural disasters, illicit arms trafficking, trafficking in persons, terrorism, intolerance (racial, ethnic and religious), and diseases
- Issues: water and sanitation, energy, health care, food security, biodiversity, and shelter
- Regions: small island countries, and least developed countries
- Instruments: capacity-building, technology-transfer, new partnerships, partnerships, dialogue, development of human resources, education and training, financial means, and good governance

(5) Multilateralism is the future
- Democratic and accountable international and multinational institutions
- Strengthening of multi-lateralism
- Monitoring of SD

(6) Making it happen
- Involving major groups
- Commitment to SD

3.3.3. Content

As a whole, the JPI is designed as a framework for action to implement commitments originally agreed to at UNCED.

The text of the JPI, made available at the end of the WSSD on 4 September, contained 170 paragraphs in 11 chapters. A list of these chapters is given in Box 5, together with the core ideas underlying them.

Overarching the discussion of the individual items, was the debate on the application of the Rio Principles 7 (common but differentiated responsibilities) and 15 (precautionary approach). While many countries assumed these principles were acquired in the SD discussion, some countries, leaded by the US, repeatedly re-questioned these basics of the Rio Declaration.

The introduction to the JPI has links with earlier UN activities, the Rio (UNCED) Conference in particular. It focuses on the meagre implementation of Agenda 21 to date.

Chapter II is on poverty eradication. Article 7(a), which is the centre-piece of this chapter, states "to halve, by the year 2015, the proportion of the world's people whose income is less than $1 a day and the proportion of people who suffer from hunger and, by the same date, to halve the proportion of people without access to safe drinking water". These targets are not original, however, as they are already mentioned in the "Millennium Declaration". Lack of originality and consacration of the WB doctrine on poverty are characteristic of this chapter. The main new element is the establishment of a world solidarity fund. In general, this chapter seeks to clarify the complex links that exist between poverty and SD, but it does so less explicitly than Agenda 21.

Changing unsustainable patterns of consumption and production is the subject-matter of chapter III. To promote fundamental changes to the ways in which societies worldwide produce and consume, the JPI recommends greater use of selected instruments (e.g. life-cycle analysis, consumer awareness and information, eco-efficiency, and cleaner production and practices) as well as attention to particular sectors (e.g. energy, transport, waste and chemicals). For chemicals, the JPI calls explicitly for the ratification and implementation of the Rotterdam Convention on Prior Informed Consent Procedures for Certain Hazardous Chemicals and Pesticides in International Trade, and the Stockholm Convention on Persistent Organic Pollutants. However, although the JPI is more explicit on this prerequisite of SD than Agenda 21, which it brings to a sharper focus, it does not bring new ideas for discussion. The Plan thus reflects the limited progress made so far in this area of environmental research (ICS PAC, 2002). Renewable energy, energy subsidies and chemicals and health, which are central in this chapter, were among the major areas of disagreement in Johannesburg.

Chapter IV, which is the most extensive, is on "Protecting and managing the natural resource base of economic and social development". Focusing on the important role ecosystems play in providing essential resources and services, the following are considered in this chapter:

– *Drinking water:* In line with the Millennium objectives of access to safe drinking water, Articles 25–29 reflect the high priority the WSSD gave to water,

Box 5. The Johannesburg Plan of Implementation (JPI) for Sustainable Development: the core ideas.

I. Introduction
- Commitment to the Rio principles of Agenda 21 and programme for further implementation of Agenda 21.
- Further implementation of UNCED.
- Involvement of major groups and partnerships.
- Prerequisites of SD: peace, security, stability.

II. Poverty eradication
- To be based on the Millennium Declaration goal of halving by 2015 the proportion of world population whose income is less than 1 US dollar a day.
- Actions to be taken should include sustainable use of biomass and energy.
- To strengthen the contribution of industrial development through specific actions.

III. Changing unsustainable patterns of consumption and production
- Unsustainable patterns of production and consumption are a main obstacle to SD in all countries, especially developed countries.
- A ten-year work programme should make this shift possible. A move to cleaner production and eco-efficiency is crucial to this.
- Main actors in this process include private companies and authorities.
- Focal areas include: transport, waste and chemicals.

IV. Protecting and managing the natural resource base of economic and social development
- Anchors on the increasing impact on ecosystems.
- Calls to launch a programme of action on *water* and to develop integrated water resources management.
- Asks for special attention for oceans, seas, islands and coastal areas, including sustainable fisheries and to advance the implementation of the global programme of action for the protection of the marine environment against land-based activities.
- An integrated, inclusive approach to be adopted for dealing with disasters caused by floods, droughts, climate change, etc.
- Call to be a party to the Kyoto protocol and to facilitate the Montreal protocol.
- Promote sustainable agriculture and address food security.
- Strengthen the convention on desertification.
- Give particular attention to mountain ecosystems.
- Promote sustainable tourism development.
- Address biodiversity currently being lost at an unprecedented rate.
- Sustainable forest management is an essential goal.
- Actions for a sustainable use of minerals, metals and mines.

(Continued)

Box 5. Continued

V. Sustainable development in a globalising world
- The challenge is to use the growing integration of economies and societies around the world to enhance SD.
- To this end a wide array of actions should be taken, ranging from open, equitable, rule-based, predictable and non-discriminatory finance and trade systems, through capacity-building in developing countries to regional trade and co-operation agreements.

VI. Health and sustainable development
- There is an urgent need to address the causes of ill health, including environmental causes and their impact on development.
- To this end 15 actions are provided to strengthen the capacity of health care systems to deliver basic health services and to reduce environmental threats to health.
- Particular attention to be paid to HIV/AIDS, malaria and tuberculosis.

VII. Sustainable development of small island developing States
- Efforts of the small islands developing states to implement SD are being increasingly constrained by the interplay of adverse factors clearly underlined in Agenda 21 and other agreements.
- Actions should be targeted, among others, on transfer of technology, capacity-building, coastal biodiversity, fresh water, sustainable tourism, disaster prevention and mitigation of effects, and affordable and environmentally sound energy.

VIII. Sustainable development for Africa
- Implementation of SD in Africa is hindered by poverty, conflicts, inadequate investment, and HIV/AIDS.
- Action for SD in Africa should focus on supporting the vision of the New Partnership for Africa's Development (NEPAD) to promote science and technology development in African centres of excellence, development of education, contribution of the industry (mining in particular), environmental impact assessment, energy, affordable transport, aforestation and reforestation.
- Develop and strengthen health systems addressing Ebola, HIV/AIDS, tuberculosis, malaria, trypanosomiasis, displacement of people, management of conflicts, and rehabilitation of destroyed environments.
- Pay attention to potable domestic water resources, management of chemicals, sustainable tourism, biodiversity and urban development.

IX. Other regional initiatives
- Latin America and the Caribbean: concrete actions on major environmental problems and foster south-south cooperation.
- Asia and the Pacific: acknowledge population pressure and act upon capacity-building for SD; poverty; cleaner production and energy; land management and biodiversity; fresh water resources; oceans, coastal and marine resources; atmosphere and climate change.

- West Asia: scarce water and limited fertile land resources. Focus on poverty; relief of debt burden; natural resources; programmes to combat desertification; integrated coastal zone management; water and land pollution.
- Economic Commission for Europe: different levels of economic development. Implement existing agreements relevant to Agenda 21.

X. Means of implementation
- Relate SD and trade, for example by reducing or eliminating tariffs on non-agricultural products.
- Promote, facilitate and finance environmentally sound technologies and know-how.
- Build capacity for science and technology (e.g. by establishing partnerships).
- Mobilise resources to enhance education for SD.
- Enhance and accelerate human, institutional and infrastructure capacity-building
- Assure access to environmental information and information needed to implement SD.

XI. Institutional framework for sustainable development
- Develop national and international institutions to implement Agenda 21.
- Address the role of the UN institutions: General Assembly, Economic and Social Council, CSD, and the international institutions WTO, GEF, UNDP, UNEP, UN-Habitat, UNCTAD.
- Strengthen institutional arrangements and frameworks at regional and national levels.
- Enhance partnerships for SD and promote the participation of major groups.

especially drinking water. Activities at the Water Dome reflected concern over water and related issues.

– *Oceans, seas, islands and coastal areas:* This section pays ample attention to sustainable (management of) fisheries. Although there is already an impressive number of conventions and international agreements in this area, the degradation of marine ecosystems continues nevertheless. It is argued that new international legal regimes and actions are needed to reverse this trend. Interestingly, the JPI also pays attention to both environmental and health impacts of radioactive waste and global reporting and assessment of the state of the marine environment.

– *Disasters:* Assessment of risks of disasters resulting from floods and draughts, and their management, is explicitly addressed. The text calls for, among others, early warning systems and information networks for disaster management.

– *Climate change:* Articles 38 and 39, which are on climate change, show a remarkably feeble attention to what is probably the most important global environmental problem today. It is all the more remarkable when seen against the background of the Kyoto Protocol and related agreements that have been under intense international discussion. In contrast to this weak attention of the JPI for climate change, China, India and Canada announced at the WSSD that they would ratify the protocol. A call was also made by a number of countries, including Brazil and South Africa, to prepare a post-Kyoto agreement with more stringent emission reduction targets.

- *Agriculture:* In the section on sustainable agriculture and rural development, the JPI calls for enhancing access to existing markets and developing new markets for value-added agricultural products. It also calls for the use of traditional and indigenous agricultural systems, but not for genetically modified organisms.
- *Desertification:* The JPI calls for strengthening the implementation of the 1994 Convention to Combat Desertification, in particular by providing financial resources. It is suggested that the GEF should provide the necessary funds.
- *Mountain ecosystems:* As in Agenda 21, mountain ecosystems are acknowledged as being especially fragile and vulnerable.
- *Sustainable tourism:* This new item, not included in Agenda 21, is considered in Article 43 of the JPI. It is linked to a number of recent events, including the International Year of Eco-tourism (2002), the UN Year for Cultural Heritage (2002), and the World Eco-tourism Summit (2002). The focus is on partnerships, education and training, host communities, and small and medium-size local enterprises.
- *Biodiversity:* The section on biodiversity is based on the finding that currently biodiversity is being lost at unprecedented rates due to human activities. The overall goal is to achieve a significant reduction in the current rate of loss by 2010. To this end a call is made for more effective and extensive implementation of the Biodiversity Convention and related international regimes (e.g. Cartagena Protocol on Biosafety, Bonn Guidelines on Genetic Resources). Special attention is given to, among others, hot spots, alien species, traditional knowledge, and the Global Taxonomy initiative. The JPI calls for negotiations on an international regime to promote and safeguard fair and equitable sharing of benefits accruing from the utilisation of genetic resources. Biodiversity loss and fish stocks were among the most sensitive issues to be debated in Johannesburg.
- *Forests and trees:* These are addressed in the context of a call for sustainable forest management. To this end political commitment, partnerships, domestic law enforcement, and community-based forest management systems are considered essential. Remarkably, the JPI makes no mention of Rio's "Authoritative Statement of Forest Principles".
- *Mining:* It is acknowledged that while mining is essential to modern life, it has serious impacts on both the environment and mining communities. Life-cycle approach, stakeholder participation, and sustainable mining practices are advocated for improving matters.

Chapter V is on globalisation, meaning growing integration of economies and societies around the world. It seeks to strike a delicate balance between the opportunities globalisation offers, and the challenges it poses for SD. In order to make globalisation inclusive and equitable, the JPI recommends multilateral trading and financial systems, also at the regional level, with particular attention to the following: developing countries and countries with economies in transition; implementation of the Doha Declaration and the Monterrey Consensus; corporate responsibility and accountability; and narrowing of the north-south gap.

Unlike other issues discussed at the WSSD, discussions on globalisation (which was not explicitly on the Rio agenda) turned out to be both difficult and unsatisfactory. Difficult because most of the text was still to be agreed when the WSSD had begun. And unsatisfactory because it was much too feeble with regard to the environmental consequences of globalisation. When it came to the environment, the JPI only calls for examining the relationship between trade, environment and development. Another symptom of this weakness is that the JPI fails to invoke the precautionary principle, enshrined in the Rio Declaration, in dealing with potential problems occurring at the resources–environment–trade–development interface.

Chapter VI, which deals with health and SD, begins with the premise that there is an urgent need to address the causes of ill health, including environmental causes. It calls for attention to be paid to HIV/AIDS, malaria, tuberculosis, and respiratory diseases caused by air pollution. Calling for attention to cardiovascular diseases, cancer, diabetes, injuries, violence and mental health disorders, it recommends the phasing out of leaded fuels and paints and rethinking of ways in which societies generate and consume energy. It also calls for the provision of basic health services for all, the WHO Health for All Strategy, partnerships, research, public health education and training, and the development and use of effective traditional medicine. And it urges special attention to safe water, sanitation and adequate food. This chapter refers to two quantitative targets:

- Reduction, by the year 2015, of mortality rate of children under the age of 5 by two-thirds and maternal mortality rate by three-quarters (as in the Millennium Declaration), in both cases taking the rate prevailing in 2000 as the base-line; and
- Reduction of HIV prevalence among young men and women within the age range of 15–24 by 25% in the most affected countries by 2005, and globally by 2010 (UN Declaration of Commitment on HIV/AIDS).

Unlike Agenda 21, the JPI has a regional outlook too. In particular, it refers to the sustainable development of SIDS and Africa. Chapter VII is on SIDS that are considered to be especially vulnerable and whose prospects for SD are under the mounting threat of sea level rise caused by climate change. To overcome these threats, the JPI recommends greater emphasis to be placed on sustainable management of fisheries, coastal areas, freshwater, waste and pollution, sustainable tourism, and hazard and risk management. And it advocates that funding needed for these should be provided by the GEF, among others.

Chapter VIII is on the sustainable development of Africa. It gives this impressive list of recommendations for pursuing the sustainable development of that continent:

- Poverty should be replaced by sustained economic growth and SD.
- Arresting or reversing desertification in accordance with the 1994 Convention to Combat Desertification.
- Developing and strengthening healthcare systems.
- Integrated water resources management.
- Sustainable agricultural production and food security.
- Narrowing income disparity.
- Sustainable tourism.
- Implementation of the Habitat agenda.

The JPI practically covers the entire spectrum of obstacles on the way to the realisation of SD in Sub-Saharan Africa. However, some of the recommendations may be a little too enthusiastic to be useful or beneficial. For example, "adventure tourism" which may destroy biodiversity instead of protecting it. Support for the New Partnership in Africa's Development (NEPAD), which is a commitment of African leaders to the people of Africa, is noticeable in the JPI recommendations.

Chapter IX focuses on Latin America and the Caribbean, Asia and the Pacific, West Asia, and the Economic Commission for Europe. However, the focus, which is primarily on regional agreements and their implementation, is significantly less sharp than that on Africa or the SIDS.

Most of the discussions in Johannesburg centred on means of implementation (Chapter X), and on the issues of finance and trade in particular. Acknowledging that greater funds would be needed to reach internationally agreed goals, the section on *finance* calls for:

- Facilitating greater flows of foreign direct investment.
- Substantial increase in overseas development aid (ODA).
- Make use of existing financial mechanisms and institutions including the GEF.
- Reducing unsustainable debt burden.

For reducing the debt burden of developing countries, the JPI advocates *debt-for-sustainable-development-swap* as an innovative mechanism. The section on trade repeatedly refers to the Doha, Monterrey and WTO commitments and focuses on trade liberalisation which, it is claimed, will bring economic growth. However, it is one of the least balanced sections of the JPI, for it does not clarify the underlying mechanisms linking trade, environment and sustainable development.

The section on *transfer and diffusion of environmentally sound technologies*, which is closely related to the section on research and development (R&D), recommends networking, partnerships, and appropriate legal and regulatory frameworks. The section on R&D advocates greater collaboration between scientists in the social and natural sciences, as well as between scientists and policy-makers. It recommends integrated scientific assessment, science for environmental management and protection, and more research on cleaner production and production technologies. In short, the JPI makes an open call for interdisciplinary R&D for sustainable development.

The section on *education* reaffirms the Millennium Declaration goal to ensure that by 2015 all children would be able to complete full primary schooling. And that by then boys and girls should have equal access to all levels of education. It refers to the goals of the Dakar Framework for Action on Education for All, including elimination of gender disparity in primary and secondary education by 2005. Perhaps more importantly, it calls for general education for sustainable development, and a decade of education for sustainable development starting in 2005.

The section on *capacity-building* advocates establishing "centres of excellence" in developing countries, and for the provision of technical and financial assistance for them.

Finally, the section on *environmental information* emphasises access to environmental information, better statistical and analytical services, indicators of sustainable development and GIS, and remote sensing and related technologies. Although it repeatedly mentions the importance of environmental impact assessment, it failed to include strategic environmental assessment, as suggested by the Bali draft of the JPI.

The last chapter, Chapter XI, is on institutional frameworks for SD. Although it calls for the adoption of new measures to strengthen institutional arrangements for sustainable development at all levels (local, national, regional, international), it is to a large extent an introspection of the UN and its own SD framework. Interestingly, the institutional aspect is linked to good governance and to efforts to reform international financial structures and arrangements.

At the institutional level, the role of the UN Economic and Social Council is stressed. It is recommended that the council should undertake, among other things, periodic assessment of SD themes. Since Rio the CSD, established under Agenda 21, has been the topic of much discussion. The JPI has proposed an enhanced role for it that includes monitoring of progress of implementation, initiatives and partnerships (but not the key roles of following-up and evaluating partnerships). Although the roles of other international institutions such as the WTO, GEF, UNDP, UNEP, UNCTAD and Habitat are discussed in a separate section, the CSD and its role are mentioned or referred to throughout the JPI.

The JPI calls for the establishment of an effective, transparent and permanent inter-agency mechanism to coordinate the ocean and coastal issues within the UN system. At the regional level, it recommends that institutions should contribute to SD and pay particular attention to NEPAD and the Programme of Action for the Sustainable Development of the SIDS. At the national level, it recommends that states should make progress on national strategies for SD and promote both public participation and sustainable development councils. It recommends, furthermore, that the role of local authorities should be enhanced, and acknowledges that each country has the primary responsibility for achieving its own SD.

Box 6. WSSD Terminology.

At the 3rd WSSD PrepCom (25 March to 5 April) attention turned to the role of "partnerships" between different actors (local, national and global governmental institutions; NGOs; the private sector; community organisations; etc.) in delivering implementation of agreed sustainable development objectives. The general terminology used for these is:

Type I actions: These are negotiated by governments through the PrepCom process. They are divided into "Type 1A" – dealt with in the proposed WSSD Political Declaration (a short declaratory document to be endorsed by the Heads of Government). And "Type 1B" – to be addressed in the Johannesburg Programme of Action.

Type II partnerships: They are defined as a "series of implementation partnerships and commitments involving many stakeholders. . . .These would help to translate the multilaterally negotiated and agreed outcomes into concrete actions by interested governments, international organisations and major groups" (Opening Statement by the Chairman of the 3rd WSSD PrepCom). A procedure for registering Type 2 partnerships was discussed during the Bali Prepcom (see www.johannesburgsummit.org for details). However, little has so far been done to elaborate the workings of such partnerships, or about how to sustain them after the WSSD.

The final section of Chapter XI is on major groups and partnerships. Although the Type II partnerships are a major outcome of the WSSD, the JPI is most economical with information on it. The very last article calls for the participation of young people in SD.

3.4. TYPE II PARTNERSHIPS FOR SUSTAINABLE DEVELOPMENT

Partnerships for SD, conceptualised during the third PrepCom, are the third main outcome of the WSSD. The concept was mooted in response to this question: how best could civil society make project-wise contribution to the implementation of SD? However, there is at present a good deal of confusion over the concept of such partnerships and their *modus operandi*. Information given in Box 6 throws some light on the vocabulary.

To clarify the concept, a "consensus" was reached at PrepCom 4 in Bali on a set of guidelines for Type II partnerships. According to those guidelines, such partnerships should:

– Achieve further implementation of Agenda 21 and seek to reach the Millennium Declaration goals.
– Complement globally agreed Type I actions and not substitute government commitments.

- Be voluntary in nature and not subject to negotiation.
- Be participatory, with ownership shared by the partners.
- Be new initiatives, or, in the case of ongoing initiatives, demonstrate added value in the context of the Summit.
- Integrate economic, social and environmental dimensions of SD.
- Be international (global, regional or sub-regional) in scope and reach.
- Have clear objectives and set specific targets and timeframes for their achievement.
- Have a system of accountability, including arrangements for monitoring progress.

The UN Department of Economic and Social Affairs (DESA) screened the proposals submitted to the UN during the period between the Bali PrepCom and the WSSD. On 28 August a list of 128 Type II partnership projects was announced. They were organised in different groups including:

- Information for decision-making; science and education.
- Finance, trade and technology-transfer.
- Sustainable production and consumption patterns.
- Urbanisation and local authorities.
- Other areas, including the cultural dimension of SD.

In a special session, held at the main conference centre, further details of the above partnerships were given. Notable partnerships included the following projects:

- Capacity 2015 of UNDP.
- UNESCO's Encyclopaedia of Life Support Systems (EOLSS).
- Water and energy partnerships of the EU.
- International Youth Dialogue on SD of the Global Youth Network.

The UN call for further development of partnerships. It is, however, reticent about how they are to be followed-up.

The proposed partnerships are fundamentally an excellent idea. For, if one is serious about implementing SD, it makes much sense to operate under a framework that allows civil society to make its contribution. However, ever since the idea was launched, serious doubts have been expressed, notably the following:

- Worry of the NGOs that these partnerships may substitute governmental obligations (SDIN, 2002).
- Authorities fear that they might lose control over their SD policies and programmes.
- Implementation of SD is not a core activity of business or industry, neither of most of the other major groups.

Although efforts were made on different occasions to mitigate these concerns, suspicions remained, and the level of participation of the international community in the first phase of Type II partnerships was a good deal lower than anticipated.

At present the best established partnerships are those organised by international organisations such as UNESCO, UNDP, WMO, etc. They have well established and extensive international partnership networks for executing SD or related projects. These partnerships might well dominate the scene for years to come, unless other major groups organise their own partnerships without delay. In particular, scientists engaged in research on SD have much to gain from Type II partnership networks.

4. Discussion

4.1. THE WSSD

The UN Secretary General Kofi Annan's address to the WSSD was commensurate with the expectations placed on it.

> We have tempted fate for most of the past two hundred years, fuelled by breakthroughs in science and technology and the belief that natural limits to human well-being has been conquered. Today we know better and have begun to transform our societies, albeit haltingly. With some honourable exceptions, our efforts to change course are too few and too little. The question now is whether they are also too late. In Johannesburg, we have a chance to catch up. Together, we can and must write a new and more hopeful chapter in natural and human history.
>
> (From the opening address of Kofi Annan to the WSSD)

Have we been capitalising on this chance to catch up? Is a new chapter in human history really being written? No doubt the outcomes of the WSSD have their strong points. The Political Declaration, and certainly the JPI, contributes in good measure to make SD more specific than hitherto. As for the rest, and the verdict of history, only time will tell.

Selected examples of the new agreements reached and initiatives announced at the WSSD are listed in Table II. Limited to the WEHAB discussions, this list is by no means exhaustive. It is indicative, nonetheless, of the progress made at the WSSD.

In the JPI there are time-limited (time-bound in the UN jargon) targets (or commitments) that are to be realised by the year indicated. A list of nine important targets of this kind is given in Table III. Specification of time limits is significant. Because, although there is no legal obligation to meet the targets by the specified time limit, these limits have nevertheless been negotiated by the parties concerned. And to that extent they impose a moral obligation on the parties to comply. Clearly, non-compliance by the negotiated time limit would indicate occurrence of problems concerned with implementation.

The list of Table III provokes the following pertinent comments, among others:

– Why should the world wait until 2010 for loss of biodiversity to be ended? Why should economic concerns over biodiversity be allowed to transcend environmental imperatives for another eight years? Even more fundamental is the question: "Will it be possible to halt biodiversity loss by 2010 given the current economic imperative?".

TABLE II. WEHAB commitments and initiatives stemming from the WSSD.

Subject/area	Commitment	Initiative(s)
Water and sanitation	• Halve the proportion of people without access to sanitation by 2015. This matches the goal of halving the proportion of people without access to safe drinking water by 2015.	• The EU announced its "Water for Life" initiative to meet the goals of water and sanitation, primarily in Africa and Central Asia. • The USA announced its $970 million investment in water and sanitation projects during the next three years. • The UN received 21 other initiatives on water.
Energy	• Greater access to modern energy services; greater energy efficiency; and greater use of renewable energy. • Support for the NEPAD objective to ensure access to energy of at least 35% of the people of Africa within 20 years.	• The EU announced a $700 million partnership initiative on energy. • The USA announced that it would invest up to $43 million in 2003. • The UN received 32 partnership proposals on energy.
Health	• By 2020, chemicals should be used and produced in ways that do not harm human health and the environment. • Enhance cooperation to curb air pollution.	• The USA announced its planned expenditure of $2.3 billion on health in 2003. • The UN received 16 partnership proposals on health.
Agriculture	• The GEF will consider including the Convention to Combat Desertification as a focal area for funding. • In Africa, development of food security strategies by 2005.	• The USA announced its planned investment of $90 million in sustainable agriculture in 2003. • The UN received 17 partnership proposals on agriculture.
Biodiversity	• Ending biodiversity loss by 2010. • Restoring fisheries to their maximum sustainable yield by 2015. • Establishing a representative network of marine protected areas by 2012.	• The USA announced its planned investment of $53 million in forests in the next 3 years. • The UN received 32 partnership proposals on biodiversity.

– As will be seen from Table III, only three of the nine selected WSSD targets – number 2, 5 and 7 – are new. The rest can be traced back to existing, internationally agreed documents such as Agenda 21, the Rio+5 Process, the Millennium Declaration, and so on. The WSSD served only to confirm and lift some of them to a higher level of international agreement in response to lack of progress to date in implementing SD.

During negotiations some of the countries repeatedly visited elements that had already been agreed upon, most notably the Precautionary Principle which is enshrined in the Rio Declaration as a fundamental principle for SD. In spite of the fact that since Rio this principle has been enshrined in international agreements and diplomacy, consensus on it could not be reached in Bali. And so, further discussion on the principle had to be postponed for Johannesburg, where reference to it in the negotiated documents was agreed upon only after intense discussions.

TABLE III. Time-limited targets in the Plan of Implementation of the WSSD, with indication of their reference in other fora.

No.	Plan of implementation	Other fora
(1)	...to halve, by the year 2015, the proportion of people unable to access or afford *safe drinking water* and the proportion of people who do not have access to *basic sanitation*, ... (Article 8)	To halve, by the year 2015, the proportion of the world's people whose income is less than one dollar a day and the proportion of people who suffer from hunger and, by the same date, to halve the proportion of people who are unable to reach or to afford safe drinking water. (UN Millennium Declaration)
	Launch a programme of actions ... to achieve the Millennium development goal to halve, by the year 2015, the proportion of people who are unable to reach, or afford, *safe drinking water* as outlined in the Millennium Declaration and the proportion of people without access to *basic sanitation*, ... (Article 25)	
(2)	...to achieve by 2020 that *chemicals* are used and produced in ways that cause minimum significant adverse effects on human health and the environment, ... (Article 23)	
(3)	To achieve sustainable *fisheries* ... maintain or restore stocks to levels that can produce the maximum sustainable yield with the aim of achieving these goals for depleted stocks on an urgent basis and where possible not later than 2015 ... (Article 31(a))	Singly and through implementing regional international agreements eliminate over-fishing by 2010. (OECD Environmental Strategy 2001)
(4)	A more efficient and coherent implementation of the three objectives of the (Biodiversity) Convention and the achievement by 2010 of a significant reduction in the current rate of loss of *biological diversity* ... (Article 44)	11 ...to halt biodiversity loss, which is taking place at an alarming rate, at the global, regional, sub-regional and national levels by the year 2010. 15(d) ...reconfirm the commitment to have instruments in place to stop and reverse the current alarming biodiversity loss at the global, sub-regional and national levels by the year 2010. (CBD COP6 Ministerial Declaration)
(5)	Effectively reduce, prevent and control *waste* and pollution and their health-related impacts by undertaking by 2004 initiatives aimed at implementing the Global Programme of Action for the Protection of the Marine Environment from Land-based Activities in small island developing States. (Article 58(e))	
(6)	Achieve (in Africa) significantly improved *sustainable agricultural productivity* and food security ... to halve by 2015 the proportion of people who suffer from hunger .... (Article 67)	To halve, by the year 2015, the proportion of the world's people ... who suffer from hunger ... (UN Millennium Declaration)
(7)	African countries should be in the process of developing and implementing *food security strategies*, within the context of national poverty eradication programmes, by 2005. (Article 67(a))	

TABLE III. (*Continued*).

No.	Plan of implementation	Other fora
(8)	Eliminate *gender disparity in* primary and secondary *education* by 2005 . . . and at all levels of education no later than 2015 . . . (Article 120)	7(v) Eliminating gender disparities in primary and secondary education by 2005, and achieving gender equality in education by 2015 . . . (UNESCO Dakar Framework for Action, 2000) To ensure that by the same date (2015), children everywhere, boys and girls alike, will be able to complete a full course of primary schooling and that girls and boys will have equal access to all levels of education. (UN Millennium Declaration)
(9)	States should take immediate steps to make progress in the formulation and elaboration of national strategies for sustainable development and begin their implementation by 2005. (Article 162(b))	24(a) By the year 2002, the formulation and elaboration of national strategies for SD . . . (Programme for the Further Implementation of Agenda21, Rio + 5, 1997)

The WSSD was remarkable for failing to address a number of pressing issues, notably that of climate change. Judging by events such as unprecedented floods in Mozambique, south-east Asia and Europe, delayed monsoons in south-east Asia, and drought in South-Africa, climate change is no longer a myth or a distant possibility. It is already here. Yet, its treatment at the WSSD was disdainful if not surreal. The issue was not even on the formal agenda, and neither was any measure to combat it such as carbon tax.

4.2. EVALUATION OF WSSD OUTCOMES

An objective evaluation of the WSSD outcomes is facilitated by comparing them with those of the UNCED in Rio.

The Rio Declaration provided an authoritative set of principles for SD. Although there was a call in the early days of planning for the WSSD to come up with an Earth Charter in line with the Rio Declaration, it did not materialise. Also, it is very likely that the Political Declaration of the WSSD will not have an enduring impact, neither has it the intellectual sophistication or authority that the Rio Declaration still commands.

A comparison of the JPI with Rio's Agenda 21 is less straightforward, however, not least because the former is concerned with the implementation of the latter. Indeed, implementation of Agenda 21 was the buzzword in Johannesburg, and, not surprisingly, the JPI makes frequent references to items in Agenda 21. Even so, progress can be said to have been made at the WSSD in the sense that the JPI sets new targets, some of which are quantified and to be met within specified time limits, for implementing Agenda 21. Although the chapter on globalisation is new to the SD debate, Rio's SD paradigm still remains intact. According to many who attended, the ambience of the WSSD certainly lacked the excitement, enthusiasm and high expectation generated at Rio. Perhaps this was to be expected, in line with the implementation character of the Johannesburg plan.

The concept of WSSD's Type II partnerships is also new. In spite of the fact that their precise details and *modus operandi* are still to be worked out, as it would appear, and that an opportunity was missed in Johannesburg to strengthen them, they are still likely to turn out to be the enduring legacy of the WSSD. Certainly, these partnerships confirm that progress towards SD can only be made with the active and wider participation of the major groups within society. In practice this means that the current model of policy advice, dominated by business, industry and labour unions, will in future give way to one characterised by wider societal participation. To this extent Johannesburg can be said to have been a milestone in democratising the approach to SD.

Interestingly, the WSSD did not produce any convention, presumably because the Rio Conventions are as adequate and valid today as they were ten years ago. The JPI opened only few avenues for new international environmental regimes. The protection of the marine environment is probably the most nearby one.

5. Conclusion

What would be the verdict of history on the WSSD? By all accounts, its main objective was to seek pragmatic ways and means for implementing Agenda 21, which was negotiated and agreed in Rio ten years ago, more effectively than hitherto rather than to produce a plethora of yet more documents. Notwithstanding their lack of both originality and intellectual rigor compared to Agenda 21, the main WSSD documents – the Political Declaration and the JPI – set out reasonable strategies for realising the Agenda 21 objectives. They indicate how current patterns of resource consumption and pollution can be made less unsustainable. To this extent the WSSD may be said to have been a success. However, there is still confusion over both configuration and *modus operandi* of the much-vaunted Type II partnerships, in particular over how, whether, and the extent to which they might be able to deliver sustainability in a rapidly globalising world in which arguably the poor nations are destined to remain poor, or become even poorer, while the rich become richer.

In the interests of both present and future generations, and before the earth's natural environment is irredeemably degraded and its natural resources depleted by humankind's relentless pursuit of "development", clearly the onus is on political leadership and responsible international structures. Otherwise it is hard to see how even a modest degree of global SD could ever be achieved. To do so, the outcome of Johannesburg provides a sense of direction to make the world less unsustainable. The main question for the years to come remains however: "Do we have the will to move to a sustainable world?".

References

Bigg, T.: 2002, 'Partnerships and WSSD: brave new world or blind valley', *The Future is Now* **3**, 60–72.

CDB, UN, UNEP – Convention on Biological Diversity, United Nations, United Nations Environmental Programme: 2001, *Handbook of the Convention on Biological Diversity*, London, UK, Earthscan.

Engfeldt, L.-G.: 2002, 'Some highlights of the Stockholm – Rio – Johannesburg journey', in *Stockholm Thirty Years On – Proceedings from an International Conference 17–18 June 2002*, Ministry of the Environment, Stockholm, Sweden, pp. 52–54.

ICS PAC – International Coalition for Sustainable Production and Consumption: 2002, *Waiting for Delivery – A Civil Society Assessment of Progress Towards Sustainable Production and Consumption*, Rockville, MD, Integrative Strategies Forum, 81pp.

Johnson, S.P.: 1992, *The Earth Summit. The United Nations Conference on Environment and Development (UNCED)*, London, UK, Graham and Trotman/Martinus Nijhoff.

Orlando, B.: 1998, 'The Kyoto protocol: a framework for the future', *SAIS Rev.* **15**, 105–120.

SDIN – Sustainable Development Issues Network: 2002, *Paper No. 1, Taking Issue: Questioning Partnerships* (http://www.sdissues.net/SDIN/docs/Takingissue-No1.pdf).

UN – United Nations: 2002, *World Summit on Sustainable Development. Plan of Implementation*. Advanced Unedited Text, 4 September 2002.

Upton, S.D.: 2002, 'Challenges for international governance on sustainable development on the road ahead', in *Stockholm Thirty Years On – Proceedings from an International Conference, 17–18 June 2002*, Ministry of the Environment, Stockholm, Sweden, pp. 57–58.

WCED – World Commission on Environment and Development: 1987, *Our Common Future*, New York, Oxford University Press.

CHAPTER 2

POVERTY REDUCTION AND SUSTAINABLE DEVELOPMENT

FRANCINE MESTRUM
Université Libre de Bruxelles
(E-mail: Mestrum@xs4all.be)

Abstract. The prominent place of the chapter on poverty in the Johannesburg Plan of Implementation (JPI) is totally in keeping with the priority given to poverty reduction in the development thinking of the international community of today. The Johannesburg process did not lead to any new insights or new commitments in the fight against poverty. Section one sets out a factual comparison of the poverty chapters in Rio's Agenda 21(AG21) and in the JPI. Section two reviews the conceptual links between poverty reduction and sustainable development, since poverty is used both as a dependent and as an independent variable. This analysis shows a shift in the function of growth as related to environmental protection. Section three explores the "naturalization" of development thinking in its economic and social dimensions and shows how this affects the policy options for social protection. I also explain how social and environmental sustainability have become elements of risk management and how are both aimed at conflict prevention and enhanced growth. Finally, in section four three lines of action are suggested to enhance the emergence of a socially meaningful sustainable development agenda that, ideally, would make poverty reduction strategies redundant.

Key words: Development, poverty, sustainability, United Nations, World Bank

Abbreviations: AG21 – Agenda 21; CEPAL – UN Economic Commission for Latin America; HIPC – Highly Indebted Poor Countries; ILO – International Labour Organisation; IMF – International Monetary Fund; JPI – Johannesburg Plan of Implementation; LDC – Least Developed Countries; MDG – Millennium Development Goals; ODA – Official Development Assistance; PNUD – Programme des Nations Unies pour le Développement (UNDP); PPP – Purchase Power Parity; PRS – Poverty Reduction Strategy; PRSP – Poverty Reduction Strategy Paper; UN – United Nations; UNCTAD – United Nations Conference on Trade and Development; UNDP – United Nations Development Programme; UNRISD – United Nations Research Institute on Social Development; WSSD – World Summit on Social Development (1995); – World Summit on Sustainable Development (2002); WTO – World Trade Organisation

1. Introduction

The Rio Summit on Environment and Development in 1992 was the first of an impressive series of major UN conferences that shaped the new international development agenda at the end of the 20th century. It came as no surprise that "Combating poverty" was one of the first chapters of Rio's Agenda 21 (AG21), in view of the fact that it was written two years after the first comprehensive World Bank Report on poverty and at the start of the 4th UN Decade for Development in which poverty eradication is mentioned as the first priority for development. This focus on poverty is the most remarkable new feature of the development

L. Hens and B. Nath (eds.), The World Summit on Sustainable Development, 35–55.
© 2005 *Springer. Printed in the Netherlands.*

discourse of the international organisations. In 1995, the UN organized its first conference ever on social development. In 1997 it started a "Decade" for the eradication of poverty and its General Assembly approved a new "Agenda for Development". At the start of the 21st century, the commitment to reduce poverty was solemnly confirmed in the Millennium Declaration. As a consequence, "Poverty eradication" is also the first substantial chapter of the Johannesburg Plan of Implementation (JPI).

2. From Rio to Johannesburg

Ten years after Rio, the Johannesburg Summit on Sustainable Development offers an excellent opportunity for assessing the progress made in the fight against poverty.

2.1. TRENDS IN WORLD POVERTY

Even a limited study of the academic literature on poverty reveals the huge epistemological and methodological problems of poverty assessments. There is no clear theoretical framework for poverty research and there is no unambiguous definition of poverty (Øyen, 1996). When the World Bank published its first World Development Report on poverty (World Bank, 1990), it had very little reliable and comparable data. Using an absolute poverty line of 370 US$ per year per person (or 1US$/day) (in PPP of 1985), poverty in developing countries was estimated at 1,116 billion people or 33 % of the population in the developing world. It must be emphasized that the emerging international poverty discourse of 1990 had no real empirical foundation (Tabatabai, 1996). Throughout the nineties, huge efforts were made to improve the database concerning income and consumption. In its second comprehensive report on poverty (World Bank, 2001a), the World Bank declared that its data then gave a fairly reliable view of poverty. The latest data for 1998 pointed to a substantial decrease in proportional terms: from 33 % in 1985 to 23,4 % in 1998, even if the total number of extremely poor people had slightly increased. Much of this progress was registered in China and India. In sub-Saharan Africa and Eastern Europe, poverty was on the increase. Using a poverty line of 2 US$ a day, poverty was also slowly decreasing: from 61 % in 1987 to 56,1 % in 1998 (World Bank, 2001b: 6-7). For the purpose of this paper, four brief comments may assist to put these data into perspective.

2.1.1. The international poverty line used by the World Bank is highly controversial. Very different outcomes result from looking at national poverty lines or looking at relative instead of absolute poverty. This can be illustrated by two examples taken from the statistics of the World Bank itself: 5,1 % of the population of Brazil is poor according to the international poverty assessment, whereas poverty amounts to 17,4 % when applying the national poverty line

(World Bank, 2001a: 280). In absolute terms, only 15,6 % of the people in Latin America and the Caribbean are poor. With a relative poverty line, the percentage rises to 51,4 % (World Bank, 2001a: 23-24). Many UN documents state that the majority of poor people are women. However, there are no statistics on the income poverty of women. It means that all poverty assessments are to be handled with extreme caution.

2.1.2. The methodology used by the World Bank and its subsequent modifications has also been subjected to severe criticism (Reddy & Pogge, 2002). UNCTAD recently published its own poverty assessments for the least developed countries and arrived at far less positive results. According to UNCTAD, the incidence of extreme poverty is increasing in the LDC's as a whole and amounts to 50 % for the period 1995-1999. It would mean that the number of extremely poor people in the LDC's has more than doubled over the last 30 years (UNCTAD, 2002: 111).

2.1.3. Poverty is said to be a multidimensional problem, which means that income and consumption cannot be the only criteria to be taken into account. The UNDP has devised a human development index, combining income, education and life expectancy, and a human poverty index combining health and education data, without income. Human development has dramatically improved since 1970. The proportion of people with a high level of human development has increased from 55 % in 1975 to 66 % in 1997, whilst the proportion of people with a low level of human development has decreased from 20 % to 10 % (PNUD, 1999: 25). During the past 15 to 20 years however, living standards in more than a hundred countries have decreased. This means that the incomes of more than one billion people are now lower than they were 10, 20 or even 30 years ago (PNUD, 1997: 7). Despite the positive trends in extreme poverty globally, income inequality continues to grow. The 1 % constituting the wealthiest people in the world has an aggregate income that equals the income of the 57 % of the poorest people (PNUD, 2001: 19).

2.1.4. Despite the improvements of its empirical database, the World Bank focuses more and more on the non-monetary aspects of poverty. It now does not even use the traditional social indicators but – following its wide-ranging participatory poverty assessment - defines poverty in terms of vulnerability, voicelessness and powerlessness (World Bank, 2001a).

2.2. STRATEGIES FOR POVERTY ALLEVIATION

A fairly clear basic strategy for the fight against poverty can only be found in the documents of the World Bank and the UNDP of 1990. The formal consensus reached at the social summit of Copenhagen in 1995 is differently worded and goes slightly further, but in no way does it conflict with the World Bank or the UNDP approach. The strategy consists of:

a) providing opportunities, i.e. the formation of human capital;
b) increasing the capacity of the poor to take advantage of these opportunities,
   i.e. making use of the acquired human capital (World Bank, 1990; PNUD,
   1990).

For the World Bank, "providing opportunities" means encouraging economic
growth that makes use of the labour force of the poor, while "increasing the
capacity of the poor" consists of providing basic social services such as
education, health care and family planning. The UNDP sees things the other way
round. Human development consists of making available basic social services in
order to empower individuals to increase their human capital for productive,
social and political gains within a context of economic growth. In addition to this
dual approach targeted social programmes are required to help those who cannot
participate in the market; a safety net is needed to protect those who are exposed
to shocks and to take care of the victims of the competitive struggle.

This poverty reduction strategy does not require any changes in the core
policies of the Washington consensus, even in its "augmented" form.
Nevertheless, the added-on social dimension has slowly made way for issues of
governance and trade. According to the UNDP, it was a diagnostic error to think
of poverty in terms of social protection and social expenditure (PNUD, 2000: 8,
42). In 1999, the IMF relabelled its "Enhanced Structural Adjustment Facility"
and called it "Poverty Reduction and Growth Facility". In order to benefit from
the HIPC debt relief initiative, poor countries are now required to introduce a
"Poverty Reduction Strategy Paper" (PRSP) that has to be approved by the Joint
Staff of the World Bank and the IMF. The first assessments of this PRSP-process
are now under way. Finally, as has already been mentioned, the UN Millennium
Summit approved a series of "Development Goals". The first of these is to halve
the proportion of extremely poor people by 2015 and the one but last calls for
environmental sustainability. The other targets concern child and maternal
mortality, gender equality and access to basic social services. These goals have
been confirmed at the Monterrey Conference (United Nations, 2002c). However
valuable these targets are, they fall short of the aims of the Copenhagen Plan of
Action for social development (Nations Unies, 1995), where poverty was linked
to employment and social integration. It also has to be pointed out that the World
Bank, following its re-conceptualisation of poverty, now focuses its strategy on
empowerment, opportunity and security. The income dimension is almost totally
absent from all proposed strategies. Poor people are the main agents of their
empowerment process in an enabling environment created by the governments.
Apart from the institutional requirements concerning good governance, it
primarily means economic growth thanks to the countries' integration in world
markets. Less than a year after the approval of the Millennium Development
Goals (MDG), the World Bank already stated that it was highly improbable that
they would be realized (World Bank, 2001b: 26).

2.3. AGENDA 21 AND THE JOHANNESBURG PLAN OF IMPLEMENTATION

It is against this background that the Johannesburg Plan of Implementation (JPI), and more particularly its chapter on poverty should be put to the test and be compared to Agenda 21 (AG21). Five questions will be briefly examined: 1. What do the documents tell us about world poverty? 2. How are poor people referred to? 3.What is the objective of the proposed actions? 4. Who is responsible for implementing the poverty agenda? 5. What are the ways and means of fighting poverty?

2.3.1. AG21 defines poverty as a *"complex and multidimensional problem"*, linked to insufficient development (§ 3.1; 6.1) like most of the documents issued by the major UN-conferences of the 90's. JPI does not give any definition of poverty and only mentions its income dimension in passing (§ 11c). In the chapter on Africa, it states that *"most countries have not benefited fully from the opportunities of globalisation"* (§ 62). Surprisingly, and contrary to AG21 and almost all other UN documents on poverty and development, no links are made between poverty and population growth. In JPI, as well as in AG21, the poverty problem is linked to environmental degradation, though the precise nature of the link is not explained. This aspect will be examined in section 2. While the absence of any reference to trends in world poverty in AG21 can be explained by the lack of empirical data at that time, a brief mention of the magnitude of the poverty problem could usefully have been introduced in JPI, particularly in the context of the overall aim to implement AG21. A report on poverty by the Secretary-General was introduced with the preparatory documents for the Johannesburg negotiations (United Nations, 2001b). The growing gap between the rich and the poor is briefly mentioned in AG21 (Principle 5 of the Rio Declaration and AG21 § 1.1), as well as in the Johannesburg Political Declaration (United Nations, 2002g).

2.3.2. AG21 contains numerous references to disadvantaged and vulnerable groups, naming women, children and the indigenous people as the main victims of poverty. AG21 makes a clear distinction between the urban poor (§ 6.32; 7.15) and the rural poor (§ 3.2; 14.16; 26.3iv), the latter being characterized by their dependence on natural resources for their livelihood. These same groups are referred to in JPI, though without specific links to their poverty. The largest numbers of the world's poor, according to JPI, live in Asia and the Pacific (§ 76). Poverty is also said to be a major problem for most of the African countries (§ 62).

2.3.3. Whereas chapter 3 of AG 21 is called *"Combating Poverty"*, Principle 5 of the Rio Declaration speaks of *"eradicating poverty"*. It also states that its long-term objective is to enable *"all people to achieve sustainable livelihoods"* (§ 3.4). JPI clearly calls its chapter II *"Poverty Eradication"*, but its major reference is the first MDG (§ 7a) which speaks only of halving the proportion of extremely poor people by the year 2015. It does not mention a year of reference, nor does it envisage plans to go beyond this limited ambition. The Johannesburg Political Declaration states the *"urgent need to create a new and brighter world of hope"* (§ 4) and the *"banishment forever of underdevelopment"* (§ 18). In the same way as in 1992, poverty eradication is described as an *"indispensable requirement for sustainable development"* (Principle 5 of the Rio Declaration, JPI § 7), though at the same time it is now said to be its *"overarching objective"* (Johannesburg Political Declaration § 11). The question of how poverty relates to sustainable development clearly needs further examination.

2.3.4. "Each country has the primary responsibility" for the implementation of AG 21 (§ 1.3; JPI § 7). *"The role of national policies cannot be overemphasized"* (JPI § 7). Nevertheless, both AG21 and JPI stress the importance of actions at all levels. AG21 gives a detailed overview of different responsibilities for a whole range of social agents. JPI is much less explicit on who has to do what and contrary to what is said in § 7, the role of national governments is not very clear. The Johannesburg Political Declaration (§ 32) stresses *"the leadership role of the UN"* because it is *"best placed to promote sustainable development"*. The tasks of the UN's different components and of other international organisations get ample attention in § 140 to 157. Only the last chapter of JPI gives some examples of what should be done at State level (§ 162-167). Throughout the document, different other stakeholders are mentioned, such as regional groupings, local authorities, the private sector, civil society, the poor themselves, women, indigenous people ... even children are mentioned as *"agents of change"* (§ 8d). One of the most important features of the Johannesburg process is its *"multistakeholders"* approach with the approval of a whole series of "Type II agreements" based on public-private partnerships.

In fact, some doubts may arise on who is actually speaking through the JPI and the Political Declaration. Most UN conferences start their declarations with something like *"The Conference declares..."* or *"We, heads of State and government..."*. The Johannesburg Political Declaration starts with *"We, the representatives of the peoples of the world..."* and § 26 speaks of *"we"* as *"social partners"*. One of the peculiarities of UN conferences is that they are formal gatherings of national governments issuing statements on behalf of the international community, i.e. the collectivity of national governments. The involvement of civil society groups and international – intergovernmental – institutions does not change this basic configuration. However, the importance of national governments issuing messages to themselves lies in the solemn and common commitments for future cooperative actions thereby implicitly and explicitly enhancing the role of the United Nations. The built-in ambivalence of the United Nations – an organisation of States whose Charter begins with the words *"We the peoples..."* – has recently been evolving towards a less state-centred governance approach. The Security Council undoubtedly remains a forum of national States, but former Secretary- General Boutros Boutros-Ghali has paved the way for Mr. Kofi Annan to broaden the UN constituency for all development and cooperation activities. In his report to the Millennium Summit on the role of the UN in the 21st Century (Annan, K., 2000) - appropriately titled *"We the peoples ..."* - Mr. Annan calls for the introduction of *"new principles into international relations"* (p. 6), because we no longer live in an international world but in a *"global world"* (p. 11). *"For even if the United Nations is an organisation of States ...[it] exists for, and must serve, the needs and hopes of people everywhere"* (p. 6). In his call for a *"new ethic of global stewardship"* (p. 63) Mr. Annan states that *"only governments can create and enforce environmental regulations"*. Yet, the global governance he fosters is not a *"world government"* because *"the very notion of centralizing hierarchies is ... an anachronism"* (p. 13). Rather, Mr. Annan thinks of *"loose temporary global policy networks"*, *"informal coalitions for change"*. Therefore, the United Nations *"must be opened up further to the participation of the many actors whose contributions are essential to managing the path of globalisation... civil society organizations, the private sector, parliamentarians, local authorities, scientific associations, educational institutions and many others"* (p. 13).

This is the framework in which the Type II agreements are made possible and in which, moreover, poverty eradication ceases to be an exclusive competence of national social policies. The expression *"We the peoples..."* implies a mandate for the United Nations to coordinate not only the actions of national governments, but also those of other social actors, though it has no regulatory power to do so without the intermediary of the States. But, as stated in § 2 of the Johannesburg Political Declaration, the international community of States has to become a *"global society"*. The Johannesburg Summit, then, does more than confirm and expand on the participatory approach of Rio. It encourages States to share their decision-making power with other stakeholders, and it gives the UN a role in coordinating global networks on behalf of "the peoples".

2.3.5. JPI is much less explicit than AG21 on ways and means to eradicate poverty and achieve the goal of sustainable development. The poverty chapter mentions the usual basic social services – education and health, to which energy has been added – as well as access to productive resources, income generating employment opportunities, gender equality, rural infrastructure and knowledge transfer. Similar but less developed references are made to macro-economic policies (§ 141) and the major role of trade (§ 92, 47a). The importance of economic growth is mentioned in numerous paragraphs (e.g. § 47, 62, 83, 85b, 90, 101, 138...). A reference is made to the commitments of other UN conferences of the 90's, though the Political Declaration only explicitly cites the examples of the WTO Ministerial Conference in Doha and the UN Conference of Monterrey. The Social Summit of Copenhagen of 1995 is referred to at the end of JPI (§ 140c). The Political Declaration has one reference to the ILO and its Declaration on Fundamental Principles and Rights at Work (§ 28). JPI refers twice to the ILO and gives support to its work on the social dimension of globalisation. Neither the Political Declaration nor JPI mention any specific policy for social protection. The reference to human rights in the draft Political

Declaration (United Nations, 2002e: § 24) was deleted, as well as the *"right to sovereignty and the control of natural resources"* (§ 58), despite its presence in principle 2 of the Rio Declaration. The right to development was mentioned in the Rio Declaration (principle 3), but it was deleted from the draft for the Johannesburg Political Declaration. Human rights are referred to in JPI (§ 62a, 138), whereas § 169 refers to both human rights and the right to development. No reference at all is made to the International Pact on economic, social and cultural rights (one reference in AG21 § 7.6).

A totally new agreement in JPI is the establishment of a World Solidarity Fund (§ 7b), to be financed with voluntary contributions, while *"encouraging the role of the private sector and individual citizens"*. The idea of such a Fund was first launched at the Copenhagen + 5 Special Session of the General Assembly (United Nations, 2000a), though without any mentioning of the private sector. It has since been referred to in several UN resolutions and reports. A specific proposal was made by the Secretary-General to the General Assembly in a document that also lists the rules that should govern such a Fund (United Nations, 2002a).

The 20/20 pact for allocating 20 % of ODA and 20 % of national budgets, respectively, to basic social programmes was another idea for the promotion of social development and the eradication of poverty, formally launched in Copenhagen in 1995 (Nations Unies, 1995 § 88c). It was repeated in the Report of the Secretary-General and approved at the Rio + 5 Special Session of the General Assembly (United Nations, 1997b: § 27), but it was abandoned in Rio + 10.

In short, the chapter on poverty and other poverty-related elements in the Johannesburg Political Declaration and the JPI are less strongly worded and less clear than AG21. This is surprising, since Rio + 10 adopted the acronym of the World Summit on Social Development (WSSD), which founded the hopes for a stronger focus on the social dimension of sustainable development. Despite different reports by the Secretary-General in the context of the UN Decade for the eradication of poverty, a recent UN Report on the World Social Situation (United Nations, 2001a) and several comprehensive reports on poverty by the World Bank and UNCTAD, JPI has no clear-cut concept of poverty and gives no assessment of the recent trends in world poverty. Rural poverty gets much attention, while the problem of population growth as well as the association of women and poverty have disappeared. Not a single reference is made to the Bretton Woods monitored PRSP-process nor the role it can play in linking the poverty agenda to environmental protection and sustainable development. In the context of the overall negotiations and apart from the one paragraph on the World Solidarity Fund, it seems as though poverty alleviation is taken by most participants as an obvious and uncontroversial objective. By omitting the conceptual issues and the analytical links to sustainability, a consensus is at hand. Despite the numerous documents and reports published in the 90's, it looks as if the debate on the poverty agenda has been closed with the approval of the MDGs. The World Solidarity Fund is the only new element, but it gives rise to considerable doubt, since there certainly is no shortage of adequate institutions at the global level. The risk of overlapping with UNDP or other organisations is not impossible, while the "*voluntary contributions*" from the private sector constitute an additional shift from structural social development towards charity. Of course, all commitments of AG21 and all major UN conferences of the 90's are confirmed. However, the fact that few explicit references are made to only a very select number of commitments creates the impression of a further weakening of the political will of national governments to turn social development into an essential component of sustainable development.

3. Poverty, sustainability and growth

The question concerning the links between sustainable development and poverty eradication follows from the ambivalent and frequent references to poverty eradication as "*an overarching objective*" of sustainable development – that is the desirable *outcome* of a successful development process – and "*an essential requirement*" for sustainable development – that is a *means to and end* (Johannesburg Political Declaration § 11). This matter has to be seen in the broader context of poverty eradication as related to development in general. The relationship between both concepts is less obvious than it may seem at first glance. For many researchers and practitioners in the early stages of the post-war development thinking, "*under-development*" was more or less coterminous with "*poverty*". However, at the theoretical as well as at the practical level, the solution to this perceived problem was coined "*development*" and not "*poverty eradication*". Poverty reduction was thought to be the necessary and logical outcome of a development process aimed at economic modernization and political emancipation and at bridging the gap between poor and rich countries. Although it was conceived of and theorized at the level of the UN and its regional organizations, development was a matter of national sovereignty and self-determination. Concerns for its social dimension – mainly in terms of health and education – were not absent, but proved to be extremely difficult to integrate into the development concept (Wolfe, 1981). The efforts in the 70's of the ILO to focus on a concept of "*basic needs*" and of the World Bank to launch a "*war on poverty*" were short-lived, partly due to the lack of interest of the poor countries themselves and partly due to the emerging economic and debt crisis leading to the shift to "*structural adjustment*" policies. Today, very little is left of the development project as it was promoted in the 60's and 70's. After two decades of Bretton Woods conditionality, Keynesian type development economics and inward-looking industrialization are firmly condemned and are labelled as "*errors of the past*". Slowly, the development agenda has shifted from a nation-based to a human-centred project and has finally been reduced to poverty reduction in a globalised market (Mestrum, 2002a, 2002b).

This broader context has to be taken into account in order to understand the enthusiasm with which the "sustainable development" project of the Brundtland Commission was met (World Commission, 1987). Not only did it allow for a "greening" of development, by integrating the ecological dimension, but it also offered new hope for a new and broader – "holistic" – development paradigm, away from the exclusive economic focus and away from the deflationary Washington Consensus. Even if the concept of sustainability itself has given rise to divergent interpretations, the Brundtland report was welcomed as a future-oriented effort to link economic, social, environmental and participatory dimensions of development and as a new opportunity for enhanced international cooperation. This was all the more so since environmental decline is a truly global problem, ignoring national borders and implying *"common but differentiated responsibilities"* as AG21 rightly states. Development, then, becomes an issue for both the North and the South.

Still, two questions remain unanswered. First, if sustainability is the concept that allows for an environmental dimension to be added to the existing development project, how does this additional concern affect the political, economic, social and cultural dimensions of development? Second, are there any special links between the ecological dimension of development and poverty that can explain the ambivalent position of poverty, both as a dependent and as an independent variable of development?

The answer to the first question requires an investigation of development theory and a conceptualisation of "sustainability" in respect of the different dimensions of development. The second question is of direct concern for the assessment of the Johannesburg summit and the possibilities for a successful integration of the social pillar into a sustainable development process. In fact, policy coherence is difficult to achieve when the lines between the ends and the means are blurred. In the following section, I will focus on the "why's" of poverty eradication in the context of sustainable development strategies.

3.1. GROWTH AND THE ENVIRONMENT

If we take the Brundtland Report of the World Commission on Environment and Development (1987) as a starting point, we find several explicit references to poverty and social needs:

> *"Sustainable development ... seeks to meet the needs and aspirations of the present without compromising the ability to meet those of the future"* (p. 40).
> *"What is required is ... a type of development that integrates production with resource conservation and enhancement, and that links both the provision for all of an adequate livelihood base and equitable access to resources"* (p. 40).
> *"Development countries ... endure most of the poverty associated with environmental degradation"* (p. 22).

Sustainable development, then, is the process for meeting all people's needs, for today and tomorrow. Poverty eradication will be its outcome, due to the equitable distribution of the available resources. Today's poor are the victims of insufficient or unsustainable development.

"Poverty itself pollutes the environment... Those who are poor and hungry will often destroy their immediate environment in order to survive" (p. 28).
"A world in which poverty and inequity are endemic will always be prone to ecological and other crises" (p. 44).

This implies that poverty reduction is a condition for environmental protection and thus an input to sustainable development. Today's poor are guilty of environmental decline.

Poverty, then, is at the crossroads of a two-way process. Inevitably, focusing on poverty as the independent or as the dependent variable of development will have consequences for the shaping of social and environmental policies. If poverty eradication is seen as the outcome of a sustainable development process, then it seems logical to focus on the preservation and enhancement of natural resources, as well as on their equitable distribution. In that case, anti-poverty policies might focus on human rights, inequality and the unsustainable consumption patterns of the wealthy. Indeed, it is commonly admitted that the "ecological footprint" of the world's rich minority is up to 10 times larger than that of its poor majority. Yet, if poverty is seen as an obstacle to sustainable development, then everything should be done to limit the damage caused by poor people. Policies to stop population growth are the first logical element of such an approach. Another option is to give poor people access to productive resources that are less harmful for the environment.

What do the official documents tell us about these different possibilities?

According to the Rio Declaration and AG21, poverty eradication is a condition of sustainable development (principle 5 and AG21 § 3.2.). AG21 has a chapter on population policies, stating that population growth places increasingly severe stress on the life-supporting capacities of our planet (AG21 § 5.3). The wording is rather balanced, stressing the synergetic relationship of demographic trends and sustainable development and calling for more research on the interaction. It also emphasizes gender equality and the key role of women in population policies. Nevertheless, AG21 starts with a chapter on the economic dimension of the development of poor countries. Although growth seems to be the alpha and omega of development, the chapter also asks some critical questions concerning its conceptualisation as related to wellbeing (§ 4.6; 4.11). Chapter 4 elaborates on the major responsibilities for environmental degradation and poverty of industrialized countries and their unsustainable consumption and production patterns. AG21 clearly seeks to balance different responsibilities and explicitly mentions the aim to give the poor sustainable livelihoods.

The preparatory documents for the Johannesburg summit, clearly make a similar effort to reach a balanced approach. The poverty problem is present at both sides of the sustainable development nexus. Only in some UNDP and UNEP documents does one find a clear emphasis on environmental protection as an element of human rights and human development, whereby lasting poverty eradication is seen as an outcome of sustainable development. Similarly, EU documents ask for a better integration of the environment and poverty agendas and point to the need of more empirical research (Commissie, 2001). However, two remarkable changes on the Johannesburg process have to be pointed out:

-   JPI totally omits any mention of population growth and related policies. This has been explained by the pressure exerted by conservative religious groups opposed to family planning and gender equality. UNEP and the World Bank also note that in the coming decades "*more countries [will] pass through the demographic transition*" (UNEP, 2002: 323) and world population "*is expected to stabilize by the end of this century*" (World Bank, 2003: 4). "*The next 20 to 50 years are a demographic window of opportunity*" (World Bank, 2003: 184).
-   Poverty eradication is no longer exclusively linked to environmental decline. "*The ever increasing gap between the developed and the developing world pose a major threat to global prosperity, security and stability*" (Johannesburg Political Declaration §12). "*... the poor of the world may lose confidence in their representatives and the democratic systems to which we remain committed*" (§ 15). These ideas are also put forward in the multistakeholders dialogue (United Nations, 2002b: 25): "*poverty alleviation and economic stability are crucial for environmental and social sustainability*". Sustainability, then, acquires a broader meaning as it is linked to social and economic stability.

This last point is particularly clear in the documents of the World Bank. They remind us of the economic origins of the sustainability concept. The French version of the 1992 World Development Report gives an idea of its ambiguity. Even if the Bank states to completely agree with the definition of the Brundtlant report – though seeing it more as a metaphor than as a precise concept (Banque mondiale, 1992: 8) – the report mentions "*développement soutenu*" (p. 1), "*développement soutenable*", "*Développement écologiquement viable*" and "*l'élévation durable des niveaux de bien-être*" (all on p. 8). Other documents mention "*sustained*" and "*sustainable*" (World Bank, 2002c: 15) or "*durable*" (World Bank, 2003: 23) growth.

The environment chapter of the World Bank Sourcebook on poverty reduction (Bojö et al., 2001b) is particularly interesting because it explains the different ways in which environmental protection can contribute to poverty reduction and because it points to the difficulties of measuring the impact on the environment in country-specific contexts.

The 2003 World Development Report (World Bank, 2003) elaborates on the concept of sustainability that now refers to the utilization rate of the resource base of development, be it in social, environmental or economic terms. Poverty alleviation, then, belongs to the social pillar of sustainability and is linked to "*social stress – and, at the extreme, social conflict*" (p. 14). It refers to the Bank's concept of social capital, as part of the capital stock needed for improved productivity and growth (World Bank, 2002b; Serageldin & Grootaert, 2000). In its renewed institutional approach relational and natural assets are both part of the *"broader portfolio"* to be managed by governments. *"For the assets most at risk – the natural and the social – markets cannot provide the basic coordination function of sensing problems, balancing interests, and executing policies and solutions"* (World Bank, 2003: 184). In this approach both the environment and poverty eradication (*"and other forms of conflict prevention"*) are inputs into a sustained growth process needed for enhancing well-being through time.

In the 50's and 60's, "sustained growth" was a concept coined by liberal economists to refer to growth that would not be destroyed by deficit spending or within the social chaos it engendered (Moore, 1995: 4). It now seems that for the World Bank, the original connotation of "sustainability" has not been lost.

Thirty years after the first report of the Club of Rome calling for "limits to Growth" (Meadows et al., 1972) in order to protect the environment, we have now come full circle: environment protection is said to be needed in order to preserve the growth process. The focus on poverty reduction in developing countries allows for the emphasis on the need for more growth in the absence of redistributive policies while at the same time it contributes to alleviate the burden on developed countries to change their unsustainable production and consumption patterns. It helps to explain the growing conceptual convergence of sustainability, poverty reduction and conflict prevention.

3.2. NATURALISING DEVELOPMENT THINKING

The growth versus the redistribution approach in development theory has a long history. Proposals for global re-distributional mechanisms do continue to appear in some documents from international organisations, but it is now commonly admitted that what poor countries need in the first place is economic growth. The integration of the environmental dimension into development thinking was expected to boost solidarity between the North and the South, since never before had the interdependence of all countries and peoples been so clear. This has not come about, although ecological thinking did influence the economic and social development thinking of the international organisations.

3.2.1. The laws of nature and of the economy

The post-war development project for poor countries cannot be disassociated from the specific context in which it was born. It was linked to the economic crisis of the 30's and the lessons learnt from it. It was linked to the reshuffling of power relations and the dismantling of the colonial empires. It was also the result of the frustrated ambitions of the League of Nations to achieve peace through international cooperation and social justice. Its philosophical origins are founded on theories of progress and modernity, more particularly the belief in the need and ability of human societies to shape their own environment. The modification of the natural environment was seen as a normal and necessary part of it. Thus, development means *"dominating nature"* (Nations Unies, 1951) and progress requires people to give up *"metaphysical beliefs"* which do not allow for the emergence of the individual homo oeconomicus. "Under-development" was associated with ignorance, "development" with civilization and emancipation.

In some ways, contemporary thinking on poverty still reflects this dichotomy. Poor people are said to *"live in the darkness of poverty"* (Banque mondiale, 1999: 1), *"in a state of abject poverty"* (in French: *"ils* végètent *dans la pauvreté la plus extrème")* (PNUD, 1994: 2). AG21 as well as many other UN documents constantly refer to the rural poor as being people who still *"depend"* on natural resources for their livelihood. Their lack of autonomy makes them particularly vulnerable to "shocks". The solution to their poverty, then, is an empowerment process that breaks the direct link between man and nature, in order *"to bring people into society who have never been part of it before"* (Wolfensohn, 1997).

At the same time however, and contrary to early development thinking, nature has entered economic theory. Development economics was invented as a reaction to neoclassical *"mono-economics"* that proved to be irrelevant in poor countries (Hirschman, 1984). Since Adam Smith's invisible hand did not do its work, specific economic theories were elaborated to grasp the realities and needs of poor countries. Today, mono-economics are back in favour, while definitions of development do not even mention economy anymore. For the UN *"development is a multidimensional undertaking to achieve a higher quality of life for all people"* (United Nations, 1997a) and for the World Bank economic development is defined as *"a sustainable increase in living standards that encompass material consumption, education, health and environment protection"* (World Bank, 1991: 34). Development economics has been condemned as one of the many *"errors of the past"*. With the idea of one global market and economic growth to be created by the expansion of world trade, there is no longer a need for differentiated policies. Instead of economic development, we now have *"globalisation"* and *"poverty reduction"*, both supposed to foster growth. Definitions of development no longer need economics any more, since the economy is now associated with nature, it has become an external reality that we need in order to survive, but we have learnt that we should not try to dominate it. The laws of economy are like the laws of nature, we must respect them but we cannot change them (Boutros Boutros-Ghali, 1995: 41). In the same vein, we should not try to change the world, we can only try to understand it by careful observation. This is why the so-called "interdependent" problems of today's *"global world"* are listed together as if they were all of the same nature: i.e. climate change, price fluctuations, terrorism, macro-economic shocks, epidemics, etc. Poor countries are vulnerable to the *"vagaries of global markets"*, they are hit by *"turmoil"*, *"the winds"* and *"violent hurricanes"* of globalisation. (PNUD, 1999: 2-4; PNUD, 1997: 10). The metaphors used in the globalisation discourse reveal its ideology and the belief in an immutable natural order of *"market forces"*. This is the *"drama of development"* (World Bank, 2000a: 17), since planning is now believed to be futile. Within the World Bank's renewed institutional approach, the collapse of Enron is of exactly the same order as the collapse of the cod fisheries in Newfoundland (World Bank, 2003: xiv, 50). Everything should be done to create an *"enabling environment"* allowing for the perfect functioning of markets for the benefit of all.

3.2.2. Poverty reduction and risk management

A similar line of thinking has been applied to social policies. Poverty is said to be the consequence of the same wrong ideas of the past. By not allowing markets to play their natural balancing role, poor people – and especially women – never got the right incentives and they were not allowed to contribute to and benefit from development.

In the past, social development was understood as planned social change, a process that paralleled the structural changes of the economy. Consequently, social protection was aimed at improving the living standards of the population of poor countries, by protecting them against free markets. The model to follow was the social citizenship of rich countries based on equal rights and the decommodification of certain goods, like education and health care (Midgley, 1995; Marshall, 1964). Social citizenship goes far beyond poverty reduction in that it concerns the whole population and is based on a theory of social change as a result of deliberate human actions. Thus it gives governments a central role in providing social services and in correcting or inhibiting free market forces. With poverty reduction as it is conceptualised today, the idea of social development has been eroded and has lost its transformative purpose. If free market forces have to be respected – in the same way as the laws of nature – then income guarantees and protective measures are to be banned. They may benefit some groups but they are now said to be detrimental to the poor, either because they do not reach them, or because they distort the markets. *"Market-inhibiting policies... may result in higher costs for the poor"* (World Bank, 1993: 35). Social security systems are now said to be inadequate for reducing poverty (PNUD, 1991: 55, PNUD, 2000: 42-44).

The rationale for the new social protection policies – mainly promoted by the World Bank – is based on the notion of risk. If the laws of nature and of economics cannot be changed, human societies will always remain vulnerable, though they do have the possibility - and the obligation – to protect themselves. Social protection, then, becomes an element of risk management, whether these risks are directly related to natural phenomena or to human action. Risk management is necessary in order to cope with earth quakes, the volatility of financial markets, unemployment or illness. These risks are the same for all people, though the poor are the most vulnerable. Therefore, governments have three options for their social protection policies. Ideally, risk prevention would be desirable, but *"we know"* that this is not possible at a reasonable cost or without harming growth. All governments can do is to have sound macro-economic policies and to create enabling environments. Risk mitigation is the second option and aims at alleviating the negative consequences of possible future *"shocks"*. It can imply the broadening of people's assets by enhancing their human and social capital. The third option is a set of *"coping mechanisms"*, once risk has materialized (Holzmann & Jørgensen, 2000).

The notion of risk is thus at the centre of the changing paradigm for social policies. At the end of the 19th century, it was precisely because risks and accidents had ceased to be linked to individual responsibility or to an "*Act of God*", that social insurances were created. Because of the mass poverty in the emerging industrialized societies, risks were socialized and this gave rise to the social security systems that greatly contributed to preventing the risk of income loss and poverty (Ewald, 1986). Today, risks are again accepted as natural phenomena and even have to be protected, according to the World Bank, as a factor of production. Social citizenship was based on a set of values – justice, equality, solidarity – that are meaningless in nature (Charrier, 1998). Yet the poverty reduction strategies that are now being promoted place the responsibility of income generation again upon the poor themselves, which helps to explain why the income dimension is absent from the poverty reduction strategies. Governments have to create the environment that allows markets to function and the poor to take the opportunities thus offered to them. Their basic insurance mechanism becomes the family and the local community, as the main source of social capital (World Bank, 2000a: 18). This also explains why, contrary to old development thinking, cultural traditions with their informal community-based solidarity mechanisms are no longer seen as barriers to development. While social security and social citizenship is based on a Durkheimian notion of "organic solidarity", today's poverty reduction is closer to "mechanic solidarity". Only when the social capital is eroded to the point of threatening social stability, must governments take the necessary steps to help restore it, thus preserving the sustainability of the growth process.

4. The way forward

"*Have we really come to grips with the implementation of sustainable development, do we really know what it means in operational terms?*" (Desai, 2001). The analysis of the poverty chapter of the JPI and its links to the overall sustainability agenda makes me conclude that the answer to Mr. Desai's question remains negative. The Johannesburg summit did confirm the importance of poverty eradication but it did not shed light on the linkages with the environment agenda, nor did it strengthen the social dimension of sustainable development. This does not mean that the Johannesburg process leads to a stalemate.

The most important achievement of the Johannesburg process could be its contribution to the generalized awareness that current development models are unsustainable and therefore have to be changed. Furthermore, the documents that have been approved and the contributions of different stakeholders contain useful ideas and concepts that allow for furthering and expanding the sustainability agenda. Building on the existing consensus and using the various old and new development achievements, I would like to suggest three lines along which a positive forward-looking strategy could be developed for implementing JPI while enhancing social development.

4.1. CLARIFYING THE CONCEPTUAL FRAMEWORK

Both components of *"sustainable development"* remain subject to divergent interpretations and numerous misunderstandings. *"Development"* has always been controversial and has been criticized for its ethnocentric and modernist bias. Nevertheless, the demands of poor countries still reflect the analysis made in the 60's and 70's and continue to stress the need for more equitable economic relations. The demands of the G-77 are not limited to international trade but encompass the many dimensions development has acquired during the past decades (G-77, 2000). Therefore, the Agenda for Development approved by the General Assembly of the UN (United Nations, 1997a) could serve as the main reference in which the Doha Agenda (WTO, 2001) can be integrated. Creating conceptual clarity on the different dimensions of political, economic, social and cultural development, as well as the cross-cutting themes of gender equality and environmental protection, could help to establish a framework for a meaningful contextualization of poverty reduction policies.

"Sustainability" has been welcomed into development theory as an opportunity for "greening" the agenda. However, as the concept has been applied to many other areas, such as growth, finance and social policies, its exclusive link to environmental protection has been eroded. In order to avoid it becoming a "floating signifier", *"sustainable development"* could usefully be seen as an overarching concept, encompassing the parallel processes of political, economic, social and cultural development. The UN Declaration on the *"Right to Development"* (United Nations, 1986) refers to nations, peoples and individuals as objects of development. With this reference in mind, sustainable development seems to have a greater potential for becoming a comprehensive concept than "human development". It should be noted however that the different dimensions of "sustainable development" cannot be seen as mere inputs, let alone conditionalities of development. Rather, they are parallel, interdependent and mutually reinforcing processes aimed at raising the living standards and the well-being of all people. These objectives have been laid down in the statutes of all major multilateral organisations set up after the second world war. In this context, "social sustainability" can be freed from its merely negative approach of avoiding conflict. In a more positive perspective, it would refer not only to poverty reduction, but also to a broader agenda of collective and individual empowerment and of human security as defined by the UNDP (PNUD, 1994). Conceptual clarity is not meant to give development a purely normative content. Rather, it should offer an analytical framework within which the debate can take place and choices can be made.

4.2. EXPLORING THE LINKS

In order to avoid the ideological pitfalls that have paralysed development thinking of the past decades, a broad and interdisciplinary research agenda could help to embark on exploring the empirical linkages that make policies sustainable and mutually reinforcing. By analysing specific dimensions of development practice, various UN organisations come up with pragmatic proposals for more coherent policies. The *"decent work"* agenda of the ILO, the *"Changing production patterns with social equity"* of CEPAL, the *"visible hand"* approach of UNRISD and the light shed by UNCTAD on the *"global poverty trap"* are the result of comprehensive research on the outcomes of current policies. In reference to issues of environmental and social sustainability, more knowledge is needed on the links between poverty and wealth on the one hand, and environmental degradation on the other hand. The UNDP proposal for establishing externality profiles and operationalising the concept of global public goods deserve careful research (Kaul et al., 1999). Within the context of social development, an investigation of the limits of social justice as an objective value may also be worthwhile. Finally, broadening the knowledge-base of development also allows for strengthening the scientific cooperation between the North and the South, for sharing knowledge and examining the interlocking of global and local knowledge, as well as for a more objective approach to the *"common but differentiated responsibilities"*.

4.3. AIMING AT POLICY COHERENCE

Within a comprehensive analytical framework and with better insight into the interrelatedness of the different dimensions of sustainable development, it should also be possible to find mechanisms for ensuring policy coherence. The way JPI has been worded does not in any way preclude conflicting policies that neither protect the environment nor alleviate poverty. It can lead to precautionary policies as well as to the commodification of nature. Trade-related externalities and the privatisation of basic social services are but two obvious examples that need careful monitoring. If placed within the broader context of a comprehensive sustainable development strategy, the PRSP-process could be a useful tool for monitoring policy coherence. Similar tools with appropriate indicators could be developed for rich countries. In this particular context, it would be difficult to avoid reopening the debate on economic growth. If growth has to be sustainable, than Herman Daly's concept of a *"throughput"* economy – differentiating between economic and non-economic growth – will have to be re-examined.

5. Conclusion

The Johannesburg process did not strengthen the social dimension of sustainable development. Today's poverty reduction strategy allows for an emphasis on the need for more growth, in the absence of redistributive policies. Moreover, the "social sustainability" approach paves the way for a further shift to risk management and conflict prevention and away from a rights-based social protection policy. The PRS is very different from the former approach of social citizenship, where risks were socialized. In view of the fact that the Millennium

Development Goals will not be attained within the established timeframe, other policy options should be considered. However, this implies re-thinking development and growth again. "Sustainable development" is badly in need of conceptual clarity and comprehensive research on the empirical links between its various dimensions. The Johannesburg Political Declaration and the JPI do offer useful concepts and references for enhancing this agenda. The priorities agreed upon at the major UN conferences of the 90's, together with the findings of recent research of various UN organisations can be useful inputs into this process. Thus, ideally, a project for sustainable development should make poverty reduction strategies redundant.

References

Annan, Kofi A. (2000) *"We, the Peoples"*. *The Role of the United Nations in the 21st Century*, United Nations, New York.

Banque Mondiale (1992) *Rapport sur le développement dans le monde 1992. Le développement et l'environnement*, Banque Mondiale, Washington.

Banque Mondiale (1999) *Rapport sur le développement dans le monde 1998-1999. Le Savoir au service du développement*, Editions ESKA, Paris.

Bojö, J., Chandra Reddy, R. (2001a) *Poverty Reduction Strategies and Environment. A Review of 40 Interim and Full PRSPs*, The World Bank, Washington.

Bojö, J., Bucknall, J., Hamilton, K., Kishor, N., Kraus, C., Pillai, P. (2001b) Environment, in World Bank, *Poverty Reduction Strategy Sourcebook* (www.worldbank.org/poverty).

Boutros Boutros-Ghali (1995) *An Agenda for Development*, New York, United Nations.

Charrier, B. (1998) Reflections on the Day's Discourse: Reaching for Utopia, in Serageldin, I. and Martin-Brown, J. (eds) *Ethics and Values. A Global Perspective*, The World Bank, Washington.

Commissie van de Europese Gemeenschappen (2001) *Mededeling van de Commissie aan de Raad en het Europees Parlement. Tien jaar na Rio: Voorbereiding op de Wereldtop over Duurzame Ontwikkeling in 2002*. COM (2001) 53 definitief, Brussel, 6.2.2001.

Daly, H.E. (2002) Sustainable Development: Definitions, Principles, Policies. Invited address, World Bank, April 30, Washington, DC (www.worldbank.org).

Desai, N. (2001) *Statement to the Preparatory Committee for the World Summit on Sustainable Development (New York, 30 April 2001)* (www.johannesburgsummit.org/html/documents/).

DFID, EC, UNDP, The World Bank (2002) *Linking Poverty Reduction and Environmental Management. Policy Challenges and Opportunities*, The World Bank, Washington.

Ewald, F. (1986) *L'Etat-providence*, Grasset, Paris.

G-77 (Group of 77 South Summit) (2000) *Declaration of the South Summit.Havana, Cuba, 10-14 April* (www.g77.org/summit/Declaration_G77Summit.htm).

Hirschman, A.O. (1984) A Dissenter's Confession: "The Strategy of Economic Development" Revisited, in Meier, G.M. and Seers, D. (eds) *Pioneers in Development*, published for the World Bank by Oxford University Press, Oxford.

Holzmann, R., Jørgensen, S. (2000) *Gestion du risque social: cadre théorique de la protection sociale*. Document de travail n° 0006 sur la protection sociale (http://www1.worldbank.org/sp).

Kaul, I., Grunberg, I., Stern, M.A. (eds) (1999) *Global Public Goods. International Cooperation in the 21st Century*, Published for UNDP by Oxford University Press, Oxford.

Marshall, T.H. (1964) *Class, Citizenship and Social Development*, Doubleday & Company, Inc., New York.

Mayor, F. avec la collaboration de Jérôme Bindé (1999) *Un monde nouveau*, UNESCO, Ed. Odile Jacob, Paris.

Meadows, D., Meadows, D., Randers, J., Behrens, W. (1972) *Rapport van de Club van Rome. De grenzen aan de groei,* Het Spectrum, Antwerpen.

Mestrum, F. (2002a) *Globalisering en armoede. Over het nut van armoede in de nieuwe wereldorde,* EPO, Berchem.

Mestrum, F. (2002b) Van maakbaarheid naar natuurlijke orde. De politieke economie van de internationale armoedebestrijding, *Monografieën over interculturaliteit 8,* pp.29-41.

Midgley, J. (1995) *Social Development. The Development Perspective in Social Welfare,* Sage Publications, London.

Moore, D.B. (1995) Development Discourse as Hegemony: Towards an Ideological History – 1945-1995, in Moore, D.B. and Schmitz, G.J. (eds) *Debating Development Discourse. Institutional and Popular Perspectives,* Macmillan Press Limited, London.

Nations Unies (1951) *Mesures à prendre pour le développement économique des pays insuffisamment développés.* Doc. E/1986 ST/ECA/10, 3/5/1951.

Nations Unies (1990) *Stratégie internationale du développement pour la quatrième Décennie des Nations Unies pour le développement.* Res. GA 45/199 21/9/1990.

Nations Unies (1992) *Rapport de la Conférence des Nations Unies sur l'environnement et le développement, Rio de Janeiro, 3-14 juin 1992.* Doc. A/CONF.151/26/Rev. 1(Vol. I).

Nations Unies (1995) *Rapport du Sommet mondial pour le développement social, Copenhague, 6-12 mars 1995.* Doc. A/CONF.166/9.

Øyen, E. (1996) Poverty Research Rethought, in Øyen, E., Miller, S.M., Samad, S.A. (eds.) *Poverty. A Global Review. Handbook on International Poverty Research,* Oslo, Scandinavian University Press.

PNUD (1990) *Rapport mondial sur le développement humain 1990,* Economica, Paris.

PNUD (1991) *Rapport mondial sur le développement humain 1991,* Economica, Paris.

PNUD (1994) *Rapport mondial sur le développement humain 1994,* Economica, Paris.

PNUD (1997) *Rapport mondial sur le développement humain 1997,* Economica, Paris.

PNUD (1998) *Rapport mondial sur le développement humain 1998,* Economica, Paris.

PNUD (1999) *Rapport mondial sur le développement humain 1999,* De Boeck Université, Bruxelles.

PNUD (2000) *Vaincre la pauvreté humaine. Rapport du PNUD sur la pauvreté 2000,* PNUD, New York.

PNUD (2001) *Rapport mondial sur le développement humain 2001. Mettre les nouvelles technologies au service du développement humain,* De Boeck Université, Bruxelles.

Reddy, S.G. and Pogge, T.W., (2002) How *not* to count the poor, (www.socialanalysis.org).

Sen, A.K., (2000) The Ends and Means of Sustainability, Key Note Address at the International Conference on "Transition to sustainability" of the Inter Academy Panel on International Issues, Tokyo.

Serageldin, I. and Grootaert, C. (2000) Defining Social Capital: An Integrating View, in Dasgupta, P. and Serageldin, I. (eds), *Social Capital. A Multifaceted Perspective,* The World Bank, Washington.

Tabatabai, H. (1996) *Statistics on Poverty and Income Distribution. An ILO Compendium of data,* ILO, Geneva.

UNCTAD (2002) *The Least Developed Countries Report 2002. Escaping the Poverty Trap,* New York and Geneva, UNCTAD.

UNEP (2002) *Global Environment Outlook 3. Past, present and future perspectives,* UNEP, Nairobi.

United Nations (1986) *Declaration on the Right to Development.* Res. GA 41/128, 4 December 1986.

United Nations (1997a) *Report of the Open-ended Working Group of the General Assembly on an Agenda for Development* (Doc. A/AC250/1). Res. GA 51/240.

United Nations (1997b) *Programme for the Further Implementation of Agenda 21,* Adopted by the General Assembly at its nineteenth special session (23-28 June 1997). Doc. A/RES/S-19/2, 19 September 1997.

United Nations (2000a) *Proposals for further initiatives for social development.* Twenty-fourth special session of the General Assembly entitled "World Summit for Social Development and beyond: achieving social development for all in a globalizing world", 1 July 2000 (www.un.org).

United Nations (2000b) *United Nations Millennium Declaration.* Res. GA 55/2.

United Nations (2001a) *Report of the World Social Situation 2001,* United Nations, New York.

United Nations (2001b) *Combating Poverty.* Economic and Social Council. Report of the Secretary-General. Doc. E/CN.17/2001/PC/5 (www.johannesburgsummit.org/html/documents/prepcom1.html).

United Nations (2002a) *Proposal to establish a World solidarity fund for poverty eradication. Report of the Secretary-General.* Advance unedited copy, July 2002 (www.un.org/esa/socdev/poverty/).

United Nations (2002b) *Report of the Commission on Sustainable Development acting as the preparatory committee for the World Summit on Sustainable Development.* Second Session (28 January – 8 February 2002), Annex II. Doc. A/CONF.199/PC/2.

United Nations (2002c) *Report of the International Conference on Financing for Development. Monterrey, Mexico, 18-22 March 2002.* Doc. A/CONF.198/11.

United Nations (2002d) *Global Challenge. Global Opportunity. Trends in Sustainable Development,* United Nations, New York.

United Nations (2002e) *Draft Political Declaration. The Johannesburg Commitment on Sustainable Development.* Doc. A/CONF.199/L.6, 1 September 2002.

United Nations (2002f) *Draft Plan of Implementation for the World Summit on Sustainable Development.* Advance Unedited Text, 12 June 2002.

United Nations (2002g) *The Johannesburg Declaration on Sustainable Development. From our Origins to the Future.* Report on the World Summit on Sustainable Development, Johannesburg 26 August – 4 September 2002, Doc. A/CONF.199.

United Nations (2002h) *Plan of Implementation of the World Summit on Sustainable Development.* Report on the World Summit on Sustainable Development, Doc. A/CONF.199.

Wolfe, M., (1981) *Elusive Development,* UNRISD, Geneva.

Wolfensohn, J. (1997) *The Challenge of Inclusion.* Annual Meeting Address, Hong Kong SAR, China (www.worldbank.org/html/extdr/extme/jwams972).

World Bank (1990) *World Development Report 1990. Poverty,* The World Bank, Washington.

World Bank (1991) *World Development Report 1991,* The World Bank, Washington.

World Bank (1993) *Poverty Reduction Handbook,* The World Bank, Washington.

World Bank (2000a) *World Development Report 1999/2000. Entering the 21st Century,* The World Bank, Washington.

World Bank (2000b) *Social Protection Sector Strategy: From Safety Net to Springboard,* The World Bank, Washington.

World Bank (2001a) *World Development Report 2000/2001. Attacking Poverty,* Published for the World Bank by Oxford University Press, Washington, New York.

World Bank (2001b, 4th ed.) *Poverty Trends and Voices of the Poor,* The World Bank, Washington (www.worldbank.org/poverty/mission/up1.htm).

World Bank (2002a) *Making Sustainable commitments: An Environment Strategy for the World Bank,* The World Bank, Washington.

World Bank (2002b) *Issues Paper for a World Bank Social Development Strategy,* May 2002 (www.worldbank.org/socialdevelopment).

World Bank (2002c) *The Role and Effectiveness of Development Assistance. Lessons from World Bank Experience* (http://econ.worldbank.org).

World Bank (2003) *World Development Report 2002. Sustainable Development in a Dynamic World. Transforming Institutions, Growth, and Quality of Life.* A co-publication of The World Bank and Oxford University Press, Washington, New York.

World Commission on Environment and Development (1987) *Our Common Future,* Oxford University Press, Oxford.

WTO (2001) Ministerial Declaration. Adopted on 14 November 2001. Doc. WT/MIN(01)/DEC/1 (www.wto.org).

CHAPTER 3

PRODUCTION, CONSUMPTION AND THE
WORLD SUMMIT ON SUSTAINABLE DEVELOPMENT

JEFFREY BARBER

Integrative Strategies Forum, Rockville, MD, USA
(E-mail: jbarber@isforum.org; fax:+1 301 770 6377; tel.: +1 301 770 6375)

Abstract. At the World Summit on Sustainable Development, world leaders agreed that eliminating unsustainable production and consumption is one of the three overriding objectives of sustainable development. Achieving that objective should have been a major priority for the WSSD Plan of Implementation. Increases in consumption and production over the past decade were largely responsible for the worsening environmental and social trends. Unfortunately, the negotiators of the Plan paid insufficient attention to the lessons from ten years of discussions about the concepts, the available policies and tools and their effectiveness, the impacts of those policies on developing countries, and the political commitment of countries in an era of globalisation. Despite a promising proposal for a new ten-year work programme aimed at bridging the gap implementing the Agenda 21 commitments from Rio, Summit negotiators produced barely more than a muted echo of recommendations from the past which have yet to be taken seriously enough by the world's leaders in a comprehensive intergovernmental strategy. In the ten-year review of progress to achieve sustainable production and consumption, governments quickly skipped past the critical work of examining *why* things are getting worse, avoiding the task of identifying the obstacles (which in some cases were themselves) and in turn avoiding the commitment to time-bound measurable targets. If nothing else, the World Summit on Sustainable Development demonstrated that a global strategy to achieve sustainable production and consumption will come not from a UN consensus of world leaders but from a strategic alliance of responsible governments, civil society, and others with a vision beyond the next election cycle.

Key words: Sustainable development, sustainable production, sustainable consumption, World Summit on Sustainable Development, Agenda 21, Earth Summit

Abbreviations: CSD – Commission on Sustainable Development; CSE – Centre for Science and Environment; FOE – Friends of the Earth; ICSPAC – International Coalition for Sustainable Production and Consumption; OECD – Organisation for Economic Cooperation and Development; Rio+5 – five-year review of progress on Agenda 21; Rio+10 – ten-year review of progress on Agenda 21; SPAC – sustainable production and consumption; UNGA – United Nations General Assembly; UNDP – United Nations Development Programme; UNEP – United Nations Environment Programme; WSSD – World Summit on Sustainable Development.

1. Introduction

During the past decade, sustainable production and consumption became an increasingly important category of international development policy, referred to by government and other policymaking bodies as "a key strategic approach to achieving sustainable development" (UNCSD, 1997a). "All countries should strive to promote sustainable consumption patterns", the UN General Assembly

57

L. Hens and B. Nath (eds.), The World Summit on Sustainable Development, 57–89.
© 2005 *Springer. Printed in the Netherlands.*

concluded at its 1996 Special Session review of progress since Rio, distinguishing between the responsibility of developed countries to "take the lead" and that of developed countries to "seek sustainable consumption patterns in their development process" (UNGA, 1997). The UN Development Programme (UNDP) acknowledged that "consumption patterns today must be changed to advance human development tomorrow" (UNDP, 1998: 1). "The key environmental challenge for the future", the OECD explained, "will be to continue to further increase efficiency of resource use and to reduce the pollution intensity of consumption and production" (OECD, 2002a: 27). Finally, world leaders attending the World Summit on Sustainable Development referred to changing unsustainable production and consumption patterns as one of the three "overarching objectives of, and essential requirements for, sustainable development" (WSSD, 2002).

Despite this official recognition and improvements in eco-efficiency and consumer awareness, overall efforts to reverse the growth of unsustainable production and consumption patterns have been inadequate. The imbalance between rhetoric and effective action represents one of the critical "implementation gaps" noted in the ten-year review of progress conducted as part of the preparations for the World Summit on Sustainable Development.

The World Summit on Sustainable Development played a two-fold role in progress towards this "overarching objective" of sustainable production and consumption: (1) encouraging a review and critical assessment of progress in addressing unsustainable production and consumption patterns since Rio, and (2) calling for political commitments to move from rhetoric to effective action and implementation of commitments.

Of interest are questions about the results of that review and what the negotiators of the final Plan of Implementation did with the lessons from a decade of efforts to address this issue. A critical question is how the Summit process identified and addressed the constraints and obstacles that contributed to the implementation gap. Another set of questions focus on the role played by different players in the process, e.g., governments, intergovernmental organisations, industry groups, and civil society, what they contributed in shaping the policies and practices of past and future progress.

Finally, there is the question of whether the current commitments and plans now in motion, as reflected in the Plan of Implementation and its associated initiatives and partnerships, are adequate to the task of significantly slowing and reversing the trend of growing unsustainable production and consumption, or remain curtailed by the same obstacles and taboos undermining past progress.

2. Production and consumption at Rio

2.1. A MATTER OF GRAVE CONCERN

As one of the outcomes of the 1992 Earth Summit, Principle 8 of the Rio

Declaration on Environment and Development highlights the responsibility of nations to "reduce and eliminate unsustainable patterns of production and consumption". The principle clearly points out that this task is necessary "to achieve sustainable development and a higher quality of life for all people".

Chapter 4 of Agenda 21 explicitly identifies unsustainable production and consumption patterns, "particularly in industrialised countries", as "the major cause of the continued deterioration of the global environment" (UN, 1992: para 4.3). The situation is described as "a matter of grave concern, aggravating poverty and imbalances".

In responding to this concern, the chapter identifies two broad tasks:
(1) focusing on unsustainable patterns of production and consumption, and
(2) developing national policies and strategies to encourage changes in unsustainable consumption patterns.

The first task involves (a) promoting patterns of consumption and production that "reduce environmental stress and will meet the basic needs of humanity", and (b) "developing a better understanding of the role of consumption and how to bring about more sustainable consumption patterns". The complexity of the challenge requires both a "questioning of traditional concepts of economic growth" and "new concepts of wealth and prosperity" reflected in "changed lifestyles" as well as new indicators and systems of national accounts.

The second task places responsibility squarely on national governments to create the policies and strategies needed to encourage those changes. Governments in developed industrialised countries challenged themselves to design policies and strategies to encourage the "reorientation of existing production and consumption patterns that have developed in industrialised societies". This involves the following objectives:

(a) promoting "efficiency in production processes and reducing wasteful consumption in the process of economic growth, taking into account the development needs of developing countries";
(b) developing "a domestic policy framework that will encourage a shift to more sustainable patterns of production and consumption";
(c) reinforcing "both values that encourage sustainable production and consumption patterns and policies that encourage the transfer of environmentally sound technologies to developing countries".

Among the activities recommended for achieving these objectives, Agenda 21 encourages: use of new and renewable energy sources (4.18d); recycling by industry and consumers (4.19b); reducing wasteful product packaging (4.19b); expanding environmental labelling and other environmentally related product information programmes designed to assist consumers to make informed choices (4.21); providing information on the consequences of consumption choices and behaviour (4.22a); making consumers aware of the health and environmental impacts of products, through such means as consumer legislation and environmental labelling (4.22b); reviewing and improving government procurement policies (4.23); using appropriate economic instruments to influence consumer behaviour (4.25); promoting more positive attitudes towards sustainable consumption through education, public awareness programmes and other means (4.26).

2.2. COMMON BUT DIFFERENTIATED RESPONSIBILITIES

During the Earth Summit, discussions about population growth, particularly those concerned with the rates of growth occurring in developing countries, were frequently paired with critiques of the unequal environmental pressures resulting from over-consumption by industrial countries and consumers. The argument that "one fourth of the globe's people consume 40–86 percent of the earth's various natural resources" (Durning, 1992: 50) became one of the more frequently heard points in discussions about production and consumption. Thus, the Agenda 21 objective of eradicating poverty became intertwined with the objective of achieving sustainable consumption and production, tying the question of achieving eco-efficiency to the more complicated question of economic *sufficiency*.

The principle of common but differentiated responsibilities, as defined in Principle 7 of the Rio Declaration, focuses specific attention on the imbalances in global patterns of consumption and production and the need for governments to specify their role and responsibility in establishing the proper balance. Chapter 4 then points out the need for the developed countries to "take the lead" and developing countries to include sustainable consumption in their development process. The chapter further highlights the need for policies and strategies to address unsustainability ranging from the excessive demands and lifestyles of the rich to the lack of access to food, clean water, healthcare, shelter and education by the poor. Understandably the difficulties in developing those policies and strategies are related to the difficulties in defining and agreeing upon the differentiated responsibilities.

2.3. DOMESTIC POLICY FRAMEWORKS

Agenda 21 asks each country to "develop a domestic policy framework that will encourage a shift to more sustainable patterns of production and consumption" (UN, 1992: para 4.17). This means more than calling governments' promotion of recycling or energy conservation education "sustainable consumption policy". A *national policy framework* on sustainable production and consumption implies an understanding and appreciation of the linkages involved in balancing demand- and supply-side approaches. Such a framework requires integrating various concepts such as product lifecycle, environmental space, ecological footprints, and environmental cost internalisation into a concrete inter-departmental plan with measurable targets and timetables and indicators to monitor and report on progress.

2.4. REVIEWING PROGRESS

Agenda 21 specifies a number of times the importance of monitoring and assessment in the follow-up of the implementation of Agenda 21 (UN, 1992). The authors stress that high priority be given to review "progress in achieving sustainable consumption patterns" (para 4.9) as well as "the role and impact of unsustainable production and consumption patterns and lifestyles and their relation to sustainable development" (para 4.13). Further, "due consideration" should be given to "an assessment of progress achieved in developing national policies and strategies" (para 4.26) in the overall review of the implementation of Agenda 21 (UN, 1992).

3. Progress since Rio

3.1. INTERNATIONAL PROGRAMME OF WORK

Following Rio, a number of international conferences, workshops, reports, education programs, and other activities were organised to implement the various recommendations made in Agenda 21 (ICSPAC, 2002; UNEP, 2001). In addition to ongoing discussion of production and consumption patterns by the UN Commission on Sustainable Development (CSD), many governments, intergovernmental organisations, and nongovernmental organisations engaged in numerous activities exploring questions about awareness, lifestyles, values, policies, and strategies. (See Table 1).

TABLE I. Follow-up activities on sustainable production & consumption.

1994	SORIA MORIA SYMPOSIUM (Oslo) – Organised by Norwegian Ministry of Environment
1995	OSLO MINISTERIAL ROUNDTABLE – Organised by Norwegian Ministry of Environment
1995	CLARIFYING THE CONCEPTS WORKSHOP (Rosendal) – Organised by OECD and Norwegian Ministry of Environment
1995	WORKSHOP ON POLICY MEASURES FOR CHANGING CONSUMPTION PATTERNS (Seoul) – Organised by Republic of Korea, in collaboration with Australia, UN Department for Policy Coordination and Sustainable Development (DPCSD), UNDP and OECD
1996	WORKSHOP ON SUSTAINABLE CONSUMPTION AND PRODUCTION: PATTERNS AND POLICIES (Brasilia) – Organised by Governments of Brazil and Norway
1998	WORKSHOP ON INDICATORS FOR SUSTAINABLE PRODUCTION & CONSUMPTION (New York) Organised by UNCSD
1998	ENCOURAGING LOCAL INITIATIVES TOWARDS SUSTAINABLE CONSUMPTION (Vienna) Organised by UNECE
1998	WORKSHOP ON CONSUMPTION IN A SUSTAINABLE WORLD (Kabelvag) – Organised by Norwegian Ministry of Environment
1999	FROM CONSUMER SOCIETY TO SUSTAINABLE SOCIETY: TOWARDS SUSTAINABLE PRODUCTION AND CONSUMPTION (Soesterberg) – Organised by ANPED, AKB and CRLE
1999	INTERNATIONAL EXPERTS MEETING ON SUSTAINABLE CONSUMPTION: TRENDS AND TRADITIONS IN EAST ASIA (Chejudo) – Organised by Governments of Norway and Sweden
1999	7TH SESSION OF UN COMMISSION ON SUSTAINABLE DEVELOPMENT, focus on changing production and consumption patterns (New York)
2001	WORKSHOP ON SUSTAINABLE CONSUMPTION IN LATIN AMERICA AND CARIBBEAN (Sao Paulo) – Organised by UNEP DTIE, Carl Duisberg Gesellschaft/BMZ (German Ministry for International Cooperation), Brazilian Ministry of Environment, UNESCO, Secretariat of Environment of State of Sao Paulo and its Environment Sanitation Agency (CETESB)
2002	WORKSHOP ON IMPLEMENTING SUSTAINABLE CONSUMPTION AND PRODUCTION POLICIES (Paris) – Organised by UNEP, CI, UN DESA and the governments of Denmark, Finland, Norway and Sweden

Source: ICSPAC, 2002.

Following the Earth Summit, the OECD began its exploration of the relationship between production and consumption patterns and sustainable development (OECD, 1997). The Norwegian Government hosted a series of workshops and meetings on the topic, notably the Symposium on Sustainable Consumption (UNCSD, 1994a) and the Oslo Ministerial Roundtable (Norwegian Ministry of Environment, 1995). The latter meeting established the most commonly accepted definition of *sustainable consumption* as: "the use of goods and services that respond to basic needs and bring a better quality of life, while minimising the use of natural resources, toxic materials and emissions of waste and pollutants over the life cycle, so as not to jeopardise the needs of future generations".

These and other activities contributed to the mandate by the second and third sessions of the UN Commission on Sustainable Development (UNCSD, 1994b, 1995a) to develop an international work programme on sustainable production and consumption. This was the first such global framework of programs for implementing Agenda 21 objectives. The agreed-upon programme of work (UNCSD, 1995b) consisted of five tasks:

1. Identifying the policy implications of projected trends in consumption and production patterns.
2. Assessing the impact on developing countries, especially the least developed countries and small island developing States, of changes in consumption and production in developed countries.
3. Evaluating the effectiveness of policy measures intended to change consumption and production patterns, such as command-and-control, economic and social instruments, government procurement policies and guidelines.
4. Eliciting time bound voluntary commitment from countries to make measurable progress on those sustainable development goals that have an especially high priority at the national level.
5. Revising the guidelines for consumer protection, with regard to sustainable consumption.

The first four tasks required collecting and compiling comprehensive information necessary to monitor and assess the stated trends, impacts, effectiveness of policies, and progress. Without this data, the CSD cautioned in its Third Session, "policy-making is likely to be impaired" (UNCSD, 1995a: para 39).

Some delegations and NGOs pushed, unsuccessfully, for commitments from countries to identify reasonable but specific time-bound and measurable targets, by which progress could be more easily assessed. Observing the Third Session discussions, *Earth Negotiations Bulletin* reported that "while governments have been more willing to discuss changing production and consumption patterns and the relationship between trade and the environment, there is little concrete action to report". These issues, they observed, "constitute the key indicators of sustained political will" (Earth Negotiations Bulletin, 1995).

At its Fourth Session, the CSD reported on two workshops organised to explore concepts and policy options: the OECD workshop "Clarifying the Concepts" in Rosendal hosted by the Norwegian government (OECD, 1997), and the workshop "Policy Measures for Changing Consumption Patterns" in Seoul hosted by the Republic of Korea (Republic of Korea, 1996). Reporting on the international work programme, the Commission admitted it was "mainly research oriented" and that the 1997 review of the implementation of Agenda 21 would "provide an opportunity for further directing the work programme towards a more action-oriented approach" (UNCSD, 1996).

3.2. RIO+5: MORE ACTION-ORIENTED?

Five years after Rio, the Commission on Sustainable Development reported on progress implementing Chapter 4 (UNCSD, 1997b). This was followed by a General Assembly report on its five-year work programme to further implement Agenda 21 (UNGA, 1997).

While the CSD report painted a positive picture of what had been achieved by governments and stakeholders in responding to the Chapter 4 objectives, evidence of substantive change remained abstract or anecdotal, underlining the need for clear indicators, targets and timelines, and concrete data to measure progress. "The most promising changes and developments can be observed", the CSD announced, "in the increased participation of non-governmental organisations, business, trade unions, local authorities and the academic community ... in particular, the ongoing efforts of the nongovernmental organisation and academic communities to promote sustainable lifestyles" (UNCSD, 1997b: para 24).

According to the report, progress was taking place in almost all the areas mentioned in Chapter 4 and in the international work programme – in understanding, awareness and policymaking by governments, industry, consumers and civil society organisations. Among governments, the Commission noted the leadership role taken by Australia, Brazil, the Netherlands, Norway and the Republic of Korea. Among international organisations, the Commission cited special efforts by the OECD, UNEP, UNCTAD, and UNDP. For business and industry, the World Business Council on Sustainable Development stood out in its efforts. Among nongovernmental organisations, the Commission singled out Consumers International, Friends of the Earth (FOE), and Global Action Plan (GAP) as examples of this increased participation and cooperation.

As to the many meetings and discussions taking place over the previous five years, the report claims "a consensus that the most promising and cost-effective policy strategies are those that aim at cost internalisation and improved efficiency in resource and energy use" (UNCSD, 1997b: para 6).

Yet at the end of this largely optimistic report, some serious "unfulfilled expectations" stand out (para 34). Most important is that the positive developments reported "have been largely offset by larger volumes of production". The result: "many natural resource and pollution problems persist or continue to worsen". One example is in the relentless rise of CO_2 emissions. As to integrated policy frameworks, governments tended to "shy away from additional ecotaxes and environmental regulations that intend to incorporate the cost of environmental protection into products and services" (para 37). In turn, "many governmental policies in sectors such as agriculture, economics, finance, trade, communications, tourism, energy and transport do not adequately reflect an appreciation of how they shape consumption and production patterns" (para 36).

In the 1997 General Assembly's Programme for the Further Implementation of Agenda 21, the section on changing consumption and production patterns is not so much a plan for "further implementation" as a restatement of many of the original recommendations and commitments in Agenda 21. Once again we hear that "all countries should strive to promote sustainable consumption patterns", that "developed countries should take the lead" and "developing countries should seek to achieve sustainable consumption patterns in their development process", and that this requires "enhanced technological and other assistance from industrialised countries" (UNGA, 1997). The same call heard in Agenda 21 for a "review of progress made in achieving sustainable consumption" returns, but without any recommendations on how this progress will be assessed. There is no mention of concrete, measurable targets and timetables for implementing these objectives, nor mention of appropriate indicators to assess countries' implementation efforts.

While a few policy recommendations from discussions since Rio do appear (e.g., producer responsibility, reduction and elimination of subsidies, consideration of a 10-fold long-term improvement in resource productivity and a factor-four increase in the next two or three decades), the majority of proposed "action-oriented policies" are simply echoes of Chapter 4 with no new ideas or commitments on the implied "further implementation".

In some cases important ideas from Rio are missing, such as the recommendations for countries to develop a domestic policy framework on sustainable production and consumption. Instead, the 1997 Programme mentions only that "the development and further elaboration of national policies and strategies ... are needed" (para 28). The concept of an integrative *framework* has disappeared.

Also missing from the 1997 implementation programme is the controversial but significant questioning of "present concepts of economic growth and the need for new concepts of wealth and prosperity" (UN, 1992: para 4.11). The concept of "carrying capacity" is also missing along with the Agenda 21 requirement for a "reorientation of existing production and consumption patterns that have developed in industrial societies and are in turn emulated in much of the world" (para 4.15).

Many of the civil society organisations that had been lobbying with like-minded government delegates for stronger language that would commit governments to concrete actions with targets and timetables were understandably disappointed albeit not surprised. Many also noted the absence of critical discussion and practical ideas about ways to overcome the obstacles to progress on sustainable production and consumption. Although the General Assembly admitted that "trends are worsening" and that absolute increases in consumption and production had overridden relative gains in eco-efficiency and lifestyle improvements, analysis of the obstacles was replaced by statements that "time is of the essence in meeting the challenges of sustainable development" and a renewed commitment "to ensuring that the next comprehensive review of Agenda 21 in the year 2002 demonstrates greater measurable progress in achieving sustainable development" (UNGA, 1997: para 6).

One significant precedent created by the General Assembly in its Programme for the Further Implementation of Agenda 21 was to establish production and consumption patterns and poverty as the "overriding issues" to be integrated into the future themes of the Commission's 1998–2002 work programme. Consequent CSD sessions were tasked with integrating the goal of changing production and consumption patterns in the context of agriculture, energy, transportation and other issues.

3.3. CONSUMER GUIDELINES ON SUSTAINABLE CONSUMPTION

In the International Programme of Work on Sustainable Production and Consumption, the task of revising the UN Guidelines on Consumer Protection, with regard to sustainable consumption, represents an important contribution to meeting the Agenda 21 objective of developing domestic policy frameworks. The evolution of these revised guidelines – their development, adoption, and use – offers useful insights in assessing governments' progress achieving the policy framework objective, not to mention the overall goal of promoting sustainable production and consumption patterns. The relationship between developed countries and the consumer guidelines is especially important, considering their obligation to "take the lead".

In 1985, the UN General Assembly adopted the United Nations Guidelines for Consumer Protection. The Guidelines "constitute a comprehensive policy framework outlining what governments need to do to promote consumer protection in the following eight areas: basic needs, safety, information, choice, representation, redress, consumer education and health environment" (UNCSD, 1998a). Designed for countries to use in structuring and strengthening policies and legislation for consumer protection, the Guidelines especially targeted the needs of governments of developing and newly independent countries.

A decade later, the UN Economic and Social Council requested the Secretary-General to extend the current Guidelines to the issue of sustainable consumption patterns. In January 1998, the Government of Brazil hosted the UN Inter-Regional Expert Group Meeting on Consumer Protection and Sustainable Consumption (UNCSD, 1998a), producing recommendations for revising the Guidelines discussed at the CSD's Sixth, Seventh and Eight Sessions and finally adopted by the General Assembly in December 1999.

In addition to specific policy recommendations on consumer information (e.g., on the impacts of consumption patterns and benefits of changes), education, eco-labelling, product testing, research on consumer behaviour, subsidy and tax reform, the Guidelines stressed the importance of government promotion of consumer empowerment and public participation in policy making, as well as the responsibility of developed country governments to support to support developing countries in promoting sustainable consumption and development. In particular, the revised Guidelines on sustainable consumption supported efforts to develop domestic policy frameworks on sustainable production and consumption shaped through informed partnership with all members of society.

In the report to the CSD's Sixth Session, the Experts Group co-chairs also stressed the need for "a review and revision mechanism for these guidelines ... under the aegis of the United Nations so as to assess progress in their implementation by Member States and to revise them as necessary" (UNCSD, 1998b).

3.4. RIO+10: THE IMPLEMENTATION GAP

In early 2001 at its 55th session, the UN General Assembly adopted the resolution mandating a ten-year review of progress achieved in the implementation of Agenda 21. The review process, led by the Commission on Sustainable Development and involving a wide-ranging series of national, regional and global preparatory meetings with inputs from governments, international organisations, business and industry, civil society organisations and other major groups, would culminate the following year in the World Summit on Sustainable Development. The Summit would integrate the "lessons learned" and recommendations for further action into "an integrated and strategically focused approach to the implementation of Agenda 21", addressing "the main challenges and opportunities faced by the international community in this regard" (UNGA, 2001).

In *Implementing Agenda 21*, the UN report summarising the overall results of the ten-year review of progress, the Secretary-General declared that, despite the various initiatives and achievements throughout the past decade, "progress towards the goals established at Rio has been slower than anticipated and in some respects conditions are worse than they were ten years ago". This was, he said, a "gap in implementation" (UNCSD, 2002: para 2). This gap is seen in the "fragmented approach towards sustainable development" taken by policymakers, in the "lack of mutually coherent policies" in finance, trade, investment, technology and sustainable development, and in the failure to provide the financial resources.

This gap is especially revealed in the lack of major changes in production and consumption patterns. For the Secretary-General, this situation reflects both the value systems driving the degradation of natural resources and the lack of political will to do what needs to be done. The developed countries were supposed to take the lead, but what gains they made were overridden by overall increases in consumption (UNCSD, 2002: para 83).

As to more specific Summit reports reviewing progress on changing production and consumption patterns, the Commission produced only a "brief factual overview" (UNCSD, 2001), which referred back to the more comprehensive review that took place at the CSD's Seventh Session in 1999 (UNCSD, 1999). Other organisations also contributed their assessments of progress on production and consumption (UNEP, 2002d; OECD 2002b, 2002c; ICSPAC, 2002). Klaus Töpfer summed up the situation in his Foreword to the UNEP report *Consumption Opportunities* (UNEP, 2001): "Since [Rio] progress on tracking consumption patterns, and devising the tools to change them, has been slow". With a view to the five goals of the International Programme of Work, we note the following:

3.4.1. Trends in production and consumption patterns

The most significant trend in the ten-year review is the relentless global increase of consumption and production, particularly of energy and natural resources. According to the 1998 *Human Development Report*, global consumption expenditure doubled in the past 25 years (Figure 1), reaching $24 trillion in 1998 (UNDP, 1998). This increase, linked with population growth and economic globalisation, elicits mixed responses of celebration and alarm. From one point of view such growth indicates increasing economic prosperity and wealth, ultimately providing the financial resources to pay for environmental protection and social services.

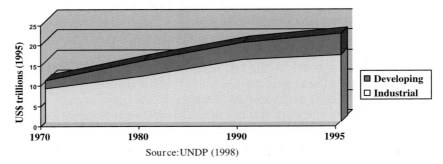

Source: UNDP (1998)

Figure 1. Global consumption expenditures in industrial and developing countries.

Another point of view laments the ecological degradation and social inequity accompanying blind economic growth. Although consumption has been increasing for the world as a whole, this is mostly concentrated among high income countries and population segments. While there is more wealth, there is also more poverty, and the gap between the two is growing (Figures 2 and 3). The report *Implementing Agenda 21* noted several areas of concern: the growing demand for water, especially in developing countries, with water use expected to increase 40 percent in the next two decades; the dramatic depletion of biological diversity, with more than 800 species already extinct due to habitat loss or degradation; a deforestation rate of 14.6 million hectares per year in tropical developing countries; destruction of coastal areas, with 27 percent of coral reefs lost due to human impacts and 32 percent threatened to follow in the next 30 years; and finally the increasing evidence of global warming, which may result in devastating changes in climate and weather, rising sea levels, and drought (UNCSD, 2002).

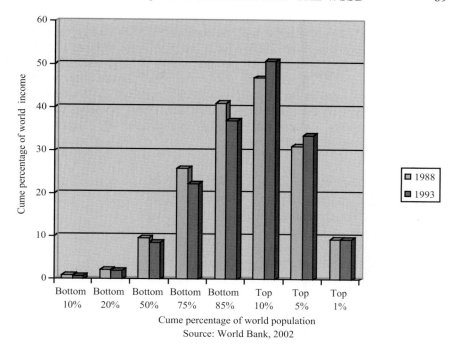

Source: World Bank, 2002

Figure 2. Gap between rich and poor.

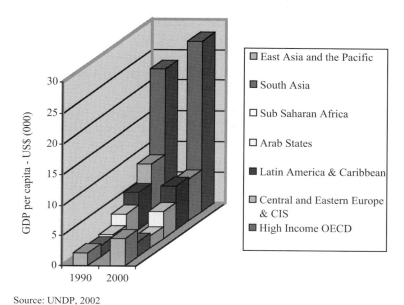

Source: UNDP, 2002

Figure 3. Regional income trends.

The 1999 CSD *Comprehensive Review of Changing Consumption and Production Patterns* already warned that "if current trends in energy and fossil fuel consumption continue, by 2010 global energy consumption and CO_2 emissions will have risen by almost 50 percent above 1993 levels". Currently, automobile use contributes 15 percent to global fossil fuel consumption and CO_2 emissions. This sector grows by 16 million vehicles per year, with one billion vehicles projected to be on the road by 2025 (UNCSD, 1999). Societies with high automobile consumption, in contrast to high population, are clearly more responsible for this rise in CO2 (Figure 4.)

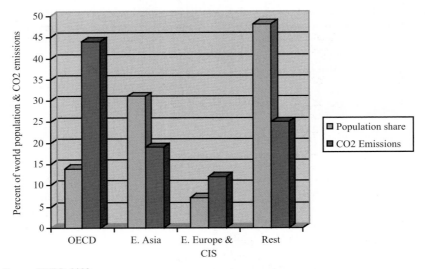

Source: UNDP, 2002

Figure 4. CO_2 emissions world-wide.

When developing policies about particular changing production and consumption patterns, the inter-linkages must be kept in mind (see Table 2). Fossil fuel use and policies are intertwined with the consumption and production of automobiles, which in turn is intertwined with the consumption and production of metals – the transportation sector using 70 percent of lead produced each year, 37 percent of steel, 33 percent of aluminium and 27 percent of copper (Worldwatch Institute, 2002: 66). The mining industry is in turn a consumer of chemicals, water, land – and motor vehicles.

TABLE II. Some global production & consumption trends.

	1970	1980	1990	2001
Cars produced (millions)	22.5	28.6	36.3	40.0
Carbon emissions from fossil fuel combustion (billion tons)	4.0	5.2	5.94	6.55
Metals mined (million tons)	621	757	820	902*
Oil spills (thousand tons)	399.9	577.9	474.4	48.6*
Global average temperature (degrees Celsius)	14.02	14.16	14.36	14.43

*Year 2002   Source: Worldwatch, 2002

The 2001 CSD report *Changing Consumption Patterns* notes that, due to agricultural expansion to meet the growing demand for food, "half of the world's wetlands area has been lost and grasslands have been reduced by more than 90 percent in some areas". Intensification of farming practices has resulted in two-thirds of the world's farmlands afflicted by soil degradation. To further meet this growing food demand, "nearly 70 percent of the world's major fish stocks are over-fished" or being fished at their biological limit. On the other hand, growing demand for fresh water, as well as electricity, has resulted in dam production and fragmentation of the world's large rivers, leaving 20 percent of freshwater species extinct or endangered (UNCSD, 2001).

3.4.2. Impacts on developing countries of changes in developed countries
Two main points stood out in reviewing the impacts of changes in industrial countries' production and consumption patterns on developing countries. First, developing countries express concern about the financial and technical burdens, especially on small and medium enterprises, from ecolabels and environment standards on their products, undermining their ability to compete on the global market. These countries fear that strategies involving product lifecycle analysis, Extended Producer Responsibility, and eco-efficiency will reduce demand for their products, especially fossil fuels, minerals and industrial raw materials. (UNCSD, 1999; UNCSD, 2001).

Second, the growing demand of consumers in industrial countries for more sustainable products, such as organic produce, open up new and growing opportunities for developing countries ready and willing to respond. The evolution of the Fair Trade movement represents one example of a niche market demand for developing country exports (UNCSD, 1999; UNCSD, 2001; Robins and Roberts, 1998).

The CSD report *Changing Consumption Patterns* raised additional points about the impact of globalisation on developing countries, particularly its capacity to "spread unsustainable lifestyles", promote increased consumption of natural resources and generation of waste, and whether global competition is undermining social and environmental policies (UNCSD, 2001: para 13).

Throughout the review process, the CSD, UNEP and many governments generally tended to avoid in-depth public discussion about the negative impacts of global media and advertising by promoting consumerism in developing countries and countries in transition. Many civil society groups directly criticised the export of western consumerism through global media and advertising, aggressively promoting unsustainable consumption values and lifestyles to the detriment of more benign traditional values and practices (Chaudhuri, 2002; Consumers International, 1999). The Centre for Science and Environment in India, for example, highlighted major marketing and advertising campaigns by tobacco companies targeting developing countries – where cigarette consumption has increased by 50 percent (CSE, 2000).

However, the silence of the Commission and UN agencies about the negative impacts of advertising is contrasted by their apparent willingness to accept the advertising industry's assertion that there are no causal links between advertising and unsustainable consumption, particularly in developing countries. Such a link is "a misperception", concluded the UNEP report on the advertising industry as part of their industry survey for the WSSD (UNEP, 2002c). Instead of exploring this controversy in more depth, the tendency has been to simply avoid friction with industry, instead "encouraging the media, advertising and marketing sectors to help shape sustainable consumption patterns" (UNGA, 1997).

3.4.3. Effectiveness of policy measures

On the policy side, governments have experimented with a range of instruments: process and product standards, ecotaxes, subsidy reform, consumer information such as ecolabels, among others. Applying the idea of product lifecycle analysis, some governments have made efforts to link these instruments into *integrated product policies (IPP)* to address the different phases of design, production, consumption and disposal. Integrated product policies also represent an important element in the cleaner production work of UNEP (UNEP, 2002a).

One ongoing policy trend has been the consistent priority given throughout by governments and the UN agencies to the concept of *eco-efficiency*. In *Implementing Agenda 21*, the Secretary-General identifies one main priority addressing the production and consumption issue, calling for "major improvements in the efficiency of resource use ... in both developed and developing countries" (UNCSD, 2002: para 224). Coined by the World Business Council for Sustainable Development, its president Björn Stigson defines eco-efficiency as "the delivery of competitively priced goods and services that satisfy human needs and bring quality of life while progressively reducing the ecological impact and resource intensity throughout the life cycle to a level at least in line with the Earth's carrying capacity" (Stigson, 1999).

However, the Rio+10 review concluded that the gains provided by this supply-side focus on promoting eco-efficiency in production are offset by trends on the demand-side – population growth and the desire for more goods and services. "The international community has not yet fully come to grips with how to address the consumption side of sustainable development", UNEP highlighted in its briefing distributed at the WSSD. According to UNEP, a "lack of awareness and understanding has led to a relatively hesitant uptake of the required policies, sometimes based on misperceptions about the economic, social and cultural consequences" (UNEP, 2002b). Ironically, the World Business Council on Sustainable Development, in its 1996 report *Sustainable Production and Consumption: A Business Perspective*, had advocated that "the definition of eco-efficiency ... is quite similar to that of sustainable production and consumption" (WBCSD, 1996: 11). For many in the business community, eco-efficiency represents an uncontroversial way to improve competitiveness, popular with both business and consumers.

"The reality is that eco-efficiency policy has so far only led to a much more efficient but still disproportionately high use of natural resources in rich industrial countries", Friends of the Earth (FOE) pointed out in their report to the WSSD (FOE, 2002a). Targeting the OECD's working programme on sustainable development, Friends of the Earth criticises it for concerning itself "almost exclusively with how a reduction in natural resource use can be achieved through economic instruments". One of the reasons why the OECD and CSD have given less attention to sustainable consumption policy is the "huge risk" that "what is ecologically necessary will not be politically feasible". The message that "we must use far fewer natural resources and must pay more" is clearly an unwelcome message to public and politicians alike. Yet the problem and challenge remain of how to effectively balance sustainable production policies with politically feasible sustainable production policies that are not seen as an attack on people's living standards.

For years, civil society organisations have called into question the overriding importance given to eco-efficiency relative to other approaches and concepts, such as the equally important but neglected concept of *sufficiency*. In 1997, the NGO Caucus on Sustainable Production and Consumption highlighted this need and stressed "moving beyond efficiency to sufficiency, promoting sustainable lifestyles and livelihoods for all" as a priority for the UN General Assembly's five-year review (SPAC Caucus, 1997). Yet the term "sufficiency" received scant attention throughout the WSSD discussions and implementation plan.

3.4.4. National and local commitments

Agenda 21 clearly states that its "successful implementation is first and foremost the responsibility of Governments" (UN, 1992: para 1.3). Thus, a major part of the ten-year review understandably should focus on governments' efforts *as well as lack of effort* to implement Agenda 21. Considering that governments, following the five-year review, committed themselves "to ensuring that the next comprehensive review of Agenda 21 in the year 2002 demonstrates greater measurable progress in achieving sustainable development" (UNGA, 1997: para 6), there is a serious problem with credibility. Yet the rhetoric of the World Summit on Sustainable Development is unabashedly filled with commitments and recommitments, seeming to ignore the public record.

Such discrepancies call for more than lists of positive initiatives, projects, and best practices – many standing out as exceptions to the rules. To win credibility a thorough review and assessment is needed, with special attention to identifying and understanding the constraints and obstacles impede and block progress.

Considering the reality of the so-called "implementation gap" and the situation in which the social and environmental trends are getting worse, we need to know why.

The actual WSSD review skirted many of these key questions, particularly where the results would involve criticism of a government or industry. The CSD report for the Summit, *Changing Consumption Patterns* discusses only the first three elements of the International Work Programme on Sustainable Production and Consumption – trends, impacts on developing countries, and policy effectiveness (UNCSD, 2001). However, the report avoids critical discussion of country commitments on production and consumption.

The Summit was not supposed to renegotiate Agenda 21 but focus on *implementation*, yet in many cases the interpretation of what is to be implemented leaves out serious parts of the original Agenda 21 programme. One of the most discussed examples concerns finance for sustainable development. Agenda 21 clearly states that "the developmental and environmental objectives of Agenda 21 will require a substantial flow of new and additional resources to developing countries" (UN, 1992: para 1.4). The reality fell far short. In turn, in the conclusion of the five-year review, governments agreed to "reconfirm the financial commitments and targets for official development assistance (ODA) made by industrialised countries at the Earth Summit, and call for intensified efforts to reverse the downward trend in ODA" (UNGA, 1997). By 2002 this recommitment also did not materialise.

Although Agenda 21 points out that "national strategies, plans, policies and processes are crucial in achieving this [implementation]", little attention is given to the lack of national sustainable development strategies throughout most of the decade, nor the continuing reluctance of some governments to engage in such effort, particularly among the most developed countries (e.g., the United States).

For the Agenda 21 commitments to promote sustainable production and consumption, most nations had little to brag about for the ten-year review. While some countries had made efforts to promote the global discussion about the concepts, policies and practices needed (e.g., Norway, Brazil, Republic of Korea), overall implementation of Chapter 4 recommendations fell far short of what was needed to make a difference.

One measure of serious effort to promote sustainable production and consumption can be found in each country's record to develop a domestic policy framework. The revised UN Guidelines for Consumer Protection, highlighting policies for sustainable consumption, were designed to provide some help to this development. Thus, after their adoption of the revised Guidelines in 1999, how countries followed up in using these Guidelines offers insight into overall national efforts to develop a domestic framework of sustainable production and consumption policies.

3.4.5. Consumer guidelines for sustainable consumption
In addition to avoiding critical discussion of country commitments in *Changing Consumption Patterns,* the CSD report also leaves out any mention of follow-up

to the revised UN Guidelines on Consumer Protection – one of the concrete achievements of the International Work Programme. The report concludes that "a broader policy framework is required to address the scale pressures of current patterns, while encouraging efficiency improvements and promoting improvements in standards of living, particularly in developing countries" (UNCSD, 2001). For some reason the Guidelines are not considered by the CSD report as contributing to that framework.

For years civil society organisations promoted the revised UN Guidelines as a tool to aid governments in developing their national policy frameworks on production and consumption. In Soesterberg, Netherlands, at the 1999 conference *From Consumer Society to Sustainable Society*, NGOs encouraged support for the Guidelines as both an indicator and tool for national efforts promoting sustainable production and consumption (ANPED, 1999). The Guidelines mentioned several policy mechanisms which NGOs agreed to be key elements for national policies and strategies, such as right to know laws for better consumer information, reform of government subsidies, responsible state procurement policies, among others.

NGOs included monitoring the development and implementation of the Guidelines as part of the SPAC Watch initiative launched in Soesterberg and adopted by the CSD NGO Caucus on Sustainable Production and Consumption to monitor progress by countries in promoting sustainable production and consumption (SPAC Caucus, 2000a). Focusing on the Guidelines was also stressed in the NGO Statement presented January 30, 2002 at the multi-stakeholder dialogue at the Second Preparatory Meeting for the WSSD (ICSPAC, 2002: 74):

"[NGOs] call upon governments to support and actively develop, implement and monitor national policy frameworks and plans of action to achieve sustainable production and consumption in partnership with civil society. The starting point for this should be the implementation of the UN Guidelines for Consumer Protection, with special emphasis on confronting barriers to change."

Governments did not immediately respond to this call from the NGOs, but the UN Environment Programme and Consumers International took an important follow-up step to the adoption of the revised Guidelines, organising a global survey from October 2001 to March 2002 of government familiarity and use of the Guidelines (UNEP, 2002d). The survey revealed that over a third of governments were not aware of the Guidelines before the survey, indicating the initial low priority given to concrete follow-up to the adoption of the Guidelines in 1999. On the other hand, the majority of governments contacted showed interest in learning more about the Guidelines. In conclusion, UNEP recommended launching at the WSSD a five-year programme "aimed at comprehensive and integrated implementation of the guidelines at national, regional and international level[s]" (UNEP, 2002d: 58). The report also recommended various activities for raising awareness of the Guidelines and building capacity among governments.

4. Production, consumption and the WSSD Plan of Implementation

On 4 September 2002, the World Summit on Sustainable Development concluded. For many civil society advocates of sustainable production and consumption, the Summit was a big disappointment. The Introduction to the WSSD Plan of Implementation clearly identifies changing unsustainable patterns of production and consumption as one of the three "overarching objectives of, and essential requirements for, sustainable development". With this in mind, one would expect the final plan for achieving this objective to build from the experience and lessons of the past, from the decade of discussion about the trends, concepts, policies and practices needed to address the problem, analysed carefully in the ten-year review of progress that was the primary object of the previous year of national, regional, and global preparatory meetings.

4.1. LESSONS FROM THE PAST

In the WSSD Plan of Implementation, Section III speaks to the objective of changing unsustainable production and consumption patterns. Despite the work that had been done on this over the years, the Section does not mention the previous Programme of Work, neither as something to build and expand upon or something requiring a whole new approach. It does not refer to any of the previous discussions about the trends, the impacts on developing countries or to the discussions about policy effectiveness and country commitments. The opening paragraph of Section III restates of Principle 7 on common but differentiated responsibilities, yet without acknowledgement of responsibility for the increasing unsustainable production and consumption patterns taking place over the past ten years. The paragraph also repeats the Agenda 21 admonition about "the developed countries taking the lead".

Despite the General Assembly's adoption of the UN Guidelines on sustainable consumption and the efforts made by UNEP, Consumers International, the NGO Caucus on Sustainable Production and Consumption and others to call attention to this important tool and achievement of the earlier Programme of Work, Section III does not even mention, much less encourage, using the Guidelines.

Neglect of the Guidelines calls further attention to the silent treatment given to the Agenda 21 commitment to develop domestic policy frameworks. The opening paragraph of Section III is symptomatic of the weakening of political will plaguing the Summit. Mentioning Principle 7 from the Rio Declaration, the text stresses that "all countries should *promote* sustainable consumption and production patterns"; however, no mention is made in the WSSD Plan of Principle 8, which more powerfully emphasises "States should *reduce and eliminate* unsustainable patterns of production and consumption" [emphasis added]. What was described in Rio as "a matter of grave concern" requiring concrete actions to ensure the "reorientation of existing production and consumption patterns" devolved in Johannesburg into a more polite and ambiguous "promotion" of recommendations made a decade ago.

Disregard for the lessons and commitments of the past, mixed with avoidance of opportunities and strategies for measurable progress towards clear objectives permeates Section III. For example, Agenda 21 asked countries to encourage greater efficiency in the use of energy and resources, including the use of new and renewable sources of energy (UN, 1992: para 4.18d); the Programme for the Further Implementation of Agenda 21 called on nations to promote "international and national programmes for energy and material efficiency with timetables for their implementation", with "consideration of a 10-fold improvement in resource productivity in industrialised countries in the long term and a possible factor-four increase in industrialised countries in the next two or three decades" (UNGA, 1997: para 28f). Despite numerous ambiguous references in the Plan of Implementation to promoting renewable energy, the proposal by the EU and other countries to set a target and timetable to increase to ten percent the share of new renewable energy by 2010 did not survive the assaults by the United States and OPEC delegations (Parmentier, 2002).

Several voiced their disappointment with the Plan. "The final document consists only of repackaged soft targets", complained the Centre for Science and Environment, "sometimes even more diluted than previous agreements" (CSE, 2002). Oxfam International described the Summit as "a triumph for greed and self-interest, a tragedy for poor people and the environment" (Oxfam, 2002); Greenpeace called it "a disaster in its official conclusions" (Greenpeace, 2002a); for Friends of the Earth the Summit was "a betrayal" (FOE, 2002b). "World leaders fail consumers", Consumers International concluded in its press release (Consumers International, 2002).

4.2. TOWARDS A TEN-YEAR PROGRAMME OF WORK

When first announced, the European Union's proposal for a ten-year programme of work on sustainable production and consumption drew much attention and enthusiastic support by sustainability advocates. As initially proposed, this idea represented to many a solid commitment by governments to an institutional vehicle which could deliver the necessary targets, timetables, monitoring and assessment processes missing from the text, providing at least one concrete implementation mechanism.

Unfortunately, in the WSSD's final Plan of Implementation this idea was also watered down and rendered ambiguous (Barber and Danada, 2002). During the last days of the Summit, negotiators unhappy with solid commitment replace the initial phrase "*develop* a ten-year work programme on sustainable production and consumption" with "*encourage and promote* the development of a ten-year framework of programs in support of regional and national initiatives to accelerate the shift towards sustainable consumption and production ..." [emphasis added](WSSD, 2002: para 15. Rather than helping develop a collective strategy and plan, the only requirement is for nations to show a positive attitude towards countries willing to act on their own. Although "all countries should take action, with developed countries taking the lead", the document fails to identify any mechanisms to coordinate, monitor or evaluate such actions or leadership.

As to implementing Agenda 21's call to give high priority to reviewing progress in achieving sustainable consumption patterns, the WSSD text loosely calls on governments to "identify specific activities, tools, policies, measures and monitoring and assessment mechanisms ... bearing in mind that standards applied by some countries may be inappropriate and of unwarranted economic and social cost to other countries ..." (para 15a). It does not mention work already done developing indicators for measuring changes in production and consumption patterns (UN, 1998), of how such work could or should be continued to support the next ten years of efforts.

The text does cite certain actions as part of encouraging and promoting the ten-year "framework" including: policymaking applying the polluter-pays principle (Principle 16 of the Rio Declaration); improving products and services by reducing their environmental and health impacts (e.g., using life-cycle analysis); developing awareness-raising programmes, particularly among youth and "relevant segments"; developing consumer information tools; and increasing eco-efficiency.

Ironically, Agenda 21 recommended these and other actions ten years ago as elements for national policies and strategies. However, in contrast to Agenda 21's recommendation to assess progress achieved in developing these national policies and strategies, and in contrast to the aim of developing a plan to further implement Agenda 21, the WSSD delegates simply restated some of these initial recommended activities, weighed down by qualifiers and conditions.

Another example of this erosion process is the Plan's treatment of consumer information tools. Agenda 21 stressed the need to "encourage the emergence of an informed consumer public" by "providing information on the consequences of consumption choices and behaviour so as to encourage demand for environmentally sound products and use of products" (4.22). The WSSD text, however, puts more emphasis on qualifying and limiting such efforts, insisting on such tools being adopted *on a voluntary basis*. There is no place here for mandatory mechanisms such as the community's right to know or legal protections against false and misleading advertising or any reference to Principle 10 access to information. Initially the paragraph on consumer information included a reference to *ecolabelling*, encouraged in Rio but in Johannesburg simply inspiring the added qualifier to "not be used as disguised trade barriers". By the Summit's end, ecolabelling disappeared, although the qualifier remains.

In spite of the ambiguity of paragraph 15, governments, international organisations, civil society organisations and others originally enthusiastic about the ten-year work programme will undoubtedly and voluntarily play a leading role in developing the ten-year "framework". The Summit did not specify an institutional vehicle for this framework, leaving the interpretation and implementation of this open.

Most likely, members of the European Union, UNEP, the NGO-oriented International Coalition for Sustainable Production and Consumption (ICSPAC), and others will through their own efforts lay the operational foundation stones that will shape and animate the proposed framework. Following the Summit, the European Union identified the ten-year framework for programmes on sustainable production and consumption as one of their five key targets (European Union, 2002). UNEP has also made the ten-year framework a priority, circulating at the Summit its own proposal of ideas for the work programme (UNEP, 2002b). This proposal will be discussed at the upcoming UNEP Governing Council meeting in 2003.

NGOs involved with the International Coalition for Sustainable Production and Consumption offered their ideas in the SPAC Watch report, *Waiting for Delivery* (ICSPAC, 2002). The SPAC Watch programme, one important contribution to the ten-year framework, was designed to foster collaboration and communications among civil society organisations around the world in monitoring and advocating progress towards sustainable production and consumption (Barber, 2002a).

4.3. CORPORATE RESPONSIBILITY AND ACCOUNTABILITY

Since Rio, two new concepts have entered the official UN lexicon and discussion about sustainable development: globalisation and corporate accountability. In the WSSD Plan for Implementation, *corporate accountability* appears in a number of places, notably in Section III on production and consumption (paragraph 18) and in Section V on sustainable development in a globalising world (paragraph 49).

The concept of corporate *responsibility* appears throughout Agenda 21, mostly used to promote arguments for self-regulation and voluntary approaches by industry, in contrast to "command and control" efforts to improve regulations and compliance. However, during the Rio+5 process, arguments highlighting the inadequacy of corporate responsibility arose, stressing the need to pair this concept with corporate accountability – referring to the legal obligation of a company to do the right thing (although there are efforts to reduce the meaning to corporate reporting). "Just as individuals in society require both morals and laws to guide their behaviour", argued the NGO Taskforce on Business and Industry during Rio+5, "responsibility and accountability are both necessary to guide corporate conduct" (Barber, 1997). While corporate responsibility is behaviour that is encouraged, corporate accountability is behaviour that is required. In the 1997, the Programme for the Further Implementation of Agenda 21 introduced the concept of corporate accountability paired with responsibility (para 133e(iii)).

In the WSSD Plan of Implementation, promotion of corporate responsibility and accountability appears in several places, although not without controversy and debate. In Section V, the heated debate on operationalising this "through the full development and effective implementation of inter-governmental agreements and measures ..." continued up to the last day of the conference (Third World

Network, 2002a). For many NGOs, this paragraph was "one of the most significant outcomes of the Johannesburg Earth Summit" (FOE, 2002c).

In the section on production and consumption, the text on corporate responsibility and accountability (para 18) remains lopsided. While calling on governments to "enhance corporate environmental and social responsibility and accountability", the four actions identified are voluntary approaches (encouraging voluntary initiatives by industry; encouraging dialogue between enterprises and communities; encouraging financial institutions to incorporate sustainable development considerations; and developing workplace-based partnerships and programmes). While corporate accountability mechanisms could play a major role in encouraging and ensuring sustainable production and consumption, none are mentioned. Even the reference to corporate reporting cites only a voluntary initiative, neglecting the accountability mechanisms called for in Principle 10 (i.e., access to information, redress and remedy) or Principle 13 ("States shall develop national law regarding liability and compensation for the victims of pollution") in the Rio Declaration. Again, the WSSD missed a possible step forward.

Extended Producer Responsibility (EPR) or product take-back, which encourages manufacturer investment in eco-efficiency, is a key policy that should have received attention in this section (Dutta, 2002). Although EPR has been highlighted throughout the various discussions and research of the International Work Programme, not to mention ongoing work by OECD, UNEP and others, the WSSD neglected even mentioning it. In many ways EPR is one of the important policy mechanisms linking corporate responsibility and accountability. The 1991 German Packaging Ordinance is an early practical example of this policy, emphasising the producer's responsibility for the impacts of the product once it reaches the disposal phase of its lifecycle, creating incentives to the producer to engage in more efficient design and recycling methods to reduce the product's environmental impacts (OECD, 1998).

4.4. CLEANER PRODUCTION AND ECO-EFFICIENCY

The UN Environment Programme has been successfully operating its Cleaner Production Programme since 1989, in partnership with a growing network of organisations worldwide. The community of cleaner production centres is generally well-regarded as helping countries and enterprises in building capacity in cleaner production methods, to increase eco-efficiency and reduce waste and pollution. Paragraph 16 in the Plan of Implementation encourages governments to support, invest in and provide incentives for investment in cleaner production programmes and centres. UNEP's plans and proposals for further development of its cleaner production work also play a key role in their proposed contribution to the ten-year framework.

4.5. OTHER PROPOSED ACTIONS

Having apparently abandoned the Agenda 21 priority of developing national policy frameworks, other parts of Section III encourage governments to "integrate the issue of production and consumption patterns into sustainable development policies, programmes and strategies" (paragraph 17) and "take sustainable development considerations into account in decision-making" (paragraph 19). Yet many governments apparently still have difficulty understanding the nature and significance of that "issue". What could have evolved as a useful integrative approach to address overriding crosscutting issues fell victim to what WSSD Secretary-General Nitin Desai described as "a fragmented approach towards sustainable development" (UNCSD, 2002: para 4).

Other actions mentioned in Agenda 21 and the Programme for the Further Implementation of Agenda 21 received little more than token mention: internalisation of environmental costs and the use of economic instruments, public procurement policies, capacity-building and training, and environmental impact assessment.

4.5.1. Internalisation of environmental costs

The Programme for the Further Implementation of Agenda 21 described the objective of internalising environmental costs of "vital importance" (para 28a). To do this, it pointed out, required two actions – both missing from Section III: (a) "shifting the burden of taxation onto unsustainable patterns of production and consumption", and (b) "a socially responsible process of reduction and elimination of subsidies to environmentally harmful activities" (UNGA, 1997). The WSSD Plan for this objective mentions no more than a vague reference to "economic instruments" – again without targets, timetables or monitoring.

Earlier CSD reports already indicated "little progress" and difficulties implementing economic instruments to internalise environmental costs; that "governments shy away from additional ecotaxes and environmental regulations" (UNCSD, 1997b: para 37). However, the calls at that time for more analysis and attention to policy effectiveness were essentially disregarded at the WSSD, along with the point that "cost internalisation and eco-efficiency approaches are most effectively and efficiently implemented in combination with *specific time-bound targets and objectives*" [emphasis added](UNCSD, 1997b: para 10).

4.5.2. Subsidy reform

Government subsidies that ultimately encourage unsustainable production and consumption patterns represent a difficult political obstacle undermining and often blocking progress. It thus made sense for the 1997 General Assembly to recommend reducing and eliminating destructive subsidies as part of the production and consumption work agenda. Initially, overall subsidy reduction was included in Section III in the draft Plan of Implementation at the end of the Bali preparatory meeting, but removed in Johannesburg.

Reference to subsidy reduction does appear in other parts of the Plan, such as paragraph 20p and 20q (energy), 31f (illegal fishing), and 92c and 97b (trade). However, removing subsidy reform from Section III also discourages attention to the fact that government subsidies have been one of the driving forces encouraging unsustainable production and consumption. Section III thus abandoned the priority of developing an integrative approach and time-bound strategy to address this cross-sectoral problem of socially and environmentally destructive subsidies (e.g., to develop mechanisms providing greater public information and education needed to support reform efforts). The WSSD Plan appears more concerned with trade barriers than eliminating destructive consumption and production patterns. Note also that the proposal to include specific targets and timetables for phasing out fossil fuel subsidies was considered and supported, but ultimately deleted from the Plan.

Removing subsidy reform from the sustainable production and consumption work agenda reinforces the fragmented approach to sustainable development; it also takes attention away from the role of governments in encouraging unsustainable production and consumption (SPAC Caucus, 2000b; Benekom, 2002). Once again, this action suspiciously looks like avoidance of responsibility, further reducing the vision and strategies required for effective policymaking on production and consumption.

The CSD acknowledged the strong political opposition to subsidy reduction, that "the major beneficiaries of such subsidies are generally privileged and politically influential groups, which makes subsidy removal politically difficult" (UNCSD, 1999: para 40). The CSD then called for further work "identifying effective measures" for removing subsidies. This further work could and should have been part of the WSSD Plan for addressing unsustainable production and consumption patterns.

Speaking directly to problem of political opposition to subsidy reform, ICSPAC pointed out in its contribution to the WSSD that "civil society groups, committed to the public interest, can bring a greater level of transparency and directness to the search for solutions, especially when that search is blocked by powerful special interests" (Barber, 2002b: 10). NGOs can help mobilise public support for subsidy reform in situations where government's hands are tied.

4.5.3. Public procurement policies

Section III calls on governments to "promote public procurement policies that encourage development and diffusion of environmentally sound goods and services" (WSSD, 2002: para 19c). Although OECD, UNEP, NGOs and a range of governments have experimented with green procurement policies and analysed the lessons, as well as organised networks to collaborate and share their experience, the WSSD text does little more than echo the recommendation made a decade ago at Rio. There was no attempt to identify reasonable targets and timetables nor any effort to relay a sense of what progress has or has not been achieved since Rio and what could be done by WSSD participants to improve effectiveness.

4.6. ROLE OF TRADE AND INVESTMENT

Ironically, but not surprising, Section III on changing unsustainable patterns of consumption and production makes no mention of the ways trade and investment policies have contributed to unsustainable production and production patterns. While calling for increasing investment in cleaner production and eco-efficiency, the text is silent on the social and environmental impacts of the increasing trade and investment flows associated with globalisation. However, the text makes a point to warn governments that consumer information tools "should not be used as disguised trade barriers" and that internalisation of environmental costs should be done "without distorting international trade and investment" (WSSD, 2002: para 16, 15e, 19b).

In turn, Section V on sustainable development in a globalising world cites "new opportunities for trade, investment and capital flows ... for the growth of the world economy, development and the improvement of living standards around the world", but makes no direct connection with the resulting unsustainable production and consumption patterns nor is the objective of changing those patterns listed among the "serious challenges" (WSSD, 2002: para 47).

This silence is ironic considering not only the fact that growth of imports and exports assumes increasing production and consumption, but also that the policy agenda for sustainable production and consumption involves so many of the policies negotiated in the World Trade Organisation: subsidy reform, government procurement, labelling, environmental standards and process and production methods (PPMs), among others (UNEP/IISD, 2000: 41; WTO, 2001).

The silence is unsurprising as the topic is a political minefield, which Section III negotiators tried to quietly avoid. In Doha, with the Johannesburg Summit on the horizon, trade ministers made a point to "strongly reaffirm our commitment to the objective of sustainable development". However, their assurances that "under WTO rules no country should be prevented from taking measures for the protection of human, animal or plant life or health, or of the environment" were immediately overridden by the requirement to be "in accordance with the provisions of the WTO Agreements". (WTO, 2001: para 6). In Johannesburg, delegates engaged in heated debate over similar language in Section X on the means of implementation. Calling for governments to "continue to enhance the mutual supportiveness of trade, environment and development with a view to achieving sustainable development", the text added "while ensuring WTO consistency".

Fortunately, opposition raised by Ethiopia and Norway against this conditionality was followed in kind by the G77 and European Union, resulting in deletion. Inclusion would have, as Third World Network put it, "bound the hands of countries in all future multilateral negotiations in any area" giving the WTO "a superior status for eternity" (Third World Network, 2002b).

The political status of changing unsustainable patterns of production and consumption, one of the three "overarching objectives" of sustainable development, thus remains intertwined with the WTO's treatment of sustainable development – not as a *framework* within which trade policy contributes to the goal of improving the quality of life for everyone, but as "an objective" pursued in accordance with WTO provisions (WTO, 2001).

As the debate on how to mediate between multilateral environmental agreements and WTO rules is transferred to future meetings of the WTO, CSD and other gatherings, so does the need to better articulate the role and priority of sustainable production and consumption in trade and investment policy.

5. Conclusions

In his Foreword to Agenda 21, Maurice Strong, then Secretary-General for the 1992 UN Conference on Environment and Development, described the situation at that time whereby "industrial countries continue to be addicted to the patterns of production and consumption which have so largely produced the major risks to the global environment". (UN, 1992:1). Unfortunately, Strong's description of the addiction to unsustainable production and consumption remains valid for the world of 2002.

Ten years after the Earth Summit, the Secretary General for the World Summit on Sustainable Development cited the increase of unsustainable production and consumption patterns as one of the primary factors undermining progress towards sustainable development since the Earth Summit. At the same time, world leaders at the WSSD agreed that eliminating unsustainable production and consumption is one of the three main objectives of sustainable development.

One would thus expect to see states give much higher priority to designing a global strategy and plan, with reasonable targets, timetables and monitoring processes, to support national implementation of the Chapter 4 objectives. Unfortunately, many of the factors undermining major advances in such policies and strategies during the past decade prevented such outcomes at the Summit.

Despite an acknowledged worsening of social and environmental trends, due to the relentless global increase of production and consumption, all the discussions, conferences, research and policymaking over the past ten years were inadequate in changing those trends. Underlying the implementation gap was not so much a lack of political will but deliberate, stubborn resistance to the "reorientation" that is necessary. As Consumers International put it, "The world leaders are in a state of unsustainable procrastination" (Consumers International, 2002). On the one hand, the current President Bush continued to maintain his father's famous refusal in Rio to negotiate the American way of life – despite Colin Powell's insistence that the United States is committed to sustainable development. On the other hand, the OPEC nations continue to band together within the G77 to ensure their source of finance for development remains unthreatened. Likewise, the advertising industry spends hundreds of billions of dollars promoting around the world a consumer culture based on the American way of waste and wants. Part of the industry's investment flows to trade associations and political lobbyists committed to blocking or minimising any regulations or constraints governments may be thinking about imposing. At the Summit, however, the advertising industry was welcomed as a partner in helping promote sustainable consumption values; criticism and talk of regulation verged on official taboo. Another taboo was military production and consumption, especially comparisons between the huge amounts spent for defence and the relatively little for sustainability and human security. Like other forms of addiction, the addiction to unsustainable consumption and production patterns is sustained through a large array of defence mechanisms – denial, rationalisation, avoidance, deception, token efforts.

"Rather than dwelling on the problems, what we must ask ourselves is *why* they persist", WSSD Secretary-General Nitin Desai urged in his address to the Summit when it opened in Johannesburg. The purpose of the World Summit, he explained, was "to tackle what has stood in the way of us making progress, and what can we do in order to get action, to get results" (Desai, 2002). With the Summit now over, that purpose and the question of *why* become more important than ever.

Before the Summit, ICSPAC produced a list of obstacles blocking progress towards sustainable production and consumption (ICSPAC, 2002: 10):

- Resistance by governments in developing national policy frameworks;
- Continued promotion of consumerism by the mass media and advertising;
- Erosion of accountability by corporations;
- Political influence of industries whose profits depend on unsustainable consumption;
- Political reluctance of government and intergovernmental organisations to criticise and more directly address cases where industry plays a negative role and influence in the problem;
- Lack of understanding of forces driving unsustainable production and consumption;
- Limited and unequal resources available to civil society for public education and political advocacy, compared with larger marketing and public relations budgets for industry promoting consumerism;
- Lack of public awareness of sustainable development as an alternative;
- Where there is awareness, the belief that sustainable consumption means a reduction in living standards and quality of life – rather than improvement.

A year before the Summit the General Assembly passed a resolution asking the CSD to "identify major constraints hindering the implementation of Agenda 21, propose specific time-bound measures to be taken and institutional and financial requirements, and identify the sources of such support" (UNGA, 2001: para 15c). Nevertheless, world leaders at the WSSD produced few time-bound measures, made no major new commitments to increasing ODA, and paid minimal critical attention towards the constraints and obstacles to eliminating unsustainable production and consumption patterns.

Perhaps part of the problem is the consensus-based system of the United Nations itself, requiring agreement among the full body of member governments, despite all their differences. Often such a process, after all the deal-making and compromises, results in the lowest-common denominator, a repackaging of past promises not kept, with some of the most important new ideas and commitments watered-down, traded away or rendered ambiguous.

For those still committed to developing "a plan of implementation focused on targets, timetables, goals and activities which can lead to concrete results", as Secretary-General Desai stressed, this may have to wait for a more informal alliance of sustainability advocates coming together to develop the ten-year framework of programmes on sustainable production and consumption. However, to achieve progress will require a willingness to move beyond the taboos and the fog of rhetoric, to discuss and understand the various obstacles and find strategies to overcome them. Above all, sustainability calls for persistence.

References

ANPED – Northern Alliance for Sustainability.: 1999, *From Consumer Society to Sustainable Society: Towards Sustainable Production and Consumption*, report from conference, 31 January – 1 February, 1999, Northern Alliance for Sustainability, Amsterdam.

Barber, J. (ed): 1997, *Minding Our Business: The Role of Corporate Accountability in Sustainable Development*. NGO Taskforce on Business and Industry, Washington, DC.

Barber, J.: 2002a, 'SPAC Watch: A civil society initiative to monitor progress towards sustainable production and consumption', in *Waiting for Delivery: A Civil Society Assessment of Progress Toward Sustainable Production and Consumption*, International Coalition for Sustainable Production and Consumption, Rockville.

Barber, J.: 2002b, 'Civil society and progress towards sustainable production and consumption', in *Waiting for Delivery*, International Coalition for Sustainable Production and Consumption, Rockville.

Barber, J. and Danada, I.: 2002, 'Sustainable production and consumption lose out at the summit', in *Taking Issue*, 3 September 2002, Sustainable Development Issues Network, Johannesburg.

Benekom, S.: 2002, 'Reforming subsidies, some strategic observations', in *Waiting for Delivery*, International Coalition for Sustainable Production and Consumption, Rockville.

Chaudhuri, R.: 2002, 'Advertising reform and sustainability: a developing country perspective', in ICSPAC, *Waiting for Delivery*, International Coalition for Sustainable Production and Consumption, Rockville.

Consumers International.: 1999, *Easy Targets: Survey of Television Food and Toy Advertising to Children in Four Central European Countries*, Consumers International, London.

Consumers International.: 2002, 'World leaders fail consumers', Press release, 3 September 2002, Johannesburg.

CSE – Centre for Science and Environment.: 2000, 'The shifting smoke of tobacco', in *Down to Earth*, December 2000, Centre for Science and Environment, New Delhi.

CSE – Centre for Science and Environment.: 2002, 'The world after', in *Down to Earth*, 30 September 2002, Centre for Science and Environment, New Delhi.

Desai, N.: 2002, 'Opening address to the World Summit on Sustainable Development', 26 August 2002, Johannesburg.

Durning, A.: 1992, *How Much is Enough? The Consumer Society and the Future of the Earth*, Worldwatch Institute, W.W. Norton and Company, New York.

Dutta, A.: 2002, 'Extended producer responsibility', in *Waiting for Delivery*, International Coalition for Sustainable Production and Consumption, Rockville.

Earth Negotiations Bulletin.: 1995, 'A summary report on the 1995 session of the United Nations Commission on Sustainable Development', vol. 5, no. 42, 1 May 1995, International Institute for Sustainable Development, Winnipeg.

Energy Caucus.: 2002, 'Energy issues at WSSD – total failure', Press release, 4 September 2002, NGO Energy and Climate Caucus, Johannesburg.

European Union.: 2002, 'Now we must turn World Summit agreement into concrete results', Press release, 4 September 2002, Johannesburg.

FOE – Friends of the Earth.: 2002a, *Sustainable Production and Consumption: A Global Challenge*, Friends of the Earth, Amsterdam.

FOE – Friends of the Earth.: 2002b, 'Earth Summit: betrayal ...', Press release, Friends of the Earth, 4 September 2002, Johannesburg.

FOE – Friends of the Earth.: 2002c, 'US plots to undermine key summit outcome', Press release, Friends of the Earth, 3 September 2002, Johannesburg.

FOE – Friends of the Earth.: 2002d, 'US wrecks earth summit', Press release, Friends of the Earth, 4 September 2002 Johannesburg.

Greenpeace.: 2002a, 'US jeered, Summit denounced', Press release 4 September 2002, Johannesburg.

Greenpeace.: 2002b, 'Who to blame ten years after Rio? The role of the USA, Canada and Australia in undermining the Johannesburg Summit', Greenpeace International, Amsterdam.

ICSPAC – International Coalition for Sustainable Production and Consumption.: 2002, *Waiting for Delivery: A Civil Society Assessment of Progress Towards Sustainable Production and Consumption*, International Coalition for Sustainable Production and Consumption, Rockville.

Norwegian Ministry of Environment.: 1995, *Oslo Ministerial Roundtable: Conference on Sustainable Production and Consumption*, Government of Norway, Oslo.

OECD – Organisation for Economic Cooperation and Development.: 1997, *Sustainable Consumption and Production: Clarifying the Concepts*, Organisation for Economic Cooperation and Development, Paris.

OECD – Organisation for Economic Cooperation and Development.: 1998, 'Extended producer responsibility: Case study on German Packaging Ordinance', May 1998, Organisation for Economic Cooperation and Development, Paris.

OECD – Organisation for Economic Cooperation and Development.: 2002a, *Working Together Towards Sustainable Development: The OECD Experience*, Organisation for Economic Cooperation and Development, Paris, p. 27.

OECD – Organisation for Economic Cooperation and Development.: 2002b, *Towards Sustainable Household Consumption? Trends and Policies in OECD Countries*, Organisation for Economic Cooperation and Development, Paris.

OECD – Organisation for Economic Cooperation and Development.: 2002c, *Ensuring Sustainable Consumption: Decoupling Environmental Pressures from Economic Growth*, Organisation for Economic Cooperation and Development, Paris.

Oxfam.: 2002, 'Crumbs for the poor', Press release, 3 September 2002.

Parmentier, R.: 2002, 'Lessons from Johannesburg: What is the future for UN summits?' Speech, 10 September 2002.

Republic of Korea.: 1996, *Report of the Workshop on Policy Measures for Changing Consumption Patterns*, Seoul.

Robins, N. and Roberts, S.: 1998, *Consumption in a Sustainable World*, Report of the workshop held in Kabelvag, Norway, 2–4 June 1998, International Institute for Environment and Development, London.

SPAC Caucus.: 1997, 'Report from the NGO Sustainable Production and Consumption Caucus: Preparations for UNGASS +5', see ICSPAC website at http://isforum.org/spac/csdcaucus.

SPAC Caucus.: 2000a, 'Update', November 2000, NGO Caucus on Sustainable Production and Consumption. See http://isforum.org/spacwatch/2000-NovSpacWatchUpdate.htm.

SPAC Caucus.: 2000b, 'Progress on eliminating subsidies', NGO Caucus on Sustainable Production and Consumption, 23 February 2000, New York.

Stigson, B.: 1999, 'What is eco-efficiency? The WBCSD perspective', speech at Roundtable on Eco-efficiency, 19 March, 1999, World Business Council for Sustainable Development, Sydney.

Third World Network.: 2002a, 'Implementation plan passed, drama on corporate accountability', Press release, 4 September 2002, Johannesburg.

Third World Network.: 2002b, 'Efforts for WTO supremacy over all future accords fails', Press release, 6 September 2002, Johannesburg.

UN – United Nations.: 1992, *Agenda 21: Programme of Action for Sustainable Development*, United Nations, New York.

UN – United Nations.: 1998, *Measuring Changes in Consumption and Production Patterns: A Set of Indicators*, Department of Economic and Social Affairs, United Nations, New York.

UNCSD – United Nations Commission on Sustainable Development.: 1994a, Appendix, 'Summary report: The symposium on sustainable consumption', Oslo, 19–20 January 1994, in *General Discussion on Progress in the Implementation of Agenda 21, Focusing on the Cross-Sectoral Components of Agenda 21 and the Critical Elements of Sustainability*, United Nations Commission on Sustainable Development, United Nations, New York.

UNCSD – United Nations Commission on Sustainable Development.: 1994b, *Report of the Commission on Sustainable Development on its Second Session*, 16–27 May 1994, United Nations, New York.

UNCSD – United Nations Commission on Sustainable Development.: 1995a, *Changing Consumption and Production Patterns: Report of the Secretary General*, Commission on Sustainable Development, Third Session, 11–28 1995, United Nations, New York.

UNCSD – United Nations Commission on Sustainable Development.: 1995b, *Report on the Third Session*, 11–28 April 1995, United Nations Commission on Sustainable Development, United Nations, New York.

UNCSD – United Nations Commission on Sustainable Development.: 1996, *Report on the Fourth Session*, 18 April–3 May 1996, United Nations Commission on Sustainable Development, United Nations, New York.

UNCSD – United Nations Commission on Sustainable Development.: 1997a, *Overall Progress Achieved Since the United Nations Conference on Environment and Development: Report of the Secretary General*, United Nations Commission on Sustainable Development, United Nations, New York, par. 38.

UNCSD – United Nations Commission on Sustainable Development.: 1997b, *Overall Progress Achieved Since the United Nations Conference on Environment and Development: Changing Consumption Patterns*, United Nations Commission on Sustainable Development, Fifth session 7–25 April 1997, United Nations, New York.

UNCSD – United Nations Commission on Sustainable Development.: 1998a, *Consumer Protection and Sustainable Consumption: New Guidelines for the Global Consumer*, Background paper for Inter-Regional Expert Meeting on Consumer Protection and Sustainable Consumption, 28–30 January 1998, Sao Paulo, United Nations Commission on Sustainable Development, New York.

UNCSD – United Nations Commission on Sustainable Development.: 1998b, *Consumer Protection: Guidelines for Sustainable Consumption: Report of the Secretary-General*, United Nations Commission on Sustainable Development, United Nations, New York.

UNCSD – United Nations Commission on Sustainable Development.: 1999, *Comprehensive Review of Changing Consumption and Production Patterns: Report of the Secretary-General*, United Nations Commission on Sustainable Development, Seventh Session 19–30 April 1999, United Nations, New York.

UNCSD – United Nations Commission on Sustainable Development.: 2001, *Changing consumption patterns: Report of the Secretary-General*, Commission on Sustainable Development acting as the Preparatory Committee for the World Summit on Sustainable Development, 30 April–2 May 2001, United Nations, New York.

UNCSD – United Nations Commission on Sustainable Development.: 2002, *Implementing Agenda 21: Report of the Secretary-General*, United Nations, New York.

UNDP – United Nations Development Programme.: 1998, *Human Development Report 1998*, Oxford University Press, p. 1.

UNEP – United Nations Environment Programme.: 2001, *Consumption Opportunities: Strategies for Change*, United Nations Environment Programme, Geneva.

UNEP – United Nations Environment Programme.: 2002a, *Cleaner Production Global Status Report 2002*, United Nations Environment Programme, Paris.

UNEP – United Nations Environment Programme.: 2002b, 'Proposal for a work programme on promoting sustainable consumption and production patterns', Briefing note distributed at WSSD, 22 August 2002, United Nations Environment Programme, Paris.

UNEP – United Nations Environment Programme.: 2002c, *Industry as a Partner for Sustainable Development: Advertising*, European Association of Communications Agencies/World Federation of Advertisers, UK.

UNEP – United Nations Environment Programme.: 2002d, *Tracking Progress: Implementing Sustainable Consumption Policies*, United Nations Environment Programme, Paris.

UNEP/IISD – United Nations Environment Program/International Institute for Sustainable Development.: 2000, *Environment and Trade: A Handbook*, International Institute for Sustainable Development, Winnipeg.

UNGA – United Nations General Assembly.: 1997, *Programme for the Further Implementation of Agenda 21*, United Nations General Assembly Special Session, New York, United Nations, par. 28.

UNGA – United Nations General Assembly.: 2001, 'Resolution 55/199 adopted by the General Assembly: Ten-year review of progress achieved in the implementation of the outcome of the United Nations Conference on Environment and Development', 5 February 2001, United Nations, New York.

WBCSD – World Business Council for Sustainable Development.: 2001, 'Shaping a deal for the Johannesburg 2002 Summit', Speech by Claulde Fussler at High Level Seminar on Globalisation, Sustainable Development and the EU, 22–23 October 2001, World Business Council for Sustainable Development, Brussels.

Worldwatch Institute.: 2002, *Vital Signs 2002*, W.W. Norton & Company, New York.

WSSD – World Summit on Sustainable Development.: 2002, *Plan of Implementation*, World Summit on Sustainable Development, United Nations, New York.

WTO – World Trade Organisation.: 2001, *Ministerial Declaration*, Ministerial Conference, Fourth Session, 20 November 2001, Doha.

CHAPTER 4

WATER FOR SUSTAINABLE DEVELOPMENT IN AFRICA

DENNIS D. MWANZA

Water Utility Partnership for Capacity Building in Africa, Abidjan, Cote d'ivoire
(e-mail: ddmwanza@wupafrica.org; fax: C225 21 758656/7; tel.: C225 21 240828/13)

Abstract. Water is a precious yet non-renewable resource. Yet in Africa, the same water can be a source of life and death. Water is not only the most basic of need but also at the centre of sustainable development and essential for poverty eradication. Water is intimately linked to health, agriculture, energy and biodiversity. Without progress on water, reaching other Millennium Development Goals (MDGs) will be difficult if not impossible. The fight against poverty will remain a pipe dream.

A lot of activities have been undertaken with the aim of highlighting the importance of water, linking water with sustainable development and indeed developing strategies for resolving the ever-increasing problems of water. These include the adoption of the Africa Water Vision in The Hague, Netherlands in March, 2000.

In order to address the many problems of water in Africa especially related to the coordination of the increasing number of initiatives in the Water sector in Africa, the African Water Task Force (AWTF) was established.

As part of developing solutions to the African water crisis, the AWTF held a regional conference in Accra Ghana. Some of the emerging issues from the Accra Conference are highlighted in the Accra declaration.

This paper highlights the linkages between water and sustainable development, water and poverty and the many facets that relate to water. It mainly addresses issues of water from the African perspective. A number of key events that have taken place and which have served as a basis for many policy pronouncements have been given.

The last section concentrates on what happened to water at the World Summit for Sustainable Development (WSSD) held in Johannesburg in September 2002.

Key words: Africa water task force, Africa water vision, integrated water resources management, poverty, sanitation, sustainable development, water supply.

Abbreviations: ADB – African Development Bank; AMCOW – African Ministerial Conference on Water; AWTF – African Water Task Force; FFA – Framework For Action; GDP – Gross Domestic Product; IGWA – Intergovernmental Agency for Water in Africa; MDG – Millennium Development Goals; NEPAD – New Partnership for Africa's Development; NGO – Non-Governmental Organization; OAU – Organization of African Unity; SADC – Southern African Development Community; UNEP – United Nations Environment Programme; UNICEF – United Nations International Children and Education Fund; UNSIA – United Nations System wide Initiative for Africa; WASAI – Water and Sanitation African Initiative; WEDC – Water Engineering and Development Centre; WHO – World Health Organization; WSSD – World Summit for Sustainable Development; WASH – Water, Sanitation and Hygiene; WSSCC – Water Supply and Sanitation Collaborative Council.

1. Introduction

It is no doubt that Water is a precious yet non-renewable resource. Yet in Africa, the same water can be a source of life and death. Water is not only the most

---

Readers should send their comments on this paper to: BhaskarNath@aol.com within 3 months of publication of this issue.

L. Hens and B. Nath (eds.), The World Summit on Sustainable Development, 91–111.
© 2005 *Springer. Printed in the Netherlands.*

basic of need but also at the centre of sustainable development and essential for poverty eradication. Water is intimately linked to health, agriculture, energy, and biodiversity. Without progress on water, reaching other Millennium Development Goals (MDGs) will be difficult if not impossible. Programmes on poverty alleviation which are at the centre of most Development programmes will not achieve much without emphasis on the issues of water. Unfortunately, there is a low priority assigned to water by countries as evidenced by the decrease of external support for this sector, by the reduction of investments by International Financial Institutions, by the low priority in national budgets, and by the absence of water as a central feature in major regional programmes. And yet, Africa today has the lowest water supply and sanitation coverage in the world. More than 1 in 3 Africans do not have access to improved water supply and sanitation facilities. Current coverage levels stand at 62% for water supply and 60% for sanitation. The reality is that the absolute number of people without access to water services is increasing and between now and the year 2020 the number will increase from 300 million to 400 million (WHO/UNICEF, 2000).

The World Summit for Sustainable Development (WSSD) held in Johannesburg, South Africa was not an end in itself. It was part of a process dating back to the 1992 Dublin Principles on water resources management and to Agenda 21 of the proceedings of the United Nations Conference on Environment and Sustainable Development that took place in Rio de Janeiro in 1992. In response to these two international events, the World Water Council and the Global Water Partnership were formed. The latter was established to operationalize and promote the Dublin Principles in the form of Integrated Water Resources Management (IWRM).

The aim of this paper is to show the linkages between water and sustainable development, water and poverty and the many facets that relate to water and the impact of the WSSD.

2. The African water vision

In March 2000, the Second World Water Forum was held at The Hague. The high point of this Forum was the formulation of a World Water Vision for 2025 (World Water Council, 2000). Along with other regions of the world, Africa participated in this Forum and developed its own water vision (quoted in Box 1), the Africa Water Vision for 2025 along with a Framework for Action (FFA) for its implementation. The Vision was endorsed during the African Caucus meeting at the Second World Water Forum, and obtained wide support among Africans and its development partners. The FFA provides for the utilization of IWRM as the basis for its implementation. Later in the year, in the Millennium Declaration of 2000, African Heads of State established goals for access to water supply and sanitation services (UN, 2000).

The Africa Water Vision was also adopted by two Africa-level United Nations entities concerned with water in Africa. One is the water cluster of UNSIA (the United Nations Special Initiative for Africa); and the other is Intergovernmental Agency for Water in Africa (IGWA), the Inter-agency Group on Water for Africa,

---

Box 1. The Africa Water Vision for 2025 (World Water Council, 2000).

The African Water shared vision is for:

An Africa where there is an equitable and sustainable use and management of water resources for poverty alleviation, socio-economic development, regional cooperation and the environment

It is a vision for an Africa in which:

1. There is a sustainable access to safe and adequate water supply and sanitation to meet basic needs of all
2. There is sufficient water for food and energy security
3. Water for sustaining ecosystems and biodiversity is adequate in quantity and quality
4. Water resources institutions have been reformed to create an enabling environment for effective and integrated management of water in national and transboundary water basins, including management at the lowest appropriate level
5. Water basins serve as a basis for regional cooperation and development, and are treated as natural assets for all within such basins
6. There is an adequate number of motivated and highly skilled water professionals
7. There is an effective and financially sustainable system for data collection, assessment and dissemination for national and transboundary water basins
8. There are effective and sustainable strategies for addressing natural and man-made water resources problems, including climate variability and change
9. Water is financed and priced to promote equity, efficiency and sustainability
10. There is political will, public awareness and commitment among all for sustainable water resources management, including the mainstreaming of gender issues and youth concerns and the use of participatory approaches.

---

also established by the United Nations. At the last IGWA meeting held in March 2001 in Niamey, Niger, two important decisions were taken. First, it was decided that The African Development Bank (ADB) and the United Nations Economic Commission for Africa (ECA) would organize a series of workshops and training programmes on IWRM and the African Water Vision. Second, the delegate from the United Nations Environment Programme (UNEP) informed the meeting that UNEP would take the lead to organize African ministers responsible for water behind the African Water Vision and also to provide a political voice and advocacy for water in Africa.

3. The African Water Task Force

In pursuit of these decisions, actions were taken to launch the African Ministe-rial Conference on Water (AMCOW). In parallel to this, the first of a series of planned workshops on IWRM was held in Accra in September 2001. This was followed immediately by a meeting in Abidjan convened by the ADB on the promotion of water resources development in Africa. At the Abidjan meeting it was noted that for the Second World Water Forum the ADB, the ECA and the Organization of African Unity (OAU) coordinated efforts to produce the African Water Vision. However, no such entity or mechanism was in place to coordinate African inputs at the WSSD scheduled held in Johannesburg (South Africa) dur-ing August/September 2002, and at the Third World Water forum scheduled to be held in Kyoto in March 2003. Given the perceived urgency of such a mech-anism, it was at this meeting that an agreement was reached to establish the African Water Task Force (AWTF) to help coordinate activities designed to priori-tise water issues at the WSSD and facilitate African participation in the third World Water Forum. The Netherlands government provided financial resources through the African Development Bank, which undertook to coordinate, the activities of the Task Force.

At its first meeting held in November 2001, the AWTF became aware of NEPAD (The New Partnership for Africa's Development). This was an initiative of African Heads of State that aimed at putting Africa on the path of sustainable development, thereby averting the risk of Africa being marginalized in the fast moving global-ization process. Given the traditional low priority normally accorded to water by government, it was decided that NEPAD should be used as a platform for mov-ing forward the proposed agenda on the Africa Water Vision and the promotion of IWRM. Accordingly, it was decided that the position paper for the proposed con-ference should be designed to show African political leaders how water can help in the attainment of sustainable development as envisaged in NEPAD.

4. The Accra Conference

A major activity in the water sector in Africa was the Regional Conference on Water and Sustainable Development in Africa held in Accra, Ghana during April 15–17, 2002. The Conference provided a good opportunity to review and reori-entate actions within the water sector in Africa. It brought together a total of 200 people from 42 African countries including policy makers, sector profession-als, representatives from financing institutions and bilateral donors, civil society, researchers, etc. The conference also included participants from outside of the African continent (Sonou, 2002).

The key issues emanating from the Conference are captured in the 'Accra Declaration' (quoted in Box 2) on water as a tool for sustainable development in Africa.

Box 2. The Accra Declaration on water as a tool for sustainable development in Africa (Sonou, 2002).

Water can make the difference to Africa's development!

We have concluded that water can make an immense difference to Africa's development if it is managed well and used wisely. Given clear policies and strategies and real commitments to implementation, we can use water to help eradicate poverty reduce water-related diseases and achieve sustainable development in Africa.

Specific action programmes are required to address the huge challenge of ensuring that the proportion of Africans without access to basic water supply and sanitation is reduced by 50% by 2015 and 75% by 2025, including actions to promote improved hygiene.

... Africa, particularly its poor, is especially vulnerable to water-related disasters such as droughts, floods and desertification, aggravated by the impact of climate change as a consequence of human activities outside Africa. Areas for action include:

- Development of a prevention based culture, rectifying knowledge gaps and strengthening policy and institutional capacity to assess and monitor climate and water and mitigate the effects of climate change and climate variability on water resources;
- Adopt approaches to mitigate the impact of disasters and climate change.
- Strengthen disaster management capacity and emergency preparedness.

Actions should be undertaken to increase public awareness and strengthen the political will needed for sustainable development and management of water resources. The building of human and institutional capacities is crucial for the implementation of IWRM. There is an urgent need to establish or strengthen institutions for research and information sharing.

Mobilizing the funds that are needed

Unless we address the underlying poverty of many African people, it will not be possible to sustain their access to safe water and hygienic sanitation or to create sustainable livelihoods using water. Improved household incomes are essential if cost-recovery based strategies are to mobilize the funds required.

Water is a public good used for social and economic purposes. Water service providers should aim for financial sustainability, charging the full cost to those who can afford to pay, with transparent subsidy arrangements from public funds and cross-subsidies where the poor cannot afford the full cost. Governments must ensure that resources are mobilized first from internal sources, using public

(Continued)

Box 2. *Continued*

funds for services for the poor, and from private sector funds to meet national
objectives. As much attention should be paid to environmental sustainability
and funding ongoing operations and maintenance costs as to initial investment.
A dedicated water fund for Africa should be established and the establishment
of similar funds at national and basin levels considered. Such funds could
support IWRM as well as initiatives to encourage cooperation on shared basins.

---

5. The African crisis

From the socio-economic point of view, Africa faces a crisis of endemic poverty
and pervasive underdevelopment. For many African countries, economic performance in the immediate post-colonial era was good. However, for most of Africa,
particularly, for sub-Saharan Africa, since the oil crisis of the mid-1970s, economic
performance has been poor and worsening (Mkandawire and Soludo, 1999).

One of the worse performing sectors has been agriculture. According to
Mkandawire and Soludo (1999), long-term growth prospects in Africa will depend
on how well agriculture performs. It is argued that in most countries in Africa,
agriculture will be the source of foreign exchange and savings. It will also be
an important source of inputs for industry and a major contributor to the market for some of Africa's infant industries. Yet, Africa is the only continent where
the growth in per capita food production has been lower than the growth in population. From 1980 to 1994, average Gross Domestic Product (GDP) growth rates
were lower than population growth rates as shown in Table I. More recent data
show, however, that between 1996 and 1998 there has been some economic recovery, and average GDP growth rates have exceeded population growth rates for the
first time in the past two decades. However, this recovery is deemed to be still
fragile, and there is a long way to go to achieve a sustainable turn-around.

6. Responding to the crisis: the role of water

Several economic instruments are being deployed to address this crisis. The success of these efforts will, however, depend heavily on the availability of sustainable
water resources. At the same time, success in these efforts is needed to ensure the
sustainable flow of funds for water resources development. One example of this is
the link between water and poverty. Due to poverty, access to adequate water and
sanitation is low in Africa. Yet, due to the inadequate access to safe water supply
and sanitation, there is a high incidence of communicable diseases that reduce vita-
lity and economic productivity on the continent. In effect, 'half the work of a sick
peasantry goes to feed the worms that make them sick.' Thus inadequate access
to water and sanitation is both a cause and a consequence of poverty. Similarly,

TABLE I. Economic performance in Africa – 1965–1998 (ADB, 1994; WB, 1991).

Indicator	Performance (%)					
	1965–73	1974–79	1980–85	1986–93	1990–94	1995–98*
Population growth rate	2.7	2.9	3.0	3.0	3.0	2.7
Growth rate of GDP (avg.)	5.7	3.5	1.8	2.5	1.9	3.75
Growth rate of per capita GDP (avg.)	3.0	0.7	−1.1	−0.5	−1.1	1.05
Growth rate of agricultural output (avg.)	2.7	3.0	1.5	2.7	2.1	3.4
Growth rate of manufacturing output (avg.)	7.3	6.7	5.2	2.5	1.3	2.9
Growth rate of investment (avg.)	9.6	6.9	−4.8	1.2	0.8	—
Savings-GDP (avg.)	16.2	20.9	16.3	15.6	15.3	—
Growth rate of exports (avg.)	8.2	2.6	0.4	3.0	0.6	5.25
Growth rate of imports (avg.)	7.4	6.2	−2.4	0.7	0.4	5.8

inadequate use of water resources can become a constraint to improved agricultural development and food security and, consequently, a constraint on resource availability for water resources development. It is estimated that on the African continent only 4–7% of cultivated lands is under irrigation, that is, less than 20% of irrigable land that can be irrigated. The inadequacy of investment for soil improvement is the cause of the excessive extraction of nutriments by agricultural activities (Sonou, 2002). Over the last 15 years, soil degradation has considerably affected productivity and agricultural yield on the African continent (Sonou, 2002).

The untapped potential and resultant poor agricultural productivity are given rise to by inadequate investment in areas such as irrigation, the use of fertilizers and other commercial inputs, mechanization, etc.

It is apparent that water and socio-economic development are mutually dependent on each other. They can be nodes in a vicious cycle that puts societies in a downward spiral of poor economic development and poor access to safe and adequate water supply and sanitation. Alternatively, they can reinforce each other in an autocatalytic way, leading to an upward spiral in which improved socio-economic development produces resources needed for improved water resources development that, in turn, buttresses and stimulates further socio-economic development. Where, then, does Africa begin? The Vision and the FFA were developed with an intention to provide a way of thinking about this problem and defining priorities for action. In this regard, the Dublin–Rio principles provide a guide for defining the key roles of water in socio-economic development and for sound management of water resources. The Dublin principles are as follows:

1. Fresh water is a finite and vulnerable resource, essential to sustain *life, development, and the environment.*

2. Water development and management should be based on a participatory approach, involving users, planners and policy makers at all levels.
3. Women play a central role in providing, managing and safeguarding of water.
4. Water has an economic value in all its competing uses and should be recognized as an economic good.

The Rio principles expand the fourth of the Dublin principles to underscore the need to regard water, not only as an economic good, but also as a social good. In a way, this modification merely clarifies the fourth principle to reflect the notion inherent in the first principle that one of the essential uses of water is *to sustain life*. In this Africa Water Vision, the first and fourth Dublin–Rio Principles are interpreted to mean that water has an economic value in all its uses. This means that, it should always be treated as an economic good in its competing uses for *development*. However, in its use for *sustaining life and the environment*, it should be treated not only as an economic good, but also as a social good. This distinction is important in the pricing of services for water supply and sanitation and in the formulation of policies on water allocation for sustaining life and the environment.

7. Salient features of water resources in Africa

Africa has abundant water resources in large rivers, great lakes, vast wetlands and limited but widespread groundwater. Much of this is located in the Central African sub-region. Africa has 17 rivers with catchment areas greater than $100\,000\,\text{km}^2$, and also has more than 160 lakes larger than $27\,\text{km}^2$, most of which are located around the equatorial region and sub-humid East African Highlands within the Rift Valley.

The continent has a huge potential of energy production through hydropower. In this respect, efforts are already under way to create regional power pools in sub-Saharan Africa. However, synergy needs to be established between energy and the power sectors, and between transboundary water countries in order to gain the maximum economic benefits from this potential.

This abundance notwithstanding, there are many features that affect water resources management in Africa. The most important of them are:

1. The multiplicity of transboundary water basins
2. Extreme spatial and temporal variability of climate and rainfall
3. Growing water scarcity
4. Availability of groundwater resources
5. Increasing demand
6. Water pollution and environmental degradation

7.1. MULTIPLICITY OF TRANSBOUNDARY WATER BASINS

A key issue in water resources management in Africa is the multiplicity of international water basins. Africa has about one-third of the world's major international water basins (basins $> 100\,000\,\mathrm{km}^2$). Virtually all sub-Saharan African countries, plus Egypt, share at least one international water basin (ADB, 1999). There are about 80 international river and lake basins in Africa (Hirji and Grey, 1997). The Nile basin, for instance, has 10 riparian countries; the Congo has 9, the Niger has 11 and the Zambezi has 9. The Volta has 6, and the Chad has 8. Then there are countries through which several international rivers pass. One extreme case is Guinea, which has 12 such rivers.

Water interdependency is accentuated by the fact that high percentages of total flows in downstream countries originate from outside their borders. For example, almost all of the total flow in Egypt originates outside its borders. In Mauritania and Botswana, the corresponding figures are 95% and 94%, respectively; in the Gambia it is 86%; and in the Sudan it is 77%. Despite this, very few shared waters are jointly managed at present, and in many respects, the issues of water rights and ownership of international waters remain unresolved, and national interests tend to prevail.

According to the background information for the African Water Vision, water interdependency is accentuated by the fact that high percentages of total flows in downstream countries originate from outside their borders. For example, almost all of the total flow in Egypt originates outside its borders. In Mauritania and Botswana, the corresponding figures are 95% and 94%, respectively; in the Gambia it is 86%; and in the Sudan it is 77%. Despite this, very few shared waters are jointly managed at present, and in many respects, the issues of water rights and ownership of international waters remain unresolved, and national interests tend to prevail over shared interests.

With so many international water basins in Africa, the use of water basins as a unit for water resources management is impossible without partnership and cooperation between countries sharing a common water basin. In the absence of such cooperation, the potential for conflicts among riparian countries has increased in recent years and is likely to intensify in the future, as water scarcity increases. While national and customary laws exist to deal with conflicts at the local and national levels, existing international laws are not adequate to fully address conflicts between countries and among riparian states.

In the field of international cooperation, the Southern African Development Community (SADC) Protocol on Shared Watercourse Systems represents a model for what can be achieved if countries cooperate over their shared water resources. Other models include the Nile Basin Initiative and a number of river basin authorities such as for the Niger and Lake Chad Basins. Joint water projects between countries are encouraging examples of positive regional cooperation. Examples are the Lesotho Highlands Water Project (between Lesotho and South Africa) and the Komati Basin Project (South Africa, Mozambique and Swaziland). The challenge is for immediate action to create an enabling environment for joint management of international water basins to become the norm rather than the exception.

It would appear that partnership should not be limited to countries with shared water basins. It should be extended to cooperation between sub-regional groups as well. In the field of water and sanitation, a number of initiatives have been developed, important examples of which are the Water and Sanitation Africa Initiative (WASAI) and the Africa 2000 Initiative of the WHO to expand water and sanitation services in Africa.

7.2. EXTREME SPATIAL AND TEMPORAL VARIABILITY OF CLIMATE AND RAINFALL

Extreme spatial and temporal variability of climate and rainfall on the continent is one of the significant features of water resources in Africa, with far reaching consequences for water resources management. It is a continent with great disparities in water availability within and between countries. While there are areas with plentiful supply of water, there are other areas where there is scarcity of water. For example, northern Africa and southern Africa receive 9% and 12%, respectively of the region's rainfall. In contrast, the Congo River watershed in the central humid zone, with 10% of Africa's population, has 30% of its annual runoff. Again, in the humid equatorial zone, annual rainfall is over 1500 mm, and it exceeds evaporation. In contrast, in the Saharan and Kalahari desert areas, the annual rainfall is less than 50 mm, and it is exceeded by evaporation.

In Southern Africa, the Lake Malawi basin, Southern Tanzania and northern Madagascar have become wetter in the last 30 years. This is in contrast to the situation in Mozambique, southeast Angola and western Zambia, which have become significantly drier over the same period. The extremes in variability have been greater in Tunisia, Algeria, the Nile Basin and in the extreme south of the continent. Another example of this variability is rainfall in the Sahel region during the period 1961–1990, which was 30% lower that it was during the period 1931–1960.

7.3. GROWING WATER SCARCITY

While these variations have led to growing abundance in water-rich areas of the continent, it has also led to endemic and spreading drought and growing scarcity of water in other areas. Thus in such dry lands as the Sahelian and some Southern African countries, there has been a significant decline in rainfall. What is more, the frequency of drought has been increasing over the past 30 years, resulting in significant social, economic and environmental costs, with the poor bearing most of the cost. Not surprisingly, there are growing constraints in water supply in the dry lands that occupy about 60% of the total land area of Africa.

For example, it was reported that in 1995 six countries (Libya, Tunisia, Algeria, Rwanda, Burundi and Egypt) were facing water scarce conditions (with less than 1667 m^3 of renewable water resources per capita per year). Another four countries (Kenya, Morocco, South Africa and Somalia) were reported to be facing water stress conditions or under water scarce conditions (with less than 1000 m^3/capita/year). It has been estimated that by 2025, the number of countries facing scarcity would increase from six to eleven, and the number facing water stress would rise from four to nine. Already, about one-third of the people in the Region live in drought-prone areas, and there is one country where one-sixth of the drinking water supply in one city comes from recycled sewage that has been put through very sophisticated treatment processes.

The apparent disappearance of Lake Chad in West Africa is symptomatic of the growing scarcity of water in Africa. Originally believed to have an area of about $350\,000\,km^2$, the lake was reduced to $25\,000\,km^2$ in the early 1960s. However, today, it is reduced to about $2000\,km^2$.

While the cause of this apparent shrinkage of the lake is not well understood, it is occurring in the same area where the two complementary processes of desertification and deforestation are combining to push the frontiers of desert further south in West Africa.

7.4. INADEQUATE INSTITUTIONAL AND FINANCING ARRANGEMENTS

The practice in African water resources management has been one of technocratic interventions in the hydrological cycle undertaken to augment and regulate water supplies for the purpose of meeting specific sets of human needs without due regard for the environment and multi-sectoral use of water. A wide range of institutions are, directly or indirectly, involved in this process. At the highest level, national governments set out the policies and legal frameworks within which water resources are managed. At lower levels, regional water management organizations, public and private water utilities and corporations, local government or community institutions and water user organizations are variously involved in licensing, water allocation, construction, service delivery, operation and maintenance.

A key issue related to the institutional and financing arrangements is the adequacy of the enabling environment under which water resources are managed at local, national and inter-country levels. Current institutional arrangements are often inadequate and the financing of investments is often unsustainable. There is therefore a need for institutional reform to improve performance in the water sector. Such reform should be underpinned by the adoption of the Dublin Principles. In addition, this reform should be based on cooperation and partnership between countries and between the sub-regions in the continent, with the water basin serving as the basic unit for resource management.

Fortunately, many African countries have risen to the challenges that confront them. In the field of water policy, strategy and institutional arrangements, a number of advances have been made. These include an increased awareness of, and political commitment to IWRM. There is also an increasing commitment to water policy reform and a strong trend towards decentralization of water institutions. Furthermore, there is a thrust towards financial sustainability in the water sector and a realization of the importance of treating water as an economic good, while providing a safety net for the poor.

7.5. INADEQUATE DATA AND HUMAN CAPACITY

A key limitation at national, sub-regional and continental level is the paucity of data on water resources. This limitation is linked with inadequacy in human capacity for the collection, assessment and dissemination of water resources data for development, planning and implementation of projects.

The skills for IWRM are not widely available in Africa. A massive programme for capacity building is therefore needed to produce a cadre of water professionals (both men and women) that are highly skilled in IWRM principles and practices. Under the Global Water Partnership, a capacity building associated programme is being developed to provide strategic assistance for developing the necessary skills for IWRM. The challenge is how to retain staff, once they are given the requisite capacities. It is generally recognized that even if the trained staff is retained, the skills they acquire may become atrophied from lack of use, unless an appropriate set of incentives is also introduced. A second challenge is, therefore, how to devise such incentives so that they are consistent with the aspirations of the staff and with the goals of the water sector. These are pressing challenges that call for immediate remedial actions.

7.6. AVAILABILITY OF GROUNDWATER RESOURCES

Aquifers buffer rainfall and provide reliable sources of potable water supplies in dry years. In fact, more than 75% of the African population uses groundwater as the main source of drinking water supply. This is particularly so in North African countries, such as Libya, Tunisia and parts of Algeria and Morocco and in Southern African countries like Botswana, Namibia and Zimbabwe. However, groundwater only accounts for 15% of the continent's total renewable water resources, and in South Africa, for example, it is only 9%. As a rule, the groundwater tends to occur in small sedimentary aquifers along the major rivers and in the coastal deltas and plains.

7.7. INCREASING DEMAND AND LOW INVESTMENTS

Scarcity is not entirely due to natural phenomena. It is due in part to low levels of investment in water resources projects in the face of growing demand for water in response to population growth and economic development. These determinants of scarcity are likely to increase in significance in the future with growth in economic activities both in the agricultural and in the industrial sectors. In the SADC region, for example, demand is projected to rise by at least 3% annually till 2020, a rate equal to the region's population growth rate. As a consequence of demands like this, it has been estimated that by 2025, up to 16% of Africa's population (23 million) will be living in countries facing water scarcity, and 32% (another 460 million) in water-stressed countries (World Water Council, 2000). The rising demand for increasingly scarce water is leading to growing concern about future access to water, especially where water resources are shared by two or more countries.

7.8. WATER POLLUTION AND ENVIRONMENTAL DEGRADATION

Water quality deterioration is another form of demand on available water resources. At best, it increases the cost of water resources development; at worst it increases water scarcity. The main threats to water quality in Africa today include eutrophication, pollution and proliferation of invasive aquatic plants. Eutrophication is a factor mainly in lakes. The main cause is pollution from food processing wastes and decay of invasive aquatic weeds. Most industries in Africa discharge their waste into receiving waters without treatment. Water hyacinth has already seriously affected most water bodies like Lake Victoria, the Nile River and Lake Chivero. Future threats may include pollution from agricultural activities, such as fertilizers and pesticides, and from small-scale industries dispersed in large urban areas.

A major water quality problem is salt-water intrusion, especially along the Mediterranean coast and on the oceanic islands like the Comoros that are highly dependent on groundwater resources.

8. The key challenges

While there are numerous challenges to be overcome in the water sector in Africa, it is possible to identify five broad categories. These are access to water supply and sanitation, food security, environmental sustainability, management of international waters and retention of motivated and highly skilled water professionals.

Given the poor performance of economic development in Africa, the challenge is how to finance a sustainable supply of safe and adequate supply of both water supply and sanitation services to all, especially the poor in the shortest possible time. This is an immediate challenge that should be addressed by 2005.

Another challenge is, therefore, how to develop Africa's water resources to support the needed expansion of both rain-fed and irrigation-based agricultural production to ensure food security and economic development in the Region. This is a pressing challenge that calls for immediate action, and should be fully addressed by 2015.

Unfortunately, water contamination is rife across the continent, the result of industrial pollution, poor sanitation practices, discharges of untreated sewage, solid wastes thrown into storm drains and leachates from refuse dumps. These problems are compounded by poor land use and agricultural practices. As a consequence, concentrations of waste frequently exceed the ability of rivers to assimilate them, and water-borne and water-based diseases are widespread. Environmental degradation has, however, not been limited to poor water quality management.

It is due, in part to failure to recognize the life-supporting functions of ecosystems (terrestrial and aquatic). Hence the requirements of ecosystems for water quantities and quality have not been taken into account in the overall allocation of available water resources in much of Africa. Hence the important role played by wetlands in many rural economies (for the provision of highly productive agricultural land, dry season grazing for migrant herd, fish, fuel wood, timber needs, medicines, etc.) has not been reflected in national water policies. Hence such wetlands are increasingly being threatened by poor cultivation, deforestation and overgrazing.

As stated earlier, the Dublin Principles explicitly call attention to the essential role of water not only for development, but also for life and *the environment*. The challenge is to recognize the legitimate use of water for sustaining the environment as well as the life-supporting functions of ecosystems. This should lead to the generation of a broad-based support and a legal basis for separating water resources into three categories. One part would be subjected to competing demands for economic development, one part would be reserved for sustaining the environment, and the third part reserved for meeting basic needs for sustaining life. This is a yet another challenge that should be met by 2005.

9. What did the WSSD achieve?

The WSSD provided a unique opportunity for the Water sector to secure a great prominence. Unfortunately the opportunity was mainly for a sub-sector of the Water sector. The WSSD concentrated on the issue of lack of access to safe and adequate water supply and sanitation. This is indeed a great problem in Africa however it should always be noted that water is not only about drinking water. As already stated above there are other facets to it which are equally important. Water must be considered in its entirety hence the promotion of IWRM. This was one unfortunate failure for the WSSD. The issues identified above are very important and would have needed adequate coverage during the summit.

However the success of recognizing water supply as key to sustainable development hence agreement on goals for both water and sanitation should not be underestimated. The water sector now has a basis from a political point of view to work.

The Summit recognized that the lack of action and low priority on water in many countries is not caused by a lack of agreement on the urgency and the need to take determined action. On the contrary, there is strong agreement on many of the key issues surrounding the issues of water and sanitation. However, on issues such as cost-recovery or financial systems to ensure access and availability and the role of the private sector, there is less agreement. More global coordinated action and higher priority by countries will help sort these differences more easily. One of the presenters during the WSSD plenary session mentioned that there are several estimates made on how much is required to reach the MDGs on water. One of these calculates that it would require between US$ 14 billion and $ 30 billion a year on top of the roughly US$ 30 billion a year already being spent (UN, 2002b).

A common theme in the areas of water and sanitation is the need to involve all stakeholders for a multi-stakeholder approach to water and sanitation. The need to examine the institutional frameworks that establish priorities and policies for water and sanitation is also important as many of the decisions regarding water and sanitation and the effects on people are taken in a variety of sectors and ministries. Some of the outputs of the WSSD are highlighted in Box 3. These outputs only relate to water issues.

Box 3. Selected water related outputs of the WSSD (UN, 2002b).

Poverty eradication

Halve, by the year 2015, the proportion of the world's people whose income is less than $1 a day and the proportion of people who suffer from hunger (*reaffirmation of MDGs*).

By 2020, achieve a significant improvement in the lives of at least 100 million slum dwellers, as proposed in the "Cities without slums" initiative (*reaffirmation of MDGs*).

Establish a world solidarity fund to eradicate poverty and to promote social and human development in the developing countries.

Water and sanitation

Halve, by the year 2015, the proportion of people without access to safe drinking water (*reaffirmation of MDG*).

Halve, by the year 2015, the proportion of people who do not have access to basic sanitation.

Develop IWRM and water efficiency plans by 2005.

Announcements: Water & sanitation

- The United States announced $970 million in investments over the next three years on water and sanitation projects.
- The European Union announced the "Water for Life" initiative that seeks to engage partners to meet goals for water and sanitation, primarily in Africa and Central Asia.
- The Asia Development Bank provided a $5 million grant to UN Habitat and $500 million in fast-track credit for the Water for Asian Cities Programme.
- The UN has received 21 other water and sanitation initiatives with at least $20 million in extra resources.

The Summit further recognized that the subjects of water and sanitation revolve around the issues of:

- Access, availability and affordability
- Allocation issues
- Capacity building and technological needs
- Social issues

Some brief on the above four issues is given below.

9.1. ACCESS AND AVAILABILITY

The numbers of people to reach with adequate and quality coverage are immense and either in the poor rural areas or in the marginal urban or peri-urban areas where the ability to pay for the services is more limited; while the summit recognized the need for strong partnerships between the communities, Central Governments and those charged with the responsibility of providing water and sanitation services, there was no clear programmes on how this can be addressed. Indeed these would be country programmes but the Summit would have been the best opportunities to highlight how countries should try and resolve this problem.

The need to come up with policies, including cross-subsidization schemes to help pay for the services of the poorest areas of the population; water is very essential for human life and yet it costs some money to treat it and transport it. A balance would need to be struck between the affordability levels of the consumers and the cost of providing the water. There is need for a *policy framework* that (i) recognizes the broad range of local actors from communities, Non-Governmental Organizations (NGOs) and the private sector, that are engaged in water and sanitation services, and (ii) opens the way to partnerships and innovations to adapt the services to the capacity and the preferences of the poor seen as full-fledged customers and stakeholders.

Good policies will have to be complemented by *implementation strategies* defining roles and responsibilities, setting frameworks and processes for giving the poor a voice and bringing in all stakeholders engaged in service provision. At the heart of such strategies will be *tariff systems* that make basic services accessible and affordable to the poor while at the same time safeguard the financial autonomy of the utility.

Eventually, good policies and sound strategies should lead to *stepped up level of investment*. The requirements even for a country like Cote d'Ivoire which stands at the high end of regional performance in water and sanitation services, are staggering; halving the proportion of urban households not yet connected to the network by 2015, would require annual outlays in the order of magnitude of $1\frac{1}{2}\%$ of the current GDP. Most countries will depend on external aid for a large percentage of these requirements.

The private sector has a role to play as a provider of technology, management and finance. It is most effective when there is a strong public sector assigning allocation priorities and where standards of accountability are present for all sectors.

The need for decentralized solutions to fit the needs of the local and rural communities, including with less costly technologies that use local human and capital inputs.

Poorer countries facing water scarcity will face difficulties in providing access to water and sanitation particularly where they are constrained by indebtedness.

9.2. ALLOCATION ISSUES

Water has many uses and many competing demands. These demands come not only from various sectors of the population but also from various sectors of the economy. An IWRM approach at the country, regional and local levels is key to mediate among the various demands in a rational way.

Because there are sectors of the population that are less able to pay for services, policies and strategies need to be formulated to ensure that there are differentiated pay schemes that can eliminate the present system which often has the poorest paying the highest costs for services.

Transboundary considerations need to be given greater attention and resolution through regional cooperation (including those affected by civil conflict that lead to ecosystem destruction).

The biggest user of water resources – agriculture – need to improve on water use efficiencies ('*more crop for the drop*').

Many ecosystems that are crucial for the water supply often lack constituencies. As a result, they are often degraded by human activity. The link between the conservation of ecosystems and water needs to be better recognized. There is need for public awareness on the link between conservation of ecosystems and water supply.

9.3. CAPACITY BUILDING AND TECHNOLOGICAL NEEDS

As already stated one of the major problems of the water sector in Africa is lack of or inadequate human resources capacity to follow on issues in the sector. This includes the highly qualified professionals as well as technicians. There is a need for capacity building particularly in the introduction, use and maintenance of technologies that fit the needs of local poor populations.

There is a need for capacity building and education on water management and conservation as well as on sanitation and hygiene; usually the traditional engineering training has not addressed the needs of poor populations. The training is more engineering related than community based. During a recent e-conference (WEDC, 2002) one of the conclusions was that high institutions of learning should incorporate topics related to community management in their programmes.

There are capacity building needs in support of integration and coordination among sectors and communities; this point should be underestimated.

Education, information and public awareness are essential in support of water management and conservation.

The Summit did not make a deliberate effort to address this very critical problem. Even though it cannot be expected from a summit to give details about capacity building but a mention of how capacity building would work to assist meet the demand for water is a very welcome. Governments need to develop adequate capacity building programmes for the water sector.

9.4. SOCIAL ISSUES

Water is a human right and most countries are in agreement with this policy statement. Unfortunately, there is less agreement as to how to put this into practice; this is the case especially as related to striking a balance between cost recovery and the need to ensure that there is universal access to water. The role of the private sector usually attracts more debate other than using it as a tool for efficient and better provision of the service. The debate is usually that the private sector is only interested in making profit. With the right policies in place it is possible to strike a balance between the interest of the private sector and efficient provision of water.

Women and children and vulnerable populations in general are taking the brunt of the negative impacts of the lack of action on water and sanitation – when there is scarcity of water and sanitation in any one country or region, it is not the rich who are affected but the poor.

Better institutional frameworks are needed for governance, decentralization, multi-stakeholder – and overarching framework that helps to bring all of these into a well-linked national – regional – and local levels (linking strategies and policies with actions at the local level). Governments need to put in place institutions that adequately address the needs of all segments of the population including the poor. As already stated above, inefficient institutions hurt the poor more. Water becomes more expensive as the poor end up purchasing water in small quantities but at a higher unit cost than those that are connected.

Allocation of water among sectors of the population should not be based on the ability to pay but on need and in the case of the ultra-poor with little or no purchasing power, measures should be taken to ensure that water is supplied as needed.

10. The WaterDome

All was not lost for IWRM approach to water. A special side event known as the WaterDome was organized in parallel to the WSSD. The WaterDome was a recognized side event of the WSSD.

The WaterDome was organized by the AWTF and had the following objectives:

1. Increase awareness at all levels of the population on the link between water and development.
2. Encourage political support to the water sector especially in view of NEPAD and AMCOW.
3. An opportunity for launching of major new initiatives in the water sector in Africa.

The WaterDome approached the issues of water in a wholesome manner as it was organized according to the following thematic days:

1. Water, regional integration and finance.
2. Water and Food security.
3. Water Energy and Climate.
4. Water, Health and Poverty.
5. Water and Globalization.

The WaterDome was probably the first ever event bringing water and development together. The timing of the WaterDome made it very appropriate to link with sustainable development. In the past United Nations events such as the Earth Summit in Rio in 1992, Water did not have a high prominence.

More than 10 000 visitors passed through the WaterDome during the six days of the Dome and more than 100 Ministers and Heads of international organizations visited the WaterDome.

But what did the WaterDome achieve as a contribution to the water sector in Africa?

As already mentioned above, the first achievement of the WaterDome was the linkage between water and sustainable development. As already mentioned above, this was probably the first time globally to have the issues of water discussed in the context of sustainable development. In trying to link water and development the subject of poverty was very adequately addressed.

A session specifically dedicated to the water crisis as it relates to development and poverty was undertaken. It was recognized that without addressing the problems of water poverty reduction will not be achieved and development will not be sustainable. Governments are therefore urged to ensure that water is part and parcel of their development agenda.

One element of poverty is lack of access to basic services such as water and sanitation. The WaterDome succeeded in highlighting the important linkage between water, sanitation and poverty. One major highlight was the launch of the Water, Sanitation and Hygiene (WASH) campaign by the Water Supply and Sanitation Collaborative Council (WSSCC). This is a campaign aimed at public awareness of the need to link water and hygiene.

Usually hydroelectric generation is treated as an issue not related to water. However the WaterDome highlighted this flaw and encouraged the need to always include hydroelectric generation as one major user or beneficiary of the water sector. Most huge dams in Africa were built primarily for electric generation including the Kariba dam in Zambia/Zimbabwe, the Volta dam in Ghana, etc.

The debate on institutional arrangements for the water sector continued to take pre-eminence. The institutional arrangements are a tool for delivering all the above aspects whether it is hydroelectric generation, irrigation or water supply (for drinking). There is need a for appropriate institutions to ensure that the water sector is driven in an appropriate manner.

11. Conclusion

The paper has sought to highlight the issues of water and sustainable development and the various efforts to achieve this. The WSSD held in Johannesburg, South Africa provided a unique opportunity to achieve this.

The question still remains as to whether the thirsty will have their thirst quenched. The figures quoted in many journals and conferences are not mere figures. Behind those figures lies the imminent tragedy of people dying unnecessarily due to preventable diseases.

While the Summit made a lot of strides in highlighting the already well-known problems of water, it fell short in clearly defining the strategies that will ensure that water is part of the development agenda in Africa.

The true test of the achievements of the Summit are the actions on the ground, which would be attributed to the Summit.

Indeed the question is how to ensure that the poor do not get poorer. As the Secretary General of the United Nations stated 'This Summit will put us on a path that reduces poverty while protecting the environment, a path that works for all peoples, rich and poor, today and tomorrow.'

Governments agreed on an impressive range of concrete commitments and action that will make a real difference for people in all regions of the world.

The overriding theme of the Summit was to promote action and major progress was made in Johannesburg to address some of the most pressing concerns of poverty and the environment. Commitments were made to increase access to clean water and proper sanitation, to increase access to energy services, to improve health conditions and agriculture, particularly in dry lands, and to better protect the world's biodiversity and ecosystems.

But rather than concluding with only the words of an agreed document, the Summit also generated concrete partnership initiatives by and between governments, citizen groups and businesses. These partnerships bring with them additional resources and expertise to attain significant results where they matter – in communities across the globe.

The Summit represents a major leap forward in the development of partnerships with the UN, Governments, business and civil society coming together to increase the pool of resources to tackle global problems on a global scale.

As a result of the Summit, governments agreed on a series of commitments in five priority areas that were backed up by specific government announcements on programmes, and by partnership initiatives. More than 220 partnerships, representing $235 million in resources, were identified during the Summit process to complement the government commitments, and many more were announced outside of the formal Summit proceedings.

Indeed as already stated above, the true test of what the Johannesburg Summit achieves, is the actions that are taken afterward.

References

ADB – African Development Bank: 1994, *Policy Framework for Integrated Water Resources Management,* Cote d'Ivoire, Abidjan.

Global Water Partnership: 2000, *African Water Vision*, The Hague, Netherlands.

Hirji, R. and Grey, D.: 1997, *Managing International Waters in Africa: Process and Progress*, Paper presented at the 5th Nile Conference on Comprehensive Water Resources Development of the Nile Basin: Basis for Cooperation. February 24–28, 1997, Addis Ababa, Ethiopia.

Mkandawire, T. and Soludo, C.C.: 1999, *Our Continent, Our Future: African Perspectives on Structural Adjustment* Africa World Press, Inc., Trenton, NJ 08607.

Sonou, M.: 2002, *Water and Food Security*, paper presented at the Accra Water Conference, Accra, Ghana.

UN – United Nations: 2000, *The Report of the Millennium Assembly of the United Nations*, 5 September 2000, Network, United States of America.

UN – United Nations: 2002a, WSSD, *Summaries of the Partnership Events*, 31st August 2002, Johannesburg, South Africa.

UN – United Nations: 2002b, *World Summit on Sustainable Development. Plan of Implementation.* Advanced unedited text, 4 September 2002, United Nations, NY.

WEDC – Water Engineering Development Centre: 2002, E-Conference *Educating Engineers: Matching Supply of Human Resource Development with Demand*, 30 September–1 November, Loughborough, United Kingdom.

Water Utility Partnership: 2001, *Kampala Statement of the Regional Conference on Reforms of the Water and Sanitation Sector in Africa* 26–28 February, Kampala, Uganda.

WHO/UNICEF: 2000, *Global Water Supply and Sanitation Assessment 2000 Report.*

WB – World Bank: 1999, Africa summary briefings. Live database. Washington DC.

World Water Council: 2000 *Final Report of the Second World Water Forum and Ministerial Conference*, 17–22 March, The Hague, Netherlands.

CHAPTER 5

ENERGY AND SUSTAINABLE DEVELOPMENT AT GLOBAL ENVIRONMENTAL SUMMITS: AN EVOLVING AGENDA

ADIL NAJAM[1]* and CUTLER J. CLEVELAND[2]
[1]*The Fletcher School of Law and Diplomacy, Tufts University*
[2]*Center for Energy and Environmental Studies, Boston University*
(*author for correspodence, e-mail: Adil Najam at adil.najam@tufts.edu; tel.: +1-617-627-2706; fax: +1-617-627-3005)*

Abstract. This paper presents a framework for understanding energy issues in the context of sustainable development. It posits that there are three important ways in which energy is related to sustainable development: (a) energy as a source of environmental stress, (b) energy as a principal motor of macroeconomic growth and (c) energy as a prerequisite for meeting basic human needs. These three dimensions correspond to the three dimensions of the often-used triangle of sustainable development: environmental, economic, and social. Using this framework, the paper traces how successive environmental summits at Stockholm (1972), Rio de Janeiro (1992) and Johannesburg (2002) have dealt with energy issues. It identifies a slow, surprising and important evolution of how energy issues have been treated at these global discussions. Energy has received increasing prominence at these meetings and become more firmly rooted in the framework of sustainable development. Stockholm was primarily concerned with the environmental dimension, Rio de Janeiro focused on both the environmental and economic dimensions, and the major headway made at Johannesburg was the meaningful addition of the social dimension and the linking of energy issues to the UN's Millennium Development Goals.

Key words: basic human needs, energy, human development, Johannesburg Earth Summit, Millennium Development Goals, sustainable development.

Abbreviations: Btu – British thermal units; CSD – Commission on Sustainable Development; E – Global Energy Use; GDP – Gross Domestic Product; GEF – Global Environmental Facility; GWP – Gross World Product; HDI – Human Development Index; LPG – Liquid Petroleum Gas; MDGs – Millennium Development Goals; NGOs – Nongovernmental Organizations; UNCED – United Nations Conference on Environment and Development (1992); UNCHE – United Nations Conference on the Human Environment (1972); UNEP – United Nations Environmental Programme; UNFCCC – United Nations Framework Convention on Climate Change; WCD – World Commission on Dams; WSSD – World Summit on Sustainable Development (2002)

1. Introduction

The 2002 World Summit on Sustainable Development (WSSD) was the most recent of a series of attempts to deal holistically with global environmental issues by holding high profile, multi-issue policy summits. The last 30 years of such summitry have not only yielded a rapidly expanding global environmental

Readers should send their comments on this paper to: BhaskarNath@aol.com within 3 months of publication of this issue.

L. Hens and B. Nath (eds.), The World Summit on Sustainable Development, 113–134.
© 2005 *Springer. Printed in the Netherlands.*

agenda but have also witnessed a noteworthy evolution in the policy framing of global environmental issues (Chasek, 2001; Banuri and Najam, 2002; Seyfang and Jordan, 2002).

The first of these mega-meetings, held at Stockholm, Sweden, in 1972, was called the UN Conference on the Human Environment (UNCHE) and dealt with – as its name rightly suggests – a rather small set of issues that were most directly related to the 'human environment' (Rowland, 1973). Twenty years later, a far more elaborate agenda came under discussion at Rio de Janeiro, Brazil, when the more ambitiously titled UN Conference on Environment and Development (UNCED) sought to radically expand the global agenda by moving well beyond the merely environmental and seeking to establish 'environment *and* development' as the central policymaking framework (Adede, 1992; Najam, 1995). By 2002, in Johannesburg, South Africa, the concept of 'sustainable development' which had already begun to assume salience at Rio gained further centrality not only by being incorporated into summit's title, but also by becoming the key motivator of the expanded Johannesburg agenda, which now included such issues as sanitation, HIV/AIDS and poverty eradication.

If this evolution – from a policy framework principally rooted in 'environmental' concerns to one imbedded in the broader and more integrated notion of 'sustainable development' – is to be anything more than rhetorical it should be reflected not only in the titles of the major conferences but also in how particular issues are tackled at these summits. This paper will review one such issue, energy, in terms of (a) how it relates to sustainable development at a conceptual level; and (b) whether there is any noticeable difference in how it was treated at the three major global environmental summits, in particular, at WSSD.

2. Energy and sustainable development: conceptual connections

Smil (1994) has argued convincingly that a direct correlation between changes in energy use – both source and converters – and advances in human well-being is one of the dominant features of human history. Here we refer to the variety of fuels and electricity that people use to meet their wants and needs. The efficient use of energy and supplies that are reliable, affordable and less-polluting are widely acknowledged as important, and even indispensable, components of sustainable development (WCED, 1987; Goldemberg and Johansson, 1995). Although perennial debates linger about precise definitions of sustainable development (Lélé, 1991; Murcott, 1997), there is increasing agreement amongst scholars and practitioners that sustainable development policy relates to three critical elements that need to be treated together: economic, social and environmental (Banuri et al., 1994). In identifying these essential elements, Munasinghe (1992) suggests that one might envision sustainable development in terms of an appropriate vector of economic, social and environmental attributes.

Energy is central to any discussion of sustainable development because it is central to all three dimensions (Munasinghe, 2002). In terms of the economic

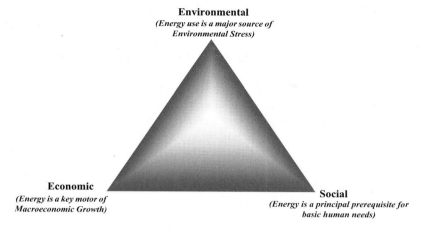

Environmental
(Energy use is a major source of Environmental Stress)

Economic
(Energy is a key motor of Macroeconomic Growth)

Social
(Energy is a principal prerequisite for basic human needs)

Figure 1. Energy and sustainable development: Deep linkages.

dimension of sustainable development, energy is clearly an important motor of macroeconomic growth. In terms of the environmental dimension, conventional energy sources are major sources of environmental stress at global as well as local levels. In terms of the social dimension, energy is a prerequisite for the fulfill-ment of many basic human needs and services, and inequities in energy provision and quality often manifest themselves as issues of social justice. Figure 1, builds on the work of Munasinghe (2002) and presents the now-familiar triangular dia-gram depicting the three essential elements of sustainable development, modified to show how the energy-dimension maps on to each of these elements. The remain-der of this section will elaborate upon the conceptual linkages between energy and each of the three dimensions of sustainable development.

2.1. ENERGY AND ENVIRONMENTAL STRESS

An important connection between energy and sustainable development concerns the environmental dimension in terms of the relationship between energy extrac-tion, processing and use, and environmental quality. This link is now well estab-lished in the scientific literature and is increasingly recognized in policy circles (IPCC, 2001). Atmospheric releases from fossil fuel energy-systems comprise 64% of global anthropogenic carbon dioxide emissions from 1850 to 1990 (Mar-land et al., 2002; Houghton and Hackler, 2001), 89% of global anthropogenic sulfur emissions from 1850 to 1990 (Lefohn and Husar, 1999), and 17% of global anthropogenic methane emissions from 1860 to 1994 (Stern and Kaufmann, 1996). Fossil energy combustion also releases significant quantities of nitrogen oxide; in the US, 23% of such emissions are from energy use (EIA, 2001). Power genera-tion using fossil fuels, especially coal, is a principal source of trace heavy metals such as mercury, selenium and arsenic. These emissions drive a range of global and regional environmental changes, including global climate change, acid deposition and urban smog.

Upstream energy sectors also have significant local impacts on the environment. Coal mining disturbs vast areas of natural habitat. In the US, for every ton of coal mined, 6 additional tons of overburden and waste are generated (Matthews et al., 2000). The exploration for and extraction of oil and natural gas can have significant impacts, particularly in sensitive ecosystems such as wetlands and tundra, and it releases hazardous and toxic wastes from drilling and field processing operations. While a potentially renewable source of energy, hydropower development can have significant environmental and social costs depending on its location and mode of development. Between 30 and 60 million people – the majority residing in China and India – have been displaced by large dams; about $500\,000\,km^2$ of land – almost the size of France – are covered by large hydroelectric reservoirs (WCD, 2001; McCully, 2001). Much of this is river valley land that supports fertile farmland and diverse forest and wetland ecosystems. Anaerobic decomposition of organic material in these reservoirs may be a significant source of methane, a potent greenhouse gas (St. Louis et al., 2000).

In terms of sustainable development, energy extraction, processing and use are major sources of environmental stress at global, regional and local levels. Although the potential of global climate change resulting from the excessive use of fossil fuels is the most dramatic and obvious of such concerns, the environmental impacts of energy use are broader than just fossil fuel use and global climate change. At a minimum, then, sustainable development policy must reflect the environmental stress resulting from the energy choices made by nations, corporations and even individuals on the global, regional and local environments.

2.2. ENERGY AND ECONOMIC GROWTH

Energy plays an equally central role along the economic dimension of sustainable development as a key driver of macroeconomic growth. At root, economic growth is a physical process: energy is used to transform materials into useful goods and services. At an aggregate level, therefore, it is not surprising that there is a strong relationship between the quantity of energy a nation uses and the size of its economy, i.e., the quantity of wealth it produces (Figure 2). Therefore it follows that the largest economies (US, China, Japan) use considerably more energy than smaller economies. It stands to reason, then, that the rise in material living standards in the poorest nations – a central goal of sustainable development – is likely to be accompanied by a substantial increase in their aggregate energy use. A similar spread appears if one compares per capita Gross Domestic Product (GDP) and energy use.

This development path is evident in the dynamics of energy and economic growth. Both Gross World Product (GWP) and Global Energy Use (E) increased steadily from the end of the second oil price shocks of 1981–1982 through 2000 (Figure 3). Although the general pattern is clear, energy use and GWP do not move in lockstep. Indeed, the E/GWP ratio declines by an average of 1% per year over this period. Analysts attribute this improvement to the shift to higher quality fuels, improvements in fuel efficiency caused in part by higher fossil fuel prices in

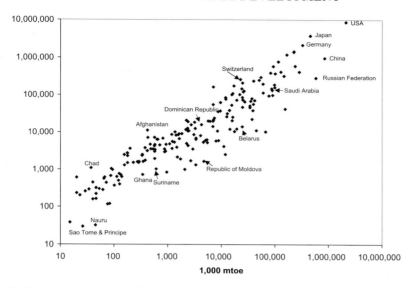

Figure 2. The international relationship between energy use (1000 million tonnes of oil equivalent) and GDP (million US dollars, 1998) (WRI, 2002).

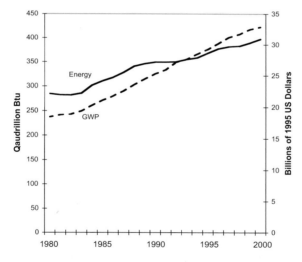

Figure 3. Global energy use (quadrillion Btu) and GWP (billion US dollars, 1995), 1980–2000 (World Bank, 2002).

industrial nations, and structural changes in the global economy (Kaufmann, 1992; Cleveland et al., 1984). Despite the decline in the E/GWP as well as national E/GDP ratios, there is a significant body of econometric research that suggests energy use and GDP are tightly linked. In particular, Granger causality running from energy use to GDP has been established for many industrialized (Stern, 2000), emerging (Hondroyiannis et al., 2002) and developing economies (Ebohon,

1996). This means that an increase in GDP is likely to require an increase in energy usage; energy, therefore, is a principal ingredient of economic growth.

Given the importance of energy, it is not surprising that energy prices have an important effect on almost every major barometer of macroeconomic performance. For example, both economic theory and the empirical evidence link rising oil prices to real GDP losses. Hamilton (1983) was the first to demonstrate this, showing that all but one of the US post-WWII recessions were preceded by rising oil prices, and that other business cycle variables could not account for the recessions. Oil prices are an important driver behind stock price movements (Sardorsky, 1999). Countries that are net importers of oil and gas tend to have a negative correlation between oil price changes and stock returns, while net exporters of oil and gas tend to have positive relationships. Energy prices are also key determinants of inflation and unemployment (Hooker, 1999). Recent work suggests that some of these relationships may have weakened in the past two decades (Brown and Yucel, 1999), although some of this debate centers around technical arguments about econometric specification and estimation techniques.

The point to be made here is that even if energy had no environmental impacts whatsoever, it would be a key issue for sustainable development policy on economic grounds alone. The fact that energy is, in fact, both a key motor of economic growth and a key source of environmental stress only makes the issue more confounding for sustainable development policy; the goal of such policy now becomes to optimize the economic virtues of increased energy use with its potential for environmental damage.

2.3. ENERGY AND BASIC HUMAN NEEDS

The connections between energy and sustainable development become all the more complex and compelling when one also focuses on the social dimension of the sustainable development equation, particularly in terms of its role in meeting basic human needs. Energy *per se* is not a need, but it is absolutely essential to deliver adequate living conditions, food, water, health care, education, shelter and employment. For example, energy availability is a key determinant of how and how much food is grown, how food is cooked, the health impacts of how food is cooked or how living spaces are heated, the time required to 'procure' household energy, and so on. The human implications of insufficient energy choices in the face of abject poverty are immediate and pressing. For example, millions of women in developing countries, particularly in Africa, have to walk long distances and spend substantial proportions of their day in gathering fuelwood, they are more susceptible to diseases of the lungs and eyes because of the energy choices they are forced to make, and they have to raise families in circumstances of extreme indoor air pollution (Wamukonya, 1995; Masera et al., 2000). Indeed, a strong relationship is evident between per capita energy use and a number of social indicators (Goldemberg and Johansson, 1995). For example, Figure 4 charts the relationship between commercial energy use and the Human Development Index (HDI). At low levels

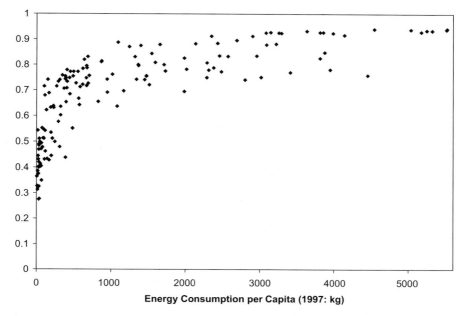

Figure 4. The international relationship between energy use (kilograms of oil equivalent per capita, 1997) and the HDI (2000) (UNDP, 2002; WRI, 2002).

of the HDI, dramatic improvements come with relatively small increases in energy use (Suarez, 1995; Reddy, 2002).

Developing nations, and particularly the poorest ones, consume far less energy per capita than developed nations. Based on this relationship, and on the energy requirements for basic household needs, Reddy (2002) estimates that about 100 watts/capita is required to achieve a reasonable quality of life corresponding to safe, clean and efficient cooking with a Liquid Petroleum Gas (LPG)-like fuel and home electrification for lighting, fans, a small refrigerator and a television. It should be noted that this 100 watts per capita is only about one tenth of the level required to support a western European living standard with modern energy sources and energy-efficient converters. Whether the number suggested by Reddy is correct or not, it is clear that some minimum amount and quality of energy is required by each person, each day, in order for him or her to meet basic human needs and sustain a decent quality of life. Indeed, one could conceive of such a minimum energy requirement as a basic human right since the inability to procure such energy can only lead to deprivation. The social dimension of sustainable development demands that the incidence of energy deprivation be determined and tackled at the human level rather than the national level, just as we seek to determine and tackle economic poverty at the human rather than the national levels (see UNDP, 1990).

Energy quality matters as well as energy quantity. Energy use is typically measured in heat units (joules, Btus, etc.). However, fuels vary in their ability to provide a service *per heat unit used* (Kaufmann, 1994; Cleveland et al., 1984). This is due to fact that the 'quality' of a fuel is a function of a host of attributes: amenability to storage, ease and cost of conversion, energy density, emissions, etc. Heat content is just one part of this picture. Based on this, coal is higher quality than wood; oil is greater than coal; and electricity higher than directly burned solid, liquid or gaseous fuels. Poor nations not only use small amounts of energy, they also tend to rely on lower quality fuels such as animal dung, agricultural wastes and fuelwood. Reliance on these fuels limits the amount of service that can be gained per heat unit used. The goal of climbing the 'energy ladder' in developing nations reflects the greater expansion of service available per heat unit of fossil fuels and electricity. An equally important ingredient of the 'quality' of energy is its direct health impacts, particularly in the case of biomass sources and their role in indoor air pollution (Ndiema et al., 1998). From the sustainable development perspective, the health dimension of energy quality would be as important as the efficiency dimension.

Just as the social dimension of sustainable development forces policymakers to look beyond aggregate development performance, it also demands that we look beyond aggregate energy performance and availability. The human development focus imposed by the social dimension of sustainable development requires us to look not simply at GDP but also at distribution of resources and opportunity across society, and particularly at that portion of the population that is most deprived (e.g. the proportion that lives at less than one dollar a day). Analogous to this, a sustainable development focus would demand that in addition to looking at the national structure of energy availability and use, policymakers pay attention to the distribution of energy within societies in terms of both quantity and quality. Once again, those most deserving of policy attention are those who are most deprived in terms of their energy options; those whose energy limitations keep them from meeting their basic human needs.

3. Global policy on energy and sustainable development

Having defined a framework for examining the conceptual relationship between energy and sustainable development, let us now examine how the issue of energy has been dealt with at the three major global environmental conferences held at Stockholm (1972), Rio de Janeiro (1992) and most recently in Johannesburg (2002). While these are not the only events of importance to have taken place in this period, they are the three most important policy conferences that have been attended by government representatives at the highest level and which have focused on a wide range of issues related to what has come to be known as sustainable development. For this reason we have not focused on a host of energy-specific conferences that have been held over this period or specific policy negotiations on energy-related issues such as global climate change. Our goal is not as much to focus on the development of energy policy, but to concentrate on the evolution of sustainable development policy in order to highlight how energy issues have featured within these discussions. Our purpose is to chart the evolution of the policy response and, in particular, to examine whether a perceptible trend of moving from a purely 'environmental' focus to a broader 'sustainable development' focus is evident in how these summits have dealt with energy issues. We will do so by examining the key documents negotiated at each of the three summits in terms of the three dimensions of energy-sustainable development linkages defined above.

3.1. STOCKHOLM, 1972

The UNCHE, held in Stockholm, Sweden, from 5 to 16 June, 1972, was a 'first' in many respects: it was the first meeting that brought the nations of the world (113 countries participated) together to discuss the environmental future of the planet; it was the first UN Conference on a single global issue; it was the first global meeting that saw a large presence and influence of non-state actors, including nongovernmental organizations (NGOs) and scholars; and it was the first meeting to seek global policy consensus on issues related to the environment.

Triggered by increasing scientific evidence of human-induced environmental degradation and a concurrent wave of growing environmental awareness in the industrialized nations of North America and Western Europe, the conference was an attempt to turn the environment into a more 'global' issue, particularly by more meaningfully incorporating the developing countries of the South into the emerging global environmental discourse. The conference turned out to be unexpectedly contentious – with most Soviet bloc countries boycotting it due to the exclusion of then East Germany and with developing countries apprehensive of the North's newfound environmental concern. However, despite the intense North–South differences, and possibly because of them, the conference stumbled towards a more authentic global agenda; one that sought to merge the North's growing concern about environmental quality and the South's long-held interest in human development. Eventually, but much later, the desire to formulate the two interests into a single composite framework would lead to the concept of sustainable development (Founex, 1972; Kay and Skolnikoff, 1972; Rowland, 1973; Pirages, 1978; Najam, 1994).

The major institutional legacy of the conference was the creation of the UN Environmental Programme (UNEP). This was accompanied by two declaratory documents – The Stockholm Declaration and the Stockholm Action Plan – ideas from which have been carried forth by subsequent summits, including at Rio de Janeiro and Johannesburg. In addition to these, a few more ritualistic declarations were also adopted: one calling for a second UN Conference on the Human Environment (which was never actually held), another establishing an annual 'World Environment Day' (which is now observed in most countries each June), and one calling for a stop to nuclear testing, particularly atmospheric nuclear testing (in fact, this was the single most hotly debated issue at the conference and inspired by the global politics of the time).

Surprisingly little was said or discussed at the conference or in any of its formal products regarding issues related to energy (see full text of Stockholm documents at http://www.unep.org). The conference Declaration, which was the major political document emerging from Stockholm, has no direct reference to the energy issue. Of the 26 'principles' laid out in the declaration, the one that can be construed to be of most relevance to energy-issues is Principle 5, which states: 'The non-renewable resources of the earth must be employed in such a way as to guard against the danger of their future exhaustion and to ensure that benefits from such employment are shared by all mankind.' A little more than one year later, energy was thrust on to the international agenda when the oil crisis nearly doubled real energy prices and plunged many national economies into recession.

The Stockholm Action Plan on the Human Environment is a more comprehensive document that includes a 'Framework for International Action' accompanied by a list of 69 specific recommendations; three of which do, in fact, deal with energy (emphasis added):

- Recommendation 57 called upon the UN Secretary-General to 'take steps to ensure proper collection, measurement and analysis of data relating to the *environmental effects* of energy use and production'.
- Recommendation 58 called for better exchange of information on energy. The recommendation is motivated by the need for 'the rationalization and integration of resource management for energy' and seeks mechanisms (such as exchange of national experiences, studies, seminars, meetings, and a 'continually updated register of research') for accessing existing information and data, particularly on 'the *environmental consequences* of different energy systems'.
- Recommendation 59 called for a 'comprehensive study to be promptly undertaken with the aim of submitting a first report, at the latest in 1975, on available energy sources, new technology, and consumption trends, in order to assist in providing a basis for the most effective development of the world's energy resources, with due regard to the *environmental effects* of energy production and use'.

Even at Stockholm itself, none of these recommendations was particularly inspiring in its scope or aspiration. They are even less so with the 30-year hindsight we now enjoy. Indeed, between them they call merely for better data collection and analysis, and the mechanisms envisaged for doing so are not particularly innovative or exciting. More importantly, even the minimal level of 'action' that the Action Plan envisaged on the energy issue never really materialized. For example, the official documents make no mention of Stockholm's implementation of the 'comprehensive study' that was sought by 1975 (Recommendation 59). The one issue related to energy that did gain wide political and policy prominence at Stockholm was atmospheric nuclear testing. This was a subject of great and heated debate at the conference, became the subject of a separate resolution, and made the International Atomic Energy Agency a frequently mentioned organization within the Stockholm Action Plan. This discussion was very much an artifact of the Cold War politics of the time and was at its root far more concerned with nuclear weapons than with energy and its role in the economy and as an agent of environmental change.

The lack of imagination or urgency on the energy issue should not be entirely surprising. Held in mid-1972, the Stockholm Conference came before the great oil shock of the 1970s (see Askari and Cummings, 1978; Allen, 1979). State delegates attending Stockholm still lived in a relatively calm world of declining real oil prices, and the possibility of spiraling energy prices or oil scarcity could not have been high on their mental maps. Similarly, global climate concerns had not yet taken root in 1972 and environmentalists were more focused on the pollution outputs of industrialization than the energy inputs for economic production. While those attending UNCHE were well aware of the many environmental implications of energy issues, these were not their most pressing priorities at that time.

What is clear from the recommendations, however, is that to the extent that energy was considered an issue of any importance, that importance derived directly from the 'environmental effects' of energy extraction, processing and consumption. While the recommendations slightly hint at the importance of energy as a motor of economic growth, their principal preoccupation is with the potential for environmental stress from the chain of energy supply and use. The dimension of energy as a prerequisite for meeting basic human needs does not figure into the equation at this conference.

3.2. RIO DE JANEIRO, 1992

The UNCED, popularly know as the Earth Summit, was the crowning moment not only of environmental summitry but of UN mega-summitry as a genre. Held at Rio de Janeiro, Brazil from June 2 to 14, 1992, it brought together more than one hundred heads of state and government, 150 nations, over 1,400 NGOs, 8,000 journalists, and nearly 35,000 participants. More than that, it caught the public imagination like no conference before or since. Some consider the summit to have failed its potential, if not its mandate in terms of the content of its substantive agreements (see, e.g. Khor, 1992; Agarwal et al., 1999). But most commentators and experts consider this Summit to have been a success in terms of elevating the global profile of environmental issues and raising awareness regarding sustainable development (Hass et al., 1992; Johnson, 1993; Gardner, 1992; Najam et al., 2002).

Held to mark the twentieth anniversary of the Stockholm Conference, the Rio Earth Summit became everything that an earlier 'Stockholm plus ten' conference, held in Nairobi in 1982, could not (see Clark and Timberlake, 1982). Indeed, it became more than even its proponents had hoped for. Instead of being the 'second' United Nations Conference on the *Human Environment*, Rio was the United Nations Conference on *Environment and Development*; putting those two terms together, which had been so much at odds at Stockholm, might itself have been Rio's most important achievement (Najam, 1995). In particular, it broadened the scope of global environmental diplomacy by adopting the notion of sustainable development, which had been advocated 5 years earlier in by the World Commission on Environment and Development (WCED, 1987) as one of its key policy frameworks (Susskind, 1994; Tolba, 1998; Chasek, 2001).

The world at Rio was, of course, very different from the world at Stockholm. In the intervening two decades the Cold War (which was the defining political framework at UNCHE) had disappeared, the level of public interest in the environment was greatly increased, environmental issues such as stratospheric ozone depletion and global climate change were now squarely on the global policy map, and energy had become a major concern for economic security in the aftermath of the oil price shocks of 1973–74 and 1980–81.

The 'products' coming out of UNCED included a political Declaration enunciating 27 principles of environment and development, a 700 page action programme called Agenda 21, a non-binding set of principles for sustainable forest management, and specific conventions on climate change and biodiversity. The institutional innovation resulting from the conference included an agreement on the operating rules for the Global Environmental Facility (GEF) and the establishment of a Commission on Sustainable Development (CSD) on the basis of an Agenda 21 recommendation. Technically, the 'official' products of UNCED were only the Rio Declaration on Environment and Development, the 'Authoritative Statement of Forest Principles', and Agenda 21; all of which were adopted by consensus (without vote) by the conference. The conventions on climate change and biodiversity were products of independent, but concurrent, negotiating processes that were opened for signatures at UNCED (full texts of all UNCED documents are available in Johnson, 1993 and at http://www.unep.org).

The Rio Declaration on Environment and Development, like its predecessor at Stockholm, had nothing specific to say about energy. Indeed, the clause about the depletion of natural resources contained within the Stockholm Declaration was dropped. While a number of the principles articulated in the Declaration could be construed to have bearing on energy, none deals with the issue directly.

Although the climate and biodiversity conventions were not direct products of the Rio process, the former is of direct relevance to the energy issue. Indeed, the UN Framework Convention on Climate Change (UNFCCC) is the nearest thing we have to a global convention dealing directly with energy concerns. Since energy production and consumption is the biggest source of anthropogenic greenhouse emissions, climate policy in the UNFCCC, and subsequently in the 1997 Kyoto Protocol, has been discussed mostly through the lens of energy policy in a wide variety of ways. Two examples, amongst many, of how climate policy becomes energy policy are the intense policy debates about the variable ability and responsibilities of different nations to change their energy consumption and production patterns (Najam and Page, 1998; Zhang, 2000; Brown et al., 2001), and the role of energy taxation as a means of emissions control (Speck, 1999; Baranzini et al., 2000; Varma, 2003).

These debates during and after Rio have been defined principally by compulsions that lie at two distinct corners of the sustainable development triangle: the environmental compulsion emanating from the ecological stresses associated with specific energy production and consumption choices, and the economic compulsion derived from the central role of energy in economic growth. The saliency that the climate issue had assumed by 1992 meant that the discussion on energy at Rio was not only more intense than it had been at Stockholm but also broader. Whereas UNCHE had been principally concerned with the role of energy as a source of environmental stress, Rio's energy concerns related to both economic and environmental dimensions.

This evolution of the energy focus was quite evident in Agenda 21, the most comprehensive of the Rio documents. Interesting, Agenda 21 (which has a total of 40 chapters) does not have a chapter on Energy. However, Chapter 9 of Agenda 21 which deals with 'Protection of the Atmosphere' serves as a *de facto* energy chapter since it focused on global climate change and related issues of fossil fuel use. In addition, the chapters on changing consumption patterns (Chapter 4), promoting sustainable human settlements development (Chapter 7), and promoting sustainable agriculture and rural development (Chapter 14) also have significant discussions on the energy issue. The vast bulk of this discussion is contextualized in the need to balance the 'environmental' and 'economic' nodes of the sustainable development triangle. What is carried over from Stockholm is a clear emphasis on the 'environmental impacts' of energy production and use (especially in terms of global climate change). New additions are the prescriptions contained in these various chapters – or the energy message of Agenda 21 – which fall into familiar categories: decrease energy consumption (see Agenda 21, Sections 4.24, 7.5), increase energy efficiency (see Agenda 21, Sections 4.18, 7.49), and develop cleaner sources of energy (see Agenda 21, Sections 9.12, 9.18).

The third dimension of the sustainable development triangle dealing with social concerns such as the role of energy as a human need do not figure as prominently in Agenda 21 as the environmental and economic dimensions. However, glimpses of such concerns do occasionally surface in the document. For example, the chapter on human settlements (Chapter 7) mentions energy as a human need at par with other needs such as water (see Agenda 21, Sections 7.27, 7.40). Section 9.9 goes the furthest by defining energy as an 'essential' component of economic as well as social development and as a prerequisite for an 'improved quality of life'. Although these references are quite general and made in passing, with very little prescriptive policy content, they do signify an important evolution from the Stockholm texts where these issues were conspicuous only by their absence.

Overall, then, one finds that UNCED did treat the energy issue very differently from UNCHE. At Stockholm energy was discussed in the most general, even cursory, fashion, and only in terms of its environmental impacts. In the years between Stockholm and Rio, concerns about the environmental stress imposed by energy production and use became more precise with the mounting evidence of global climate change. As a result, Rio was relatively more precise and prescriptive in terms of energy policy in that it went beyond simply calling for more information collection and dissemination to highlighting the need for decreasing consumption, increasing efficiency and transitioning to cleaner sources. In doing so, Rio broadened the focus from merely environmental concerns to the balance between environmental and economic concerns. However, the third node of the sustainable development triangle, the social dimension signified by energy as a human need, still remained in the shadows and peeped through the Rio documents only infrequently and rather unimpressively.

3.3. JOHANNESBURG, 2002

Few expected the 2002 WSSD to be as impressive as UNCED (Najam, 2002). Held to mark the tenth anniversary of the Rio Earth Summit and to take stock of progress on Agenda 21 in those 10 years, the run-up to Johannesburg was singularly dismal and uninspiring. The world had, once again, changed. The high hopes of a new era of global environmental cooperation that had been ignited by Rio, soon proved false. The industrialized countries of the North had remained unwilling to provide the developing countries of the South with the resources or support that had been implied at Rio, meanwhile the promise of a post-Rio harvest of global environmental treaties and implementation proved unfounded as key states, particularly but not solely the US, dragged their feet on key issues such as climate change. As a result, a malaise had set in well in advance of WSSD which was only made worse by events at the geopolitical level, where the global mood had gone sour after the tragic terrorist attack on the US and a growing sense of insecurity and violence around the world (Gardner, 2002). Held in Johannesburg, South Africa from August 26 to September 4, 2002, WSSD was different from both Stockholm and Rio in that it was not born within the optimism and high hopes that had accompanied earlier summits (Agarwal et al., 1999; Sachs, 2002).

In terms of sustainable development, the World Summit on Sustainable Development had the distinction of actually having those two magic words in its very title. However little had been achieved in the 10 years since Rio on other counts, Johannesburg was testimony to the fact that the term 'sustainable development' had gained policy acceptance. Even though some argued that the term had lost its 'edge' and was mostly being used rhetorically (Najam, 2002), the fact remained that it had also become a political necessity. For those who believed in the concept, this was a chance to put meaning into it. At best, Johannesburg was viewed as a chance to advance the agenda that had been set by Rio; at the very least, it was an attempt to keep the Rio agenda alive.

It became clear fairly early on in the Johannesburg process that WSSD would not be able to match the ambition or scope of UNCED; certainly not in terms of its products. Like Stockholm and Rio before it, the Johannesburg Summit also sought a political Declaration as its principal output. In addition, it also sought a Plan of Implementation; one that was much less ambitious in scope or scale than Agenda 21 but more extensive than the Stockholm Plan of Action. The major innovation at Johannesburg were the so-called 'Type 2' agreements. These were informal agreements involving non-state parties, sometimes amongst themselves and sometimes with individual governments. On the one hand, Type 2 agreements were a reflection of the massive change in landscape that had occurred over the previous 10 years, with NGOs and business taking a far more important role in international environmental affairs. At the same time, however, they were a reflection of the WSSD organizer's desperation and desire to get something memorable out of the summit. According to the rough count by the summit organizers, over 220 Type 2 agreements were reached at Johannesburg, signifying around US\$ 235 million in pledged resources; thirty-two of these Type 2 agreements relate to energy, accounting for US\$ 26 million in resources; the vast majority of these are programmes of technical cooperation in energy generation and conservation (http://www.johannesburgsummit.org/). It should be noted that a systematic accounting of these agreements has not yet been accomplished, and it is not yet clear how many of these agreements and how much of these resources are, in fact, new and unique. The other innovation at Johannesburg, in comparison to previous summits, relates to the fact that the Johannesburg Plan of Implementation sought agreement on actual targets and timetables rather than simple statements of intent. While it is true that in many cases (including renewable energy) such targets and timetables were not forthcoming and in others they were merely restatements of targets that had already been set (such as in access to clean water), it is also true that in a few areas (such as sanitation) meaningful headway was made in terms of reaching agreement on targets and timetables where there had previously been none (for full texts of these documents see http://www.johannesburgsummit.org/).

Looking carefully at the various products it does seem that amidst the many disappointments of Johannesburg, energy might be one of these few areas where progress was made. First, and quite strikingly, the Johannesburg Declaration is different from its predecessors from Stockholm and Rio in that it actually does have a direct reference to energy. More importantly, the Declaration (clause 18) clearly identifies energy as a human need at a par with needs such as clean water, sanitation, shelter, health care, food security and biodiversity. Although this is a declaratory clause with no enforceability, it does signify a demonstrable shift from the previous summits. It clearly defines energy as a basic human need, thereby evoking the social dimension of the sustainable development triangle more clearly than either Stockholm or Rio had done.

The Johannesburg Plan of Implementation breaks similar new ground in terms of how it deals with the energy issue. While the focus on the environmental stress-economic growth axis that had emerged at Rio is not at all lost, the Johannesburg Plan is strikingly different from its predecessors in two distinct and important ways. First, it clearly deals with energy as an issue in its own distinct right rather than as a facet of other issues. Second, it firmly adds the social dimension to the existing environmental and economic dimensions to begin dealing with the entirety of the sustainable development triangle for the very first time.

The concerns about energy in terms of environmental stress and economic growth show up very similarly to how they had surfaced at Rio. The principal arena for these concerns remains climate change and the need for a balanced approach is once again reiterated, as are the preferred mechanisms for achieving such balance: decrease energy consumption, increase energy efficiency, and transition to cleaner energy systems. This discussion is most clear in Article 20 of the Johannesburg Plan whose 23 sub-clauses relate to various environmental and economic aspects of energy in relation to sustainable development. The environmental dimension of sustainable development is most clearly and persistently manifest in the many references to the need for enhanced energy efficiency, in particular the call for establishing domestic programmes for energy efficiency (sub-clause h), the need to accelerate the development and dissemination of energy efficiency and energy conservation technologies (sub-clause i) and the call to promote and invest in research and development of such technologies (sub-clause k). The economic dimension is also strong in Article 20, parts of which call for removing market distortions including the restructuring of taxes and the phasing out of harmful subsidies (sub-clause p) and the call to support efforts to improve the functioning, transparency and information about energy markets with respect to both supply and demand (sub-clause o). One area in which Johannesburg tried, but failed, to make new headway within the environmental-economic axis relates to renewable energy. WSSD saw heated debates about setting up quantifiable targets and timetables for renewable energy use. These discussions eventually failed to yield actual timetables and targets (principally because of US opposition to them) but they did succeed in introducing more detailed language regarding energy issues than had been present in Agenda 21. The sub-clause (e) of Article 20 is, therefore, crafted in general language and calls for diversifying energy supply and 'substantially' increasing the global share of renewable energy sources.

What is new and quite intense in the Johannesburg Plan on Implementation are the repeated references to 'energy and sustainable development.' Here, the document goes beyond Agenda 21 by focusing more on the social dimension of energy and sustainable development and by concentrating on the role energy plays as a prerequisite for basic human needs including those defined in the UN's Millenium Development Goals (MDGs). The most significant of these references is made in Article 9 of the Plan of Implementation, which falls within the section on poverty eradication. New ground is covered here when the Johannesburg Plan clearly and unambiguously calls for:

> ... access to reliable and affordable energy services for sustainable development sufficient to facilitate the achievement of the millennium development goals, including the goal of halving the proportion of people in poverty by 2015, and as a means to generate other important services that mitigate poverty, bearing in mind that access to energy facilitates the eradication of poverty.

This is not only interesting but groundbreaking, because the original Millennium Development Goals (MDG) – a set of quantitative targets proposed by the UN Secretary General to the 2000 Millennium Summit as the foundation of the

United Nations' work programme from the next decades and as a framework for measuring development progress (see http://www.developmentgoals.org,
Devarajan et al., 2002) – do not mention energy at all. The case for energy as a
prerequisite for basic human needs is made, instead, by the Johannesburg Plan of
Implementation. Indeed, it is made rather convincingly. Sub-clause (a) of Article 9,
e.g., calls for improving:

> . . . access to reliable, affordable, economically viable, socially acceptable and environmentally
> sound energy services and resources, taking into account national specificities and circumstances,
> through various means, such as enhanced rural electrification and decentralized energy systems,
> increased use of renewables, cleaner liquid and gaseous fuels and enhanced energy efficiency, by
> intensifying regional and international cooperation in support of national efforts, including through
> capacity-building, financial and technological assistance and innovative financing mechanisms,
> including at the micro and meso levels, recognizing the specific factors for providing access to
> the poor.

Sub-clause (g) of the same article, elaborates the case further by calling for:

> . . . access of the poor to reliable, affordable, economically viable, socially acceptable and environ-
> mentally sound energy services, taking into account the instrumental role of developing national
> policies on energy for sustainable development, bearing in mind that in developing countries sharp
> increases in energy services are required to improve the standards of living of their populations
> and that energy services have positive impacts on poverty eradication and improve standards of
> living.

Here, then, is an example of all three dimensions of the sustainable development
triangle being invoked together and in a way that was not seen in any of the
Stockholm or Rio documents. Although this is still preliminary, it is nonetheless a
novel and welcome attempt to deal with energy fully in the context of sustainable
development by seeking policy that responds to the environmental, economic as
well as the social impulses of the concept. Policy made under the guidelines set
here would need to be evaluated not only in terms of the environmental stresses
being imposed, or effects on economic growth, but also in terms of social justice
and especially in terms of how it affects the poorest and the most vulnerable.

3.4. AN EVOLVING AGENDA

The process of evolution that seems to have taken place between the 1972, 1992
and 2002 environmental summits is depicted in Figure 5, which builds on the three
dimensions of sustainable development to illustrate how the documents emerging
from each successive conference have dealt with energy issues. Figure 5 suggests
that a rather neat evolution of the agenda has happened with the Stockholm summit
of 1972 dealing with energy issues principally as a source of environmental stress,
the 1992 Rio summit added a clear economic focus to its treatment of the subject,
while the 2002 Johannesburg Conference built upon the existing environmental
and economic focus and added the element of energy as a requisite for basic human
needs to the equation for the first time. Of course, the fact that these summits have
dealt with energy in a particular way does not imply that global energy policy has

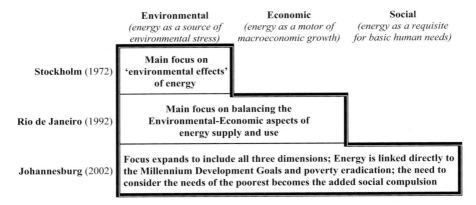

Figure 5. Energy and sustainable development: An evolving agenda.

moved in that direction automatically. The purpose of such summits is principally declaratory. However, the value of these declaratory proclamations must not be underestimated. They serve not only to advance the conceptual agenda but also tend to eventually influence the actual policies, although usually with some time lag (Susskind, 1994; Chayes and Chayes, 1995).

In terms of general conclusions, there are a number of surprises that can be highlighted:

• First, although conventional wisdom maintains that Johannesburg is a pale comparison to conferences before it (Najam et al., 2002), on the specific issue of energy it has actually made major conceptual headway by incorporating the energy issue more fully into a sustainable development framework. While the conference as a whole may not have been inspiring, on this one issue it has traversed into new and important territory.

• Second, it is rather surprising that the UN MDGs fail to identify energy as a key human development and human needs theme. However, the Johannesburg Plan of Implementation has partially filled that gap and made the argument that would have better come from the MDGs. That energy is a prerequisite for many human needs and for poverty alleviation merits explicit inclusion in the MDGs. Because of Johannesburg, this notion now has the endorsement of the comity of nations and there seems to be a strong case for incorporating energy issues explicitly within MDG programmes.

• Finally, there is a need for the practice of energy policy as well and the scholarship on energy policy to catch up with the realization imbedded in the Johannesburg Plan on Implementation regarding the importance of energy as a prerequisite for basic human needs.

• Finally, a broader argument can now be made that mega-conferences can advance global agendas in significant ways, as they have done with energy. The energy case gives cause for more optimism than many scholars invest in these conferences (Fomerand, 1996; Haas, 2002; Seyfang and Jordan, 2002). However, a corollary to be investigated further, would be whether such impact is more likely to be noticeable on specific issues rather than on the general agenda as a whole.

4. Conclusions

In focusing on the actual text of the documents negotiated at various global environmental summits and using a framework of how energy policy relates to various aspects or dimensions of sustainable development, this paper finds that there has been a slow but demonstrable evolution in how these conferences have dealt with energy. Not only has energy assumed a successively more prominent role in these global summits, but a noticeable evolution has occurred in the dimensions of energy policy that have been addressed. Moreover, this evolution has been along the trajectory of sustainable development. Although energy policy *per se* might not have made this transition as yet, these summits have given a clear signal to national and international policymakers to align energy policy more firmly to sustainable development, and to do so in more intricate ways.

In terms of evaluating the impact of the Johannesburg Earth Summit on the energy and sustainable development agenda, it seems that the one summit that had started with the worst prospects might well have achieved the most important advance in terms of conceptualizing energy policy within a framework rooted in all dimensions of sustainable development. Both the Johannesburg Declaration and the Johannesburg Plan of Implementation are remarkable in that they highlight the human need aspect of energy, in addition to the environmental and economic dimensions that had already been incorporated at Stockholm and Rio.

Acknowledgements

The authors, who are co-leaders of The Project on Human Development at the Fredrick Pardee Center for the Study of the Longer-Range Future at Boston University, thank the Pardee Center for a grant that made part of this research possible. They also thank Jenny K. Ahlen, Miquel Munoz, Claire Norris and Janice M. Poling, all graduate students at Boston University, for their assistance and comments. Finally, they wish to thank three anonymous reviewers for their very useful comments.

References

Adede, A.O.: 1992, 'International environmental law from Stockholm to Rio', *Environmental Policy and Law*, **22**(2), 88–105.
Agarwal, A., Narain, S., and Sharma, A. (eds): 1999, *Green Politics: Global Environmental Negotiations 1*, New Delhi, Centre for Science and Environment.
Allen, L.: 1979, *OPEC Oil*, Oelgeschlager, Gunn & Hain Publishers, Cambridge.

Askari, H. and Cummings, J.T.: 1978, *Oil, OECD, and the Third World: A Vicious Triangle?*, Austin, University of Texas Press.

Banuri, T. and Najam, A.: 2002, *Civic Entrepreneurship: A Civil Society Perspective on Sustainable Development*, Islamabad, Gandhara Academy Press.

Banuri, T., Hyden, G., Juma, C. and Rivera, M.: 1994, *Sustainable Human Development: From Concept to Operation*, New York, United Nations Development Programme.

Baranzini, A., Goldemberg, J. and Speck, S.: 2000, 'A future for carbon taxes', *Ecological Economics*, **32**(3), 395–412.

Brown, M.A., Levine, M.D., Short, W. and Koomey, J.G.: 2001, 'Scenarios for a clean energy future', *Energy Policy*, **29**(14), 1179–1196.

Brown, S.A. and Yucel, M.: 1999, Oil Prices and U.S. aggregate economic activity: A question of neutrality,' *Economic and Financial Review*, Second Quarter, Dallas, TA, Federal Reserve Bank of Dallas, pp. 16–23.

Chasek, P.: 2001, *Earth Negotiations: Analyzing Thirty Years of Environmental Diplomacy*, Tokyo, United Nations University Press.

Chayes, A. and Chayes, A.H.: 1995, *The New Sovereignty: Compliance with International Regulatory Agreements*, Cambridge, MA, Harvard University Press.

Clarke, R. and Timberlake, L.: 1982, *Stockholm Plus Ten: Promises, Promises?*, London, Earthscan.

Cleveland, C.J., Costanza, R., Hall, C.A.S. and Kaufmann, R.: 1984, Energy and the U.S. economy: A biophysical perspective, *Science*, **255**, 890–897.

Devarajan, S., Miller, M.J. and Swanson, E.V.: 2002, *Goals for Development: History, Prospects, Costs*, Washington DC, The World Bank.

Ebohon, O.J.: 1996, 'Energy, economic growth and causality in developing countries: A case study of Tanzania and Nigeria', *Energy Policy*, **24**(5), 447–453.

EIA.: 2001, *Annual Energy Review*, Washington, DC, Energy Information Administration.

Fomerand, J.: 1996, 'UN Conferences: Media events or genuine diplomacy?' *Global Governance*, **2**(3), 361–375.

Founex.: 1972, *Development and Environment. Report and Working Papers of Experts Convened by the Secretary General of the United Nations Conference on the Human Environment (Founex, Switzerland, June 4–12, 1971)*, Paris, Mouton.

Gardner, G.: 2002, The challenge for Johannesburg: Creating a more secure world, *State of the World 2002*, New York, W.W. Norton, pp. 4–23.

Gardner, R.N.: 1992, *Negotiating Survival: Four Priorities after Rio*, New York, Council on Foreign Relations Press.

Goldemberg, J. and Johansson, T.B.: 1995, 'Energy as an instrument for socio-economic development', in T.B. Johansson and J. Goldemberg (eds.), *Energy for Sustainable Development: A Policy Agenda*, New York, United Nations Development Programme, pp. 9–17

Haas, P.: 2002, 'UN conferences and constructivist governance of the environment', *Global Governance*, **8**(2), 82–83.

Hamilton, J.D.: 1983, 'Oil and the macroeconomy since World War II', *Journal of Political Economy*, **91**, 228–248.

Hass, P.M., Levy. M.A. and Parson, E.A.: 1992, 'Appraising the Earth Summit: How should we judge UNCED's success', *Environment*, **34**(8), 12–36.

Hondroyiannis, G., Lolos, S. and Papapetrou, E.: 2002, 'Energy consumption and economic growth: Assessing the evidence from Greece', *Energy Economics*, **24**(4), 319–336.

Hooker, M.A.: 1999, Oil and the macroeconomy revisited, *Finance and Economics Discussion Series* 1999–1943, Washington DC, Board of Governors of the Federal Reserve System.

Houghton, R.A., and Hackler, J.L.: 2001, *Carbon Flux to the Atmosphere from Land-use Changes: 1850–1990*, ORNL/CDIAC-79, NDP-050/R1, Oak Rigde, TN, Carbon Dioxide Information Analysis Center, Oak Ridge National Laboratory, U.S. Department of Energy.

IPCC: 2001, *Synthesis Report of the Third Assessment Report of the Intergovernmental Panel on Climate Change*. Cambridge, Cambridge University Press.

Johnson, S.P.: 1993, *The Earth Summit: The United Nations Conference on Environment and Development (UNCED)*, London, Graham & Trotman/Martinus Nijhoff.

Kaufmann, R.K.: 1992, 'A biophysical analysis of the energy/real GDP ratio: Implications for substitution and technical change', *Ecological Economics*, **6**(1), 35–56.

Kaufmann, R.K.: 1994, 'The relation between marginal product and price in US energy markets: Implications for climate change policy', *Energy Economics*, **16**(2), 145–158.

Kay, D.A. and Skolnikoff, E.B.: 1972, *World Eco-crisis: International Organizations in Response*, Madison, University of Wisconsin Press.

Khor, M.: 1992, 'Losers and winners at Rio', *Third World Resurgence*, **24-25** , 32–37.

Lélé, S.M.: 1991, 'Sustainable development: A critical review', *World Development*, **19**(6): 607–621.

Lefohn, A.S.H. and Husar, R.B.: 1999 'Estimating historical anthropogenic global sulfur emission patterns for the period 1850–1990', *Atmospheric Environment*, **33**(21), 3435–3444.

Matthews, E., Amann, C., Bringezu, S., Fischer-Kowalski, M., Hüttler, W., Kleijn, R., Moriguchi, Y., Ottke, C., Rodenburg, E., Rogich, D., Schandl, H., Schütz, H., van der Voet, E. and Weisz, H.: 2000, *Weight of nations: Material outflows from industrial economies*, Washington, DC, World Resources Institute.

Marland, G., Boden, T.A. and Andres, R.J.: 2002, *Trends: A Compendium of Data on Global Change. Global, Regional, and National CO2 Emissions*, Oak Ridge, TN, Carbon Dioxide Information Analysis Center, Oak Ridge National Laboratory, U.S. Department of Energy.

Masera, O.R., Saatkamp, B.D. and Kammen, D.M.: 2000, 'From linear fuel switching to multiple cooking strategies: A critique and alternative to the energy ladder model', *World Development*, **28**(12), 2083–2103.

McCully, P.: 2001, *Silenced rivers: The Ecology and Politics of Large Dams*, London, Zed Books.

Munasinghe, M.: 1992, *Environmental Economics and Sustainable Development*, (originally presented at the United Nations Conference on Environment and Development, Rio de Janeiro, Brazil), Washington, DC, World Bank.

Munasinghe, M.: 2002, 'The sustainomics trans-disciplinary meta-framework for making development more sustainable: Applications to energy issues', *International Journal of Sustainable Development* **5**(1/2), 125–182.

Murcott, S.: 1997, Sustainable development: A meta-review of definitions, principles, criteria, indicators, conceptual frameworks, information systems. Presented at the Annual Conference of the American Association for the Advancement of Sciences held at Seattle, WA, February 13–18.

Najam, A.: 1994, The case for a South Secretariat in international environmental negotiation, Program on Negotiation Working Paper 94-8, Cambridge, MA, Program on Negotiation at Harvard Law School.

Najam, A.: 1995, 'An environmental negotiation strategy for the South', *International Environmental Affairs*, **7**(3), 249–287.

Najam, A.: 2002, 'Financing sustainable development: Crises of legitimacy', *Progress in Development Studies*, **2**(2), 153–160.

Najam, A. and Page, T.: 1998, 'The climate convention: Deciphering the Kyoto protocol', *Environmental Conservation*, **25**(3), 187–194.

Najam, A., Poling, J.M., Yamagishi, N., Straub, D.G., Sarno, J., DeRitter, S.M. and Kim, E.M.: 2002, 'From Rio to Johannesburg: Progress and prospects', *Environment*, **44**(7), 26–38.

Ndiema, C.K.W., Mpendazoe, F.M. and Williams, A.: 1998, 'Emission of pollutants from a biomass stove', *Energy Conversion and Management*, **39**(13), 1357–1367.

Pirages, D.: 1978, *Global Ecopolitics: The New Context for International Relations*, North Scituate, Duxbury Press.

Reddy, A.K.N.: 2002, 'Energy technologies and policies for rural development', in T.B. Johansson and J. Goldemberg (eds.) *Energy for Sustainable Development: A Policy Agenda*, New York, United Nations Development Programme, pp. 115–136.

Rowland, W.: 1973, *The Plot to Save the World*, Toronto, Clarke, Irwin & Co.

Sachs, W. (ed): 2002, *The Jo'burg memo: Fairness in a Fragile World*, Berlin, Heinrich Boll Foundation.

Sadorsky, P.: 1999, 'Oil price shocks and stock market activity', *Energy Economics*, **21**(5), 449–469.

Seyfang, G. and Jordan, A.: 2002, 'The Johannesburg Summit and sustainable development: How effective are environmental mega-conferences?', *Yearbook of International Co-operation on Environment and Development, 2002/2003*, London, Earthscan, pp. 19–26.

Smil, V.: 1994, *Energy in World History*, Boulder, CO, Westview Press.

Speck, S.: 1999, 'Energy and carbon taxes and their distributional implications', *Energy Policy*, **27**(11), 659–667.

St. Louis, V.L., Kelly, C.A., Duchemin, E., Rudd, J.W.M. and Rosenberg, D.M.: 2000, 'Reservoir surfaces as sources of greenhouse gases to the atmosphere: A global estimate', *BioScience*, **50**, 766–775.

Stern D.I. and Kaufmann, R.K.: 1996, 'Estimates of global anthropogenic methane emissions 1860–1993', *Chemosphere*, **33**, 159–176.

Stern, D.I.: 2000, 'A multivariate cointegration analysis of the role of energy in the US macroeconomy', *Energy Economics*, **22**(2): 267–283.

Suarez, C.E.: 1995, 'Energy needs for sustainable development: Energy as an instrument for socio-economic development', in T.B. Johansson and J. Goldemberg (eds.), *Energy for Sustainable Development: A Policy Agenda*, New York, United Nations Development Programme, pp. 18–27.

Susskind, L.E.: 1994, *Environmental Diplomacy: Negotiating more Effective Global Agreements*, New York, Oxford University Press.

Tolba, M.K.: 1998, *Global Environmental Diplomacy: Negotiating Environmental Agreements for the World, 1973–1992*, Cambridge, MA, MIT Press.

UNDP: 1990, *Human Development Report 1990*, New York, United Nations Development Programme and Oxford University Press.

UNDP: 2002, *Human Development Report 2002*, New York, United Nations Development Programme and Oxford University Press.

Varma, A.: 2003, 'UK's climate change levy: Cost effectiveness, competitiveness and environmental impacts', *Energy Policy*, **31**(1), 51–61.

Wamukonya, L.: 1995,'Energy consumption in three Kenyan households: A survey', *Biomass and Bioenergy*, **8**(6), 445–451.

WCD: 2001, *Dams and development: Report of the World Commission on Dams*, London, Earthscan Publishers.

WCED: 1987, *Our common future: Report of the World Commission on Environment and Development*, New York, Oxford University Press.

World Bank: 2002, *World Development Indicators*, Washington, DC, CD-ROM, World Bank.

WRI: 2002, *World Resources 2002–03*, New York, CD-ROM, World Resources Institute.

Zhang, Z.X.: 2000, 'Can China afford to commit itself an emissions cap? An economic and political analysis', *Energy Economics*, **22**(6), 587–614.

CHAPTER 6

MANAGEMENT OF CHEMICALS
FOR SUSTAINABLE DEVELOPMENT

LARRY W. OLSON

Arizona State University, Technology Center, Mesa, 85212 Arizona, USA
(E-mail: Larry.Olson@asu.edu)

Abstract. This chapter traces the growth of global actions related to the management of chemicals and hazardous wastes since the UN Conference on Environment and Development in 1992, through the World Summit on Sustainable Development in 2002, and projections into the future as far as 2020. It is important to understand this relationship, since the groundwork for essentially all of the recommendations found in the Article 23 of the Plan of Implementation from Johannesburg is found in Chapter 19 of Agenda 21.

Significant progress has been made in understanding the risks associated with chemical exposure and in how to manage those risks to effectively reduce the threat to human health and the environment. The Plan of Implementation calls for transparency and accessibility in sharing this information with all countries and assistance to developing countries, and countries with economies in transition, in establishing the capacity for sound management of chemicals within their borders. Ratification of the Rotterdam and Stockholm Conventions is called for by 2003 and 2004, respectively. Full implementation of the new Globally Harmonised System for classifying and labelling chemicals is sought by 2008. Attention is given to risks posed by heavy metals, with a particular focus on the health and environmental effects of mercury and efforts to reduce anthropogenic releases. Finally, the Bahia Declaration and Priorities for Action beyond 2000 are used as examples of a strategic global approach to management of chemicals.

Chemistry must play a central role in reducing poverty and improving standards of living by more efficient and sustainable use of resources than is the case today as outlined in Principle 8 of the Rio Declaration. All of the actions called for in Article 23 of the Plan of Implementation are achievable and the time frames specified are reasonable. Progress to date has demonstrated the potential for effective cooperation between private industry, governments, international groups, and non-governmental organisations, yet much remains to be done, particularly in the area of Green Chemistry.

Key words: chemicals, classification, Globally Harmonised System, labelling, persistent organic pollutants, risk reduction, Rotterdam Convention, Stockholm Convention

Abbreviations: CG/HCCS – Coordinating Group for the Harmonisation of Chemical Classification Systems; COP – Conference of the Parties; DGD – Decision Guidance Document; FAO – Food and Agriculture Organisation; GHS – Globally Harmonised System; IFCS – Intergovernmental Forum on Chemical Safety; IGO – intergovernmental organisation; ILO – International Labour Organisation; INC – Intergovernmental Negotiating Committee; IOMC – Inter-Organisation Programme for the Sound Management of Chemicals; IRPTC – International Register of Potentially Toxic Chemicals; NGO – non-governmental organisation; OECD – Organisation for Economic Co-operation and Development; PIC – Prior Informed Consent; POP – Persistent Organic Pollutant; UN – United Nations; UNECE – United Nations Economic Convention for Europe; UNEP – United Nations Environmental Programme; UNIDO – UN Industrial Development Organisation; UNITAR – UN Institute for Training and Research; UNSCEGHS – United Nations Economic and Social Council's Sub-Committee of Experts on the Globally Harmonised System of Classification; WHO – World Health Organisation; WSSD – World Summit on Sustainable Development.

L. Hens and B. Nath (eds.), The World Summit on Sustainable Development, 135–150.
© 2005 *Springer. Printed in the Netherlands.*

1. Introduction

The essential commitment made at the World Summit on Sustainable Development (WSSD) in Johannesburg with respect to sound management of chemicals throughout their lifecycle, was to ensure by the year 2020 "that chemicals are used and produced in ways that lead to the minimisation of significant adverse effects on human health and the environment ..." (UN, 2002). Since over 75,000 chemicals are now used in commercial products, and only a fraction of these have undergone a rigorous risk assessment process (Brown *et al*, 2000), complying with this commitment will represent a significant challenge.

The global chemicals industry had estimated sales of US$1500 billion dollars in 1998, up from US$171 billion in 1970. This accounted for seven percent of global income, nine percent of international trade, and the employment of over ten million people. By the year 2010, the Organisation for Economic Co-operation and Development (OECD) Reference Scenario predicts a world chemical output of US$3920 billion (OECD, 2001). Perhaps more importantly, there is projected to be a shift in production, with OECD countries providing primarily higher value specialty and life science chemicals and developing countries producing more basic chemical feedstocks. Since the per capita consumption of chemicals in the developed world far exceeds that of developing countries, there is reason to expect increased demand in the future from non-OECD countries. As noted in the OECD Environmental Outlook for the Chemicals Industry report, "the tremendous growth rate in exports and imports of chemicals from and to non-OECD countries – as compared with the mature markets in OECD countries – represents a major change" (OECD, 2001: 33).

Since the amount of scientific and technical expertise in developing countries is typically less than in the industrialised world, there is an incumbent responsibility to help them address the risks associated with this chemical production (Hildebrandt and Schlottmann, 1998). The Plan of Implementation for WSSD made an explicit commitment in Article 23 to provide financial and technical assistance to developing countries to improve their capacity to manage chemicals and hazardous wastes.

Although considerably briefer than Chapter 19 of Agenda 21 (UN, 1999) pertaining to chemicals and hazardous waste, Article 23 of the Plan of Implementation from the WSSD outlined a number of specific actions that would contribute to protecting human health and the environment (UN, 2002). These are summarised below and discussed in the balance of this chapter.

(a) Promote the ratification and implementation of relevant international instruments on chemicals and hazardous waste, including the Rotterdam Convention on Prior Informed Consent Procedures for Certain Hazardous Chemicals and Pesticides in International Trade so that it can enter into force by 2003 and the Stockholm Convention on Persistent Organic Pollutants so that it can enter into force by 2004 and support developing countries in their implementation.

(b) Further develop a strategic approach by 2005 to international chemicals management based on the Bahia Declaration and Priorities for Action beyond 2000 of the Intergovernmental Forum on Chemical Safety. Emphasise cooperative actions on the part of international organisations in dealing with chemical management.

(c) Encourage countries to implement the new globally harmonised system for the classification and labelling of chemicals as soon as possible and have the system fully operational by 2008.

(d) Encourage partnerships that promote activities aimed at enhancing environmentally sound management of chemicals and hazardous wastes.

(e) Promote efforts to prevent international illegal trafficking of hazardous chemicals and hazardous wastes.

(f) Develop coherent and integrated information on chemicals, such as through national pollutant release and transfer registers.

(g) Reduce risks posed by heavy metals that are harmful to human health and the environment.

2. Setting the stage for Johannesburg: Agenda 21 and post-Rio actions

Chapter 19 of Agenda 21, entitled *Environmentally Sound Management of Toxic Chemicals, Including Prevention of Illegal International Traffic in Toxic and Dangerous Products* recognised that a modern economy requires the use of chemicals and that substantial progress has been made in developing management practices that minimise human exposures to toxic chemicals and the environmental degradation resulting from release of such chemicals. However, it also identified a number of problems related to the use of chemicals in both industrialised and developing countries. These were summarised in a proposal for six programme areas (UN, 1999):

a) Expanding and accelerating international assessment of chemical risks.
b) Harmonisation of classification and labelling of chemicals.
c) Information exchange on toxic chemicals and chemical risks.
d) Establishment of risk reduction programmes.
e) Strengthening of national capabilities and capacities for management of chemicals.
f) Prevention of illegal traffic in toxic and dangerous products.

A significant concern in 1992 was the need to develop better information on the risks associated with both occupational exposure and that of the general public to specific chemicals. Despite the tens of thousands of chemicals in some degree of commercial use, only about 1500 chemicals comprise more than 95% of total world production (OECD, 2001). Yet even for many of these high volume chemicals, there is a lack of appropriate data needed for hazard assessment, risk assessment, and risk management.

One of the first international agencies to address this concern was the International Register of Potentially Toxic Chemicals (IRPTC), established by the United Nations Environmental Programme (UNEP) in 1976 in response to the United Nations Conference on the Human Environment in Stockholm. IRPTC, now known as UNEP-Chemicals, had as its goal to link a formal network of government nominated institutions that provide information on production, distribution, release, disposal, and adverse effects of chemicals (UNEP, 2003a). The "*Inventory of Information Sources on Chemicals*" was published in 1997 and is available through UNEP's website (UNEP, 2003b).

As a response to Agenda 21, the Intergovernmental Forum on Chemical Safety (IFCS) was created in 1994 to help coordinate international work on chemicals (IFCS, 2003). It consists of governmental representatives, intergovernmental organisations (IGOs), and non-governmental organisations (NGOs). This partnership has created a unique approach to chemical safety issues and is consistent with the spirit of openness and cooperation of Agenda 21. By consolidating and harmonising risk assessment procedures, chemical classifications, and chemical management IFCS seeks to avoid duplication of effort, identify gaps in scientific understanding, and recommend priorities for national and international actions. IFCS concluded in June 1996 that there was sufficient evidence of risks from exposure to Persistent Organic Pollutants (POPs) to justify pursuing an internationally binding legal instrument. This ultimately led to the Stockholm Convention on Persistent Organic Pollutants (IFCS, 1996).

The Inter-Organisation Programme for the Sound Management of Chemicals (IOMC) was established in 1995 to promote coordination among international organisations involved in implementing Chapter 19 of Agenda 21. IOMC's current membership includes the United Nations Environment Programme (UNEP), International Labour Organisation (ILO), Food and Agriculture Organisation (FAO), World Health Organisation (WHO), UN Industrial Development Organisation (UNIDO), UN Institute for Training and Research (UNITAR), and OECD. Links to Programme Areas in Risk Assessment, Classification and Labelling, Information Exchange, Risk Reduction, Capacity Building, and Illegal International Trade provide detailed information and assistance through the IOMC website (IOMC, 2003). Among its other actions, IOMC has been responsible for coordinating the work on the Globally Harmonised System (GHS) for the classification and labelling of chemicals.

One objective of Agenda 21 was to target several hundred priority chemicals or groups of chemicals to be assessed by the year 2000, using accepted, peer reviewed methodology and taking into account the precautionary approach as outlined in Principle 15 of the Rio Declaration (UN, 1992). Developed countries were to take the lead and bear the brunt of the costs because they were primarily responsible for the introduction of these chemicals. In order to improve efficiency and minimise the cost of such assessments, increased collaboration and information exchange between governments, industry, academia, and relevant international organisations were specifically targeted. Using a common, internationally accepted framework for risk assessment and for obtaining toxicological and epidemiological data on chemical exposures was proposed as a means of improving the risk management process. Ultimately, it was hoped that this would result in phasing out the use of chemicals with unreasonable or unmanageable risks, as well as a greater emphasis on pollution prevention. An example of this action is the list of persistent organic pollutants in the Stockholm Convention (Stockholm Convention, 2003).

Hampering the safe handling and transport of chemicals in 1992, was the lack of a globally harmonised hazard classification and labelling system that could be used in all United Nations official languages, including adequate pictograms. This system should include material safety data sheets, standardisation of hazard communication terminology and symbols, and translation of information into the end-user's language. Agenda 21 had a goal of completing systems for classification and labelling by the year 2000. After more than ten years of negotiations, the Globally Harmonised System (GHS) for the Classification and Labelling of Chemicals is now available for use (UNECE, 2003).

Agenda 21 also expressed the need to develop and strengthen national capacities within developing countries for dealing with toxic chemicals. This would include a critical mass of technical staff capable of evaluating potential exposures and risk analysis within their own country. National capacity also includes adequate legislation and the ability to implement and enforce laws related to chemical safety and environmental protection. The ability to rehabilitate contaminated sites or treat exposed individuals, including emergency response efforts, is further evidence of a country's ability to manage toxic chemicals. Educational efforts to establish local expertise in interpreting relevant technical data and ready access to information, including establishing national registers and databases for chemical hazards, are key components of developing this national capacity. Agenda 21 called for support of these efforts by international organisations with the goal that by the year 2000, all countries would have in place a system for the environmentally sound management of chemicals. The emphasis in the WSSD Plan of Implementation on assisting developing countries and countries with economies in transition to improve this type of capacity is evidence that this goal has not yet been met.

Finally, Chapter 19 of Agenda 21 expressed concerns over the export to developing countries of chemicals whose use had been banned or severely restricted in industrialised or producing countries. This could include the illegal import or export of toxic and dangerous products, which were defined as those that are "banned, severely restricted, withdrawn or not approved for use or sale by Governments in order to protect public health and the environment" (UN, 1999: paragraph 66). Growing international trade in hazardous chemicals had led the UNEP Governing Council in 1987 to adopt the London Guidelines for Exchange of Information on Chemicals in International Trade focused on industrial chemicals (UNEP, 2001). This worked in tandem with the International Code of Conduct on the Distribution and use of Pesticides (FAO, 1985). The purpose of these non-binding agreements was to allow countries access to information about certain hazardous chemicals (pharmaceuticals, radioactive materials, and food additives were among those excluded) that would facilitate informed choices on importation, handling, and use (Hunter *et al*, 2002). They utilised a voluntary Prior Informed Consent (PIC) procedure to provide information on banned or severely restricted chemicals to participating countries. Together with improved monitoring to detect and prevent illegal transboundary movement of toxic and dangerous substances, and appropriate penalties, the PIC procedure was considered key to resolution of this issue.

3. Rotterdam Convention

The first deadline for action in the Plan of Implementation was to promote the ratification of the Rotterdam Convention on Prior Informed Consent Procedures for Certain Hazardous Chemicals and Pesticides in International Trade so that it can enter into force by 2003 (Rotterdam Convention, 2003). The Rotterdam Convention creates legally binding obligations, but builds upon the voluntary PIC procedures of the London Guidelines and the FAO Code by requiring exporters to obtain Prior Informed Consent of importers before commencing trade in certain hazardous substances. The list of regulated substances is expanded in the Rotterdam Convention (see Annex III). No longer requiring affirmative governmental action, the Convention includes pesticides or chemicals withdrawn voluntarily from the market. Developing countries or countries with economies in transition may propose the listing of a severely hazardous pesticide formulation in Annex III if there are problems associated with the use of the pesticide within its territory.

Labelling requirements that adequately identify risks and/or hazards to human health or the environment, as well as a safety data sheet, are also required for listed exported chemicals. A Decision Guidance Document (DGD) for each relevant chemical will be adopted by the Intergovernmental Negotiating Committee (INC) and will be provided to importing countries for each chemical covered by the PIC procedure. The DGD is intended to assist authorities in deciding whether to import or prohibit import of the listed chemical based upon toxicological, ecotoxicological, or safety information. The status of the listed chemical and its permitted uses in various countries is included in the DGD. Exporting countries are required to take appropriate measures to ensure that a listed chemical is not exported to an importing Party that has decided not to import it.

The Convention requires that each Party must establish the infrastructure and institutions necessary to provide the capacity to manage chemicals throughout their lifecycle. Towards this goal, exporting countries are required to provide technical assistance, including training, for developing countries or countries with economies in transition. Any Party may request assistance in evaluating whether or not to import a chemical.

The Convention was adopted on 10 September, 1998. It will enter into force after being ratified by at least fifty parties. As of October, 2003, forty nine states, plus the European Community, out of seventy three signatories had ratified the Convention. The Convention will thus soon enter into force, and the first Convention of the Parties will likely occur in 2004. During this period before the Convention becomes legally binding, the INC continues to meet to address and oversee issues related to the PIC procedure. The resolution on interim arrangements calls for implementation of the voluntary PIC procedures, including provisions for adding new chemicals to the list. Thus, progress in the standard of care for international trade in hazardous chemicals is occurring, even in the absence of legally binding commitments. However, of those countries that have already ratified the Convention, only 31% are in full compliance for import notifications of all chemicals listed in Annex III, and 15% have not submitted any import responses. Compliance issues will be a major part of the initial COP (IISD, 2002).

Capacity building and technical assistance to developing countries in managing imported chemicals, a key component of the Convention, has so far received little attention. Another concern is the time frame required to add a chemical to the regulated list in a world where 1500 new chemicals are introduced each year. Nevertheless, UNEP Executive Director Hans Töpfer calls the Rotterdam Convention vital and a first line of defence in protecting human health and the environment from chemical hazards (IISD, 2002).

4. Stockholm Convention on Persistent Organic Pollutants

Negotiations to restrict or ban the use of Persistent Organic Pollutants (POPs) began in response to a call by the Governing Council of the UNEP in 1997 (UNEP, 1997a). The first conference on this subject was held in Montreal in 1998 and was attended by officials from over ninety countries. At this meeting, a goal of concluding a treaty in 2000 was embraced and an initial list of twelve POPs was targeted. This list encompassed three categories: industrial chemicals, chlorinated pesticides, and unintentional by-products of combustion or the manufacturing process. Following several additional negotiating conferences, ninety two signatories agreed to the Stockholm Convention on Persistent Organic Pollutants was adopted in May 2001. 151 countries have since signed the treaty, as of October 2003.

Persistent organic pollutants can have either natural or anthropogenic sources, although most of the initially targeted compounds are man-made. They are semi-volatile substances with low water solubility and high lipid solubility that are resistant to photolytic, chemical, or biological degradation. Consequently, they have long environmental half-lives, some measured in decades, and tend to remain bound to soils or sludges. Due to their high lipid solubility, they bio-accumulate in fatty tissues of living organisms, increasing in concentration as one goes up the food chain, where they can be present at several orders of magnitude above background levels (Buccini, 2002).

Although POPs have low vapour pressures, particulate matter contaminated with POPs can be spread over great distances and across international borders. These pollutants have been detected in the remotest parts of the globe, such as the Arctic, even when they had never been used in that region (AMAP, 1997). Thus, actions banning or restricting the use of these chemicals on the part of individual countries cannot solve the problem. A global solution is required.

Making the problem worse is the fact that persistent organic pollutants are among the most toxic chemicals known, not only to human beings but also to other organisms in the environment like fish, birds, or mammals. Some of the risks associated with POPs include cancer, damage to central and peripheral nervous system, reproductive disorders including endocrine disruption, and disruption of the immune system (AMAP, 1997 and Landelac and Bacon, 1999: 574).

The Stockholm Convention seeks to eliminate or restrict the production and use of all intentionally produced POPs and to reduce and ultimately eliminate the unintentional release of POPs. However, no single solution will work for every chemical on the POP list, or for every country. Some uses may be eliminated by replacement with less toxic materials or non-chemical approaches such as integrated pest management. However, POPs that are unintentional by-products, such as dioxins and furans, will require new controls on incinerators or a change in production processes (World Bank, 2002). The issue is particularly severe in many developing countries and countries with economies in transition because of uncontrolled dumpsites dating back to the 1950s and lack of finances to pay for new technologies. Thus, part of any solution must be to provide technical and financial assistance to remediate already contaminated sites. According to Hans Toepfer, Executive Director of UNEP, "We all understand that the Convention will be asking developing countries to shoulder new responsibilities. Where they lack the means to accomplish its goals, we must find a way to work with them." (UNEP, 2000).

The Stockholm Convention contains provisions to add chemicals to the regulated list. Screening criteria include evidence of persistence, bioaccumulation, potential for long-range environmental transport, and adverse effect. A risk profile must be developed that includes information on sources, releases, environmental fate, monitoring data, exposure assessment, hazard assessment, and risk evaluation. Socio-economic considerations must also be evaluated. These include possible control measures, costs, impact on health and the environment, and disposal implications (Stockholm Convention, 2003: Annex F).

A stated goal of Johannesburg was to obtain the necessary 50 ratifications of the Stockholm Convention so that it could enter into force by 2004. As of October 2003, there were 40 Parties to the Convention. Given the number of countries in the process of ratification, the Convention is expected to go into force in 2004, with the first Conference of the Parties in 2005. COP-1 will have to deal with several controversial issues that have divided developed and developing countries including technical assistance, financing, and dispute settlement (IISD, 2003).

5. Bahia declaration and priorities for action beyond 2000

The Third Forum of the Intergovernmental Forum on Chemical Safety, meeting in Salvador, Bahia, Brazil in October 2000, made a comprehensive review of progress made since the Rio Conference. The result was the Bahia Declaration on Chemical Safety and a document entitled Priorities for Action Beyond 2000 (IFCS, 2000).

The Bahia Declaration and Priorities for Action Beyond 2000 reaffirmed the commitment of IFCS to the Rio Declaration and Chapter 19 of Agenda 21, as well as to continued research into the effects of exposure to chemicals and the development of appropriate policies and infrastructure for chemicals management in all countries. They recognised that many countries still lacked the essential capacity to effectively and safely manage chemicals. This was due, in part, to insufficient international and local resources to dispose of stockpiles of obsolete pesticides and hazardous chemicals that remain around the world and to regulatory structures for chemical safety that fall short of adequate protection human health and the environment.

A plea was made to governments, private interests, international organisations, and the public to join in a global cooperation for chemicals management, pollution prevention, and sustainable agriculture. Transparency regarding the risks associated with chemical manufacture, use, and disposal together with effective national policies, legislation, and infrastructure were seen as key to ensuring that all countries develop the capacity needed for effective chemicals management. The precautionary approach, as outlined in Principle 15 of the Rio Declaration, should be applied to reduce the risk of exposure to chemicals. This is especially important for susceptible groups, such as children, the sick, elderly, and pregnant women. Information on risks of chemical exposure should be disseminated to the public by government, industry, and non-governmental organisations. As well, the concerns of the public regarding toxic chemicals should be made known to policy makers.

Key goals, including dates for implementation, were set out in the Bahia Declaration and Priorities for Action Beyond 2000. Taken together, these actions constitute an effective blueprint for improving chemical safety world-wide. The goals are summarised as follows.

By 2001:
- The Convention on Persistent Organic Pollutants will have been adopted.

By 2002:
- Most countries will have adopted a National Profile on chemicals management using a multi-stakeholder process.
- Seventy or more countries will have implemented systems aimed at preventing major industrial accidents and systems for emergency preparedness and response.

- Poison centres will have been established in thirty or more countries that do not have such centres and strengthened in at least seventy more.

By 2003:

- The Rotterdam Convention will have entered into force.
- The Globally Harmonised System for the Classification and Labelling of Chemicals (GHS) will have been adopted.
- An effective Information Exchange Network on Capacity Building for the Sound Management of Chemicals will be operating.
- IFCS will consider recommendations for prevention of illegal traffic in toxic and dangerous products. National strategies will be developed.
- A report will have been prepared on acutely toxic pesticides and severely hazardous pesticides formulations that will include management options.
- All countries will have reported on risk reduction initiatives they have taken on other chemicals of major concern.

By 2004:

- Common principles and harmonised approaches for risk methodologies on specific toxicological endpoints will be available.
- An additional one thousand chemical hazard assessments will have been completed and made available.
- Procedures to ensure hazardous materials carry appropriate and reliable safety information will be established by most countries.
- Most countries will have integrated and ecologically sound pest and vector management strategies.
- Most countries will have established action plans for safe management of obsolete stocks of pesticides and other hazardous chemicals. At least two countries in each IFCS region will have commenced implementation of their plan.
- The Convention on Persistent Organic Pollutants will have entered into force.
- At least two additional countries in each IFCS region will have established a Pollution Release and Transfer Registry or emissions inventory.

By 2005:

- At least five countries in each IFCS region will have full arrangements in place for exchange of information on hazardous chemicals.
- Most countries will have developed national policies with targets for improving the management of chemicals.

Beyond Forum V (expected in 2005 or 2006):

- The Globally Harmonised System for the Classification and Labelling of Chemicals will be fully operational.
- Most countries in each IFCS region will have fully operational arrangements in place for the exchange of information on hazardous chemicals.

6. Globally Harmonised System (GHS) for the classification and labelling of chemicals

The Johannesburg Plan of Implementation called for the new Globally Harmonised System for the classification and labelling of chemicals to be implemented as soon as possible and be fully operational by 2008. Chapter 19 of Agenda 21 had called for such a system to be available by the year 2000, so that this is obviously an element that has lagged in its development. The official document, the product of more than a decade of work with many stakeholders, was based upon existing systems in the U.S., Canada, European Union, and the United Nations Recommendations on the Transport of Dangerous Goods. The work was coordinated by the IOMC Coordinating Group for the Harmonisation of Chemical Classification Systems (CG/HCCS). The GHS was adopted in December 2002 at the UN and can be found at the United Nations Economic Convention for Europe website (UNECE, 2003). It is intended to serve as the initial basis for global implementation and may be revised as experience dictates. The new United Nations Economic and Social Council's Sub-Committee of Experts on the Globally Harmonised System of Classification (UNSCEGHS) will maintain it. The target audience for the GHS includes consumers, workers, transporters, and emergency responders.

The Globally Harmonised System provides a common approach to defining and classifying chemicals based on their physical, health and environmental hazards. Although similar in most countries, the differences in existing laws for hazard definition are significant enough that different labels and multiple safety data sheets might be required for the same product involved in trade to different countries. This impacts the protection of human health because users may see different warning labels for the same chemical and it makes international trade more expensive due to an increased regulatory burden. For developing countries, the GHS should provide consistent and appropriate information on chemicals that they import or produce. This is the first step towards improving the national capacity for safely and effectively managing chemicals to protect human health and the environment.

The GHS also includes hazard communication elements such as labelling (including pictograms) and safety data sheets. The system applies to all hazardous chemicals and mixtures of chemicals, but the mode of hazard communication (e.g. labels, safety data sheets) may vary by product category or stage of the life cycle. Labels will not apply to pharmaceuticals, food additives, cosmetics and pesticide residues in food at the point of intentional intake.

7. Reduce risks posed by heavy metals to human health and the environment, including a global assessment of mercury and its compounds

Heavy metals are naturally occurring constituents of the earth's crust and, unlike synthetic organic molecules, they cannot be degraded or destroyed. Their natural abundance in soils, sediments, and surface or ground waters can be greatly affected, however, by human activity. In an aquatic environment, heavy metals can enter the food chain where they pose risks to both human consumers of seafood and to marine biota (UNEP, 1997b).

Anthropogenic sources of heavy metals include mining activities, smelters, combustion by-products, and consumer products. These originate from both point and non-point sources. Long-range transport of heavy metals is possible for relatively volatile heavy metals and contaminated airborne particulate matter, as well as for surface run-off containing dissolved metals or contaminated sediments.

Although a number of heavy metals, such as lead and cadmium, have been widely used in industry and have significant human health and environmental concerns, the potential adverse effects from exposure to mercury are a key concern. At the UNEP's 22nd Governing Council in February 2003, agreement was reached to focus on a global crackdown on mercury pollution (UNEP, 2003c). The decision reflected the concerns expressed in the recently released UNEP Global Mercury Report that "there is sufficient evidence of significant global adverse impacts from mercury and its compounds to warrant further international action to reduce the risks to human health and the environment from the release of mercury and its compounds to the environment." (UNEP, 2003d). Among the recommended actions of the UNEP Programme for International Action on Mercury were:

- Improve the scientific basis for understanding the fate and transport of mercury in the environment, what populations and ecosystems are at risk, and the degree of risk involved.
- Enhance risk communication about mercury, particularly to at-risk populations.
- Improve the global collection and exchange of information on mercury exposure, production, trade, disposal, and release.
- Reduce the demand for and uses of mercury and the anthropogenic releases of mercury.

Priority was given to the need of developing countries and countries with economies in transition to build their capacity to address this issue. This includes not only a national assessment of the nature and magnitude of the problem, but waste management strategies, appropriate regulatory structures, risk communication, and alternatives to commercial uses of mercury.

If this effort is successful, similar strategies can be employed for other heavy metals. Although details will differ on health effects, emission inventories, and technical solutions for disposal or product substitutions, the essential procedures outlined in the UNEP Programme for International Action on Mercury can be applied in other situations.

8. Conclusions

The goal of eliminating the adverse impact of chemicals on human health and the environment will never be completely achieved. Each year new products and new chemical formulations are introduced, global trade increases the number of people potentially affected, and the demands of providing for an ever-increasing population place additional stress on the environment. However, the steps outlined in Article 23 of the WSSD Plan of Implementation provide a realistic and achievable means of improving the way in which chemicals are used. The global framework now in place to regulate chemical use and disposal, identify and control hazards, and reduce risk to human health and the environment far exceeds even the best efforts of developed countries or private industry from forty years ago. The current challenge is to ensure the implementation of these procedures, to provide financing for regulatory oversight, and to accelerate the pace of technology transfer and capacity strengthening in developing countries.

Demonstrable progress has been made in meeting the goals outlined in Chapter 19 of Agenda 21 and Article 23 of the Plan of Implementation for the WSSD. The Rotterdam and Stockholm Conventions will soon be ratified and come into force, meeting specific objectives of the Plan of Implementation. These two instruments represent significant achievements of international diplomacy and govern the use of some of the most hazardous chemicals in commercial use, while providing for special assistance to developing countries and countries with economies in transition. With the prospect of the first Conference of the Parties meeting soon, there is even more incentive for ratification, as only Parties will have an official seat at the table. It is too early to tell how well these Conventions will operate in practice; however, there remains division about the details of enforcement and finances, primarily, although not exclusively, between developing and developed countries. During the interim period between adoption of the treaty and ratification, voluntary compliance has been spotty at best. Whatever disagreements do exist, however, relate to procedural issues and not to the fundamental need for such internationally accepted standards on handling hazardous chemicals.

The Globally Harmonised System for the classification and labelling of chemicals has also been finalised, although it has not yet been widely adopted. Implementation by the 2008 date called for in the Plan of Implementation will require changes in complex national legislation involving worker safety, transportation, and public health and will need to be driven by economic incentives related to lower costs in international trade of chemical products. For developing countries that need to improve their capacity to manage chemicals, the GHS provides a cost effective model template.

The Bahia Declaration and the Priorities for Action beyond 2000 lay out specific dates and objectives through 2005. Included in these are recommendations on considering how to prevent illegal traffic in toxic and hazardous substances and improving the process of information exchange on hazardous chemicals. The Internet and multilateral agreements that standardise data collection are targeted as key components of this process. The Bahia Declaration and the Plan of Implementation both call for national pollutant release and transfer registers.

Several new international organisations related to chemical use have been developed since Agenda 21, including the Intergovernmental Forum on Chemical Safety and the Inter-Organisation Programme for the Sound Management of Chemicals. These partnerships address Objective (c) in Article 23 of the Plan of Implementation to promote activities that enhance sound management of chemicals and hazardous wastes and aid in implementing multilateral environmental agreements.

Objective (g) of Article 23 calls for reducing the risks posed by heavy metals in the environment. Based upon the risk factors to human health and biota to increased exposure to mercury resulting from anthropogenic activity, as documented in the UNEP Global Mercury Report, the Governing Council of the UNEP has targeted mercury for an international effort to reduce the release of mercury and mercury compounds into the environment. This effort will serve as a model for future actions dealing with other heavy metals and therefore has implications beyond just the toxicity of mercury itself.

Little mention was made in Johannesburg about the proactive approach to reducing pollution and energy consumption, sometimes called Green Chemistry. Unlike the "command and control" environmental regulations of the 1970 and 80s, Green Chemistry focuses on minimising the amount of waste products produced in the manufacture of chemicals, replacing toxic chemicals with less toxic materials, using less energy, and finding renewable sources for feedstocks rather petroleum derivatives. As mentioned previously, the manufacture of base chemicals, those produced in large quantities of more than a million tons per year, is shifting increasingly to developing countries. Thus, new sustainable methods that reduce energy requirements for processing or use renewable methods of production would have a large beneficial impact in all parts of the world (Eissen, et al, 2002). Future international agendas on managing chemicals and their hazards will undoubtedly need to highlight these efforts.

References

AMAP – Arctic Monitoring and Assessment Programme.: 1997, *Arctic Pollution Issues: A State of the Arctic Environment Report*, http://www.amap.no/assess/soaer-cn.htm, downloaded 12-03-2003.

Brown, L., Renner, M. and Halwell, B. (eds): 2000, *Vital Signs 2000: The Environmental Trends that are Shaping Our Future*, Worldwatch Institute.

Buccini, J.: 2002, "Getting on Top of the POPs", *Chemicals and the Environment, 2002*, http://www.ourplanet.com/home.html, downloaded 12-03-3003.

Eissen, M., Metzger, J.O., Schmidt, E. and Schneidewind, U.: 2002, "10 Years after Rio-Concepts on the Contribution of Chemistry to a Sustainable Development", *Angew. Chem. Int. Ed.*, **41**, 414–436.

FAO – Food and Agriculture Organisation.: 1985, "International Code of Conduct on the Distribution and Use of Pesticides", FAO Conference Resolution 10/85.

Hildebrandt, B. and Schlottmann, U.: 1998, "Chemical safety – an international challenge", *Angew. Chem. Int. Ed.*, **37**, 1316-1326.

Hunter, J., Salzman, J. and Zaelke, D.: 2002, *International Environmental Law and Policy (2ⁿᵈ Edition*, Foundation Press, New York.

IFCS – Intergovernmental Forum on Chemical Safety.: 1996, "Discussion paper on response strategies for reducing/eliminating certain persistent organic pollutants (POPs)" (IFCS/WG.POPs.4), presented by the Nordic Countries and the United States at meeting of the IFCS Ad Hoc Working Group, Manila, 21–22 June 1996, http://www.chem.unep.ch/pops/indxhtms/manwg4.html, downloaded 10-03-2003.

IFCS – Intergovernmental Forum on Chemical Safety.: 2000, "Forum III Final Report" (IFCS/Forum III/23w), IFCS, Bahia, http://www.who.int/ifcs/Documents/Forum/ForumIII/f3-finrepdoc/Priorities.pdf, downloaded 10-02-2003.

IFCS – Intergovernmental Forum on Chemical Safety.: 2003, "In partnership for global chemical safety", http://www.who.int/ifcs/index_pag2.htm, downloaded 10-03-2003.

IISD – International Institute for Sustainable Development.: 2002, "Summary of the ninth session of the Intergovernmental Negotiating Committee for an international legally binding instrument for the application of the prior informed consent procedure for certain hazardous chemicals and pesticides in international trade", *Earth Negotiations Bulletin*, **15**(75), 1–12.

IISD – International Institute for Sustainable Development.: 2003, "Summary of the seventh session of the Intergovernmental Negotiating Committee for an international legally binding instrument for implementing international action on certain persistent organic pollutants", *Earth Negotiations Bulletin*, **15**(81), 1–12.

IOMC – Inter-Organisation Programme for the Sound Management of Chemicals.: 2003, http://www.who.int/iomc/.

Landelac, L. and Bacon, M.: 1999, "Will we be taught ethics by our clones? The mutations of the living from endocrine disruptors to genetics", *Bailliere's Clinical Obstetrics and Gynaecology*, **13**(4), 571–592.

OECD – Organisation for Economic Cooperation and Development.: 2001, *Environmental Outlook for the Chemicals Industry*, OECD (Environment, Health and Safety Division), Paris.

Rotterdam Convention.: 2003, "PIC – Rotterdam Convention", http://www.pic.int/index.html, downloaded 10-03-2003.

Stockholm Convention on Persistent Organic Pollutants (POPs): 2003, "Convention text", http://www.pops.int/documents/convtext/convtext_en.pdf, downloaded 11-03-2003.

UN – United Nations General Assembly.: 1992, "Report of the United Nations Conference on Environment and Development", Rio de Janeiro, 3–14 June, 1992, Annex I: Rio Declaration on Environment and Development (A/CONF.151/26 [Vol. I]), http://www.un.org/documents/ga/conf151/aconf15126-1annex1.htm, downloaded 11-03-2003.

UN – United Nations.: 2002, "Report of the World Summit on Sustainable Development: Plan of Implementation": Article 23 (A/CONF.199/20).

UN – United Nations Division for Sustainable Development.: 1999, "Toxic Chemicals", Agenda 21, Chapter 19, http://www.un.org/esa/sustdev/agenda21chapter19.htm, downloaded 10-03-2003.

UNECE – United Nations Economic Commission for Europe.: 2003, "The globally harmonised system of classification and labelling of chemicals", http://www.unece.org/trans/danger/publi/ghs/officialtext.html, downloaded 10-03-2003.

UNEP – United Nations Environment Programme.: 1997a, "UNEP Governing Council Decision 19/13c on Persistent Organic Pollutants (POPs)", http://www.chem.unep.ch/pops/gcpops_e.html, downloaded 11-03-2003.

UNEP – United Nations Environment Programme.: 1997b, "Heavy metals", http://www.unep.org/unep/gpa/pol2a10.htm, downloaded 11-03-2003.

UNEP – United Nations Environment Programme.: 2000, "Progress made in negotiating global treaty on persistent organic pollutant", UNEP News Release, 20 March 2000.

UNEP – United Nations Environment Programme.: 2001, *London Guidelines for the Exchange of Information on Chemicals in International Trade,* Amended 1989, http://www.chem.unep.ch/ethics/english/longuien.htm, downloaded 12-03-2003.

UNEP – United Nations Environment Programme.: 2003a, UNEP – Chemicals website, http://irptc.unep.ch.

UNEP – United Nations Environment Programme.: 2003b, *Inventory of Information Sources on Chemicals*, http://www.chem.unep.ch/irptc/invent/igo.html.

UNEP – United Nations Environment Programme.: 2003c, "Decisions adopted by the governing council at its twenty-second session/global ministerial environment forum", http://www.unep.org/GoverningBodies/GC22/Information_documents.asp, downloaded 11-03-2003.

UNEP – United Nations Environment Programme.: 2003d, "Global mercury report", UNEP/GC.22/INF/3, http://www.unep.org/GoverningBodies/GC22/Information_documents.asp, downloaded 11-03-2003.

World Bank.: 2002, "Overview of the Stockholm Convention on Persistent Organic Pollutants", http://lnweb18.worldbank.org/ESSD/essdext.nsf/50DocByUnid/D7CCC88D7EF2344685256B A8007165E0/$FILE/OverviewoftheStockholmConventionMarch2002.pdf, downloaded 12-03-2003.

CHAPTER 7

HEALTH: A NECESSITY FOR SUSTAINABLE DEVELOPMENT

ALEX G. STEWART[1]*, EWAN WILKINSON[2], C. VYVYAN HOWARD[3]

[1]*Locum Consultant in Health Protection and Communicable Disease Control, Cheshire and Merseyside Health Protection Team, Chester Microbiology Laboratory, Countess of Chester Health Park, Liverpool Road, Chester, CH2 1UL, UK*
*(*author for correspondence, e-mail: astewart@nwhpa.nhs.uk)*
[2]*Consultant in Public Health, Central Liverpool Primary Care Trust, Liverpool L3 6AL, UK.*
(Formerly District Health Officer, Zomba District, Malawi)
[3]*Senior Lecturer, Developmental Toxic-Pathology Research Group, Department of Human Anatomy and Cell Biology, University of Liverpool, Liverpool L69 3GE, UK*

Abstract. Health and sustainable development are mutually dependent: neither is possible without the other. Health is determined by many factors: hereditary; life-style; the level of education; work; the social community; the effects of the natural environment. The greatest health-related threat to sustainable development comes from infections (diarrhoea, respiratory infections, AIDS, tuberculosis and childhood infections and malaria) and psychiatric disorders, particularly depression, but including schizophrenia (madness). Infection and psychiatric disorders are given too little importance in the Johannesburg declaration. Development is not without problems that affect health: increased population stressing the food supplies; changing disease patterns; the marginalisation of vulnerable groups who tend to have poorer health, nutrition, education, access to help or little control of their circumstances; corruption; unjust trade and other economic decisions; inappropriate education and international aid; bias towards the provision of health services while ignoring the factors affecting health; unthinking introduction of western beliefs and practices; traffic fumes and accidents; and industrial and domestic pollution. Many of these problems are tackled in the declaration in some measure, but the declaration is unbalanced in its approach, focussing on minor issues to the detriment of important health matters. However, the processes to change in both health and development are the same and operate across society. Achieving sustainable development and improving health at the same time is attainable, but may take a major investment by the developing countries. The Western nations cannot be relied upon to contribute wisely and unselfishly to health and development in the rest of the world.

Key words: Burden of disease, demographic transition, development, economics, education, epidemiological transition, globalisation, health, pollution, vulnerable groups, Westernisation

Abbreviations: AIDS – Acquired Immunodeficiency Syndrome; HIV – Human Immunodeficiency Virus; PCB – Polychlorinated Biphenyl; POP – Persistent Organic Pollutant; PVC – Polyvinyl Chloride; WHO – World Health Organisation; WSSD – World Summit on Sustainable Development.

1. Introduction

It used to be thought that improvements in health resulted from development. Therefore, the argument went, if the development occurred, the health of the population would improve without question. This is a similar argument to the idea in economics that wealth will trickle down from the rich to the poor. The sad fact

L. Hens and B. Nath (eds.), The World Summit on Sustainable Development, 151–181.
© 2005 *Springer. Printed in the Netherlands.*

is that neither is true, the trickling down of wealth from the rich to the poor, or improvements in health arising from development. To improve health needs specific investment of time, money, people and effort.

Health is not an automatic result of any development, but is, nevertheless, very closely linked to sustainable development. Without some degree of health neither individuals nor a community will be able to achieve the goals they set in development. And without development and the resulting improvement in the standard of living of those at the margins of society, health suffers.

This chapter looks at the main causes of ill health and the links between health and sustainable development, drawing out what is known about this association (Johannesburg Declaration, section VI [Health and sustainable development] paragraph 46 - hereafter, referenced as JDVI with the appropriate paragraph number).

2. Health

What do we mean by health? It can be very hard to define. Early attempts at a definition often start as an absence of disease. It is, of course, much more than this. Some people suffering the results of a definable disease feel healthy, while other people feel unhealthy with very little pain or discomfort.

Health has been variously characterised. The most famous definition is that from the World Health Organisation:

> "Health is a state of complete physical, mental and social well being, and not merely an absence of disease or infirmity. It is a fundamental human right. The attainment of the highest possible level of health ... requires the action of many other social and economic sectors in addition to the health sector."
>
> (WHO, 1978)

If this definition misses anything, it omits the spiritual, which is very important to many people but irrelevant to others. This omission demonstrates the difficulty of defining health in a way that satisfies everyone.

Health is something we all crave, something that we seldom notice when we have it, but we are acutely aware of its absence. Good health is something that many people around the world, particularly in the developing countries, do not have. But without some degree of health the prospect of sustainable development will be harder to achieve.

Ill health removes or reduces the ability of a person and a community to give time and resources to the changes needed for development. Ill health even reduces the interest in progress since development needs effort; someone with poor health lacks the ability to make such an effort. The effort that an ill person can exert can be fully used in reducing the effects of their condition and in sustaining life, in other words: given to "just living", nothing else.

Health is determined by many factors, working at different levels. From the genetic make-up we inherit from our parents, through the social situation of the community in which we live, to the natural environment around us. This is illustrated in the health rainbow (Figure 1). Sustainable development improves health through economic, social and personal means.

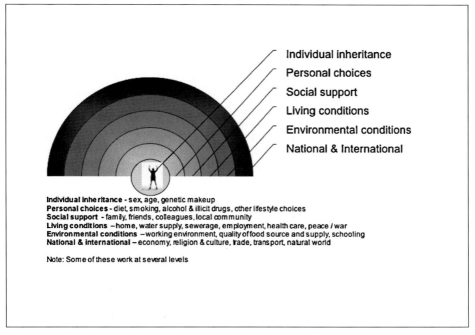

Figure 1. Main factors affecting health (Used with permission).
Source: Developed and expanded from an original illustration in Dahlgren and Whitehead (1991)

The personal factors that affect health are perhaps the clearest for many of us. Young people are affected by different diseases than older people. Young children suffer more diarrhoea, for example, while older people get more heart or lung diseases.

Some diseases are specific to each sex: women suffer from breast and ovarian disease, while men have problems with their prostate and testes. But many diseases are commoner in one sex than the other, for example, thyroid disease and depression are commoner in women in some societies, while the men may suffer more alcoholism or tobacco related diseases.

Lifestyle factors include the fact that young men are willing to take more risks than other members of society, and consequently suffer more accidents. AIDS may be caught as a consequence of a choice about the kind of partner we have sex with.

Social and community networks affect health in a number of ways. Isolated people suffer from psychiatric and emotional problems more than those who enjoy close family and social relationships. People who live in stressful relationships are more prone to heart disease than those whose relationships are gentler. Workers tend to have more trouble with blood pressure than their bosses.

The effect on health of living and working conditions is most clearly seen in the provision of water and sanitation. Access to clean water, along with good sewage disposal that does not contaminate the water supply, reduces the amount of diarrhoeal diseases in a community.

Nutrition is recognised to be important as well, since lack of good food leads to loss of weight and the problems of starvation. But working conditions, housing and education also affect health. For example, unemployment is associated with poorer mental health and higher rates of heart disease, while work may put people at risk of accidents or expose them to toxic chemicals such as lead.

It is important to note that health care services are listed as only one of the many factors that affect our health. While the health care services are important in restoring health when people fall ill, they are not major factors in keeping people healthy in the first place. Nor are they the only things necessary to improve health. Good health services do not replace clean water or a balanced diet.

The general conditions of the environment and society we live in also affect us (JD VI 47). War is clearly a cause of much ill health, and not just of injuries from arms and fighting. The disruption to food and water supplies has often caused more deaths that the war itself. But society has influences on health. A society where smoking is seen as a sign of adulthood will be full of people who suffer from chest and heart disease. A poor community where women and children only eat after the men have chosen the best parts will suffer from poor nutrition in the women and children; such poorly nourished children grow up to be unhealthy adults, so the effect lasts throughout life (JD VI 47f).

When one, or more, of the factors that affect our health (Figure 1) stress us, we may become ill. Certainly, the greater the number of different determinants that cause us stress at any one time, the more likely we are to become ill. Sustainable development, by improving the conditions of many of these determinants, is likely to result in better health for the individual and the community.

There is a need for sufficient levels of health in the members of a community for that community to be able to support sustainable development.

3. Disease and Ill Health

There are a number of questions that can be asked about health in the context of sustainable development: What are the main diseases that affect the global population? What are the results of these diseases? What impact do these diseases have on sustainable development?

There are a number of different ways of deciding which are the most important diseases, causing the most problems in the community (WHO, 2000).

The commonest way is to look at causes of death. On a global scale, the top two diseases account for over half of all deaths: first are the cardiovascular diseases (diseases of the heart and circulation): 16.6 million deaths in 2001; 29 per cent of all deaths. All infectious diseases a close second: 14.9 million; 26 per cent. The third group, cancers, lies a long way behind: 7.1 million; 13 per cent. Even if respiratory infections (JD VI 49), the largest sub-group of infections, are separated from the other infections (Table 1) it is clear that cancer comes a long way behind the other main causes of death (JD VI 47o).

TABLE I. Deaths in 2001 from various disease groups.

Cause	Both sexes [number]	% Total
Cardiovascular	16,585,393	29.3
Infectious and parasitic	10,937,452	19.3
Malignancy	7,114,896	12.6
Respiratory infections	3,947,426	7.0
Respiratory [other]	3,560,422	6.3
Unintentional injury	3,508,197	6.2
Perinatal	2,503,534	4.4
Digestive	1,987,021	3.5
Intentional injury	1,594,096	2.8
Neuropsychiatric	1,023,178	1.8
Diabetes mellitus	895,454	1.6
Genitourinary	824,719	1.5
Maternal	509,021	0.9
Congenital abnormalities	506,593	0.9
Nutritional deficiencies	476,907	0.8
Endocrine disorders	246,628	0.4
Musculoskeletal	112,737	0.2
Skin	67,128	0.1

Source: *Global Burden of Disease*, WHO, 2002.
http://www.who.int/evidence/bod

However, death is not the only measure of disease and ill health. Years of life lost because of a disease killing a young person, and years of life lived with a disability resulting from an illness or injury are two useful ways of showing how much life is lost to a particular disease or disease group.

Figures 2 and 3 give a picture of these different measures. The order of the groups of disease in both figures are by years of life lost. Figure 3 is an enlargement of the lower part of Figure 2 in order to show the variation between the various disease groups.

Death and loss resulting from common disease groups, 2001

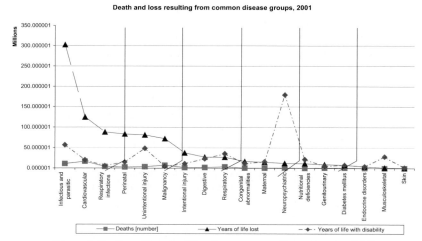

Figure 2. Death and loss resulting from common disease groups, 2001.
[y axis measures numbers in millions]
Source: *Global Burden of Disease*, WHO, 2002.
http://www.who.int/evidence/bod

Death and loss resulting from common disease groups, 2001

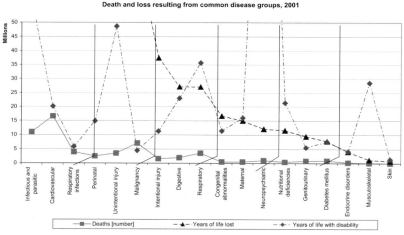

Figure 3. Death and loss 2001 – bottom of figure 2 enlarged.
[y axis measures numbers in millions]
Source: *Global Burden of Disease*, WHO, 2002.
http://www.who.int/evidence/bod

By this measure, the biggest killer is infection, which steals more than twice as many years as cardiovascular diseases. Cardiovascular diseases are the commonest killer. However, since they tend to kill older (and richer) people they are only number two by the measure of years of life lost. Because infections kill at all ages, but in particular they kill children (Gwatkin and Heuveline, 1997), they have a much greater impact on the community as a whole, as seen in the years of life lost measure. In fact, the number of years of potential life that all infections remove is over three times the number of years removed by the diseases of the heart and circulation (Table 2).

TABLE II. Years of life lost to disease.

Cause	Total number of years
Infectious and parasitic diseases	302,303,136
Cardiovascular diseases	124,358,381
Respiratory infections	88,131,330
Perinatal conditions	83,567,130
Unintentional injuries	81,182,333
Malignant neoplasms	72,220,261
Intentional injuries	37,450,831
Digestive diseases	27,110,541
Respiratory diseases	27,084,513
Congenital anomalies	16,697,204
Maternal conditions	14,951,158
Neuropsychiatric conditions	12,056,099
Nutritional deficiencies	11,557,035
Genitourinary diseases	9,459,286
Diabetes mellitus	7,793,391
Endocrine disorders	4,063,951
Musculoskeletal diseases	1,196,500
Skin diseases	795,979

Source: *Global Burden of Disease*, WHO, 2002.
http://www.who.int/evidence/bod

The infections that lead to the greatest number of years of life lost from early death (JD VI 47g) are respiratory infections (pneumonia and others), AIDS, tuberculosis (JD VI 48), the childhood infections (whooping cough, measles, tetanus), diarrhoeal diseases and malaria.

Looking at the years of life lost (Table 2), it is clear that, after all infections and cardiovascular disease, the next groups of diseases (respiratory infections, conditions around birth (perinatal), accidents (unintentional injuries) and cancers (malignant neoplasms) all cause about the same loss of life each year. This is a heavy burden, particularly since many of the conditions which form part of these disease groups, along with infections, can be prevented.

"Years of life lost" do not measure the burden of ongoing disease. "Years of life with disability" shows a very different pattern from the other measures of ill health.

By far the largest burden from disability comes, perhaps surprisingly, from the psychiatric disorders – a staggering 180 million years (Table 3), twice that of the next cause, all infections. Depression is far and away the largest problem within the psychiatric group of illnesses, leading to the greatest number of years lived with a disability (WHO, 2000) (JD VI 47g, 47o).

Table III. Years of life lived with a disability from a disease.

Cause	Total number of years
Neuropsychiatric conditions	179,204,287
Infectious and parasitic diseases	57,074,049
Unintentional injuries	48,670,752
Respiratory diseases	35,757,277
Musculoskeletal diseases	28,601,915
Digestive diseases	23,062,293
Nutritional deficiencies	21,401,331
Cardiovascular diseases	20,112,169
Maternal conditions	15,991,627
Perinatal conditions	14,854,568
Congenital anomalies	11,406,769
Intentional injuries	11,277,540
Diabetes mellitus	7,652,980
Respiratory infections	5,905,656
Genitourinary diseases	5,550,881
Malignant neoplasms	4,495,350
Endocrine disorders	4,167,893
Skin diseases	1,375,233

Source: *Global Burden of Disease*, WHO, 2002.
http://www.who.int/evidence/bod

This burden (of years lived with a disability) is not a measure of the effects of an early death, as in the previous measure, but is a measure of an ongoing problem in peoples' lives. "Years of life with disability" is perhaps the most important measure of the burden that ill health places on any community since it measures more than illness or death. It gives some indication of the scale of resources needed to cope with any particular disease or disease group.

This burden works at several levels: there is the burden on the patient of the illness from the pain or disability or even disgrace associated with that particular diagnosis. Then there is the burden of caring for the patient by the immediate family. Added to that is the loss to the local community of the time and effort that the patient and the carers could give in the fields, socially and in industry. Only the burden carried by the patient is measured by "years of life with disability" but it is easy to see that any disease that results in a large number of years spent with disability will have serious, if unmeasured, consequences for the family and the community.

So, although the cardiovascular diseases kill the greatest number, the infections are the cause of the greatest loss of years of life, while the biggest on going burden of disability are the psychiatric illnesses (JD VI 47g).

It is not an exaggeration to say that infections and depression are the greatest threat to sustainable development, burdening not just the person suffering from them but the whole community in which that person lives. Development can decrease the frequency of the infections. But depression also is affected by development. The traps of poverty, large families, repeated illness and death, which contribute to depression, can be eased, though not always eliminated, by the improved standard of living that development brings.

It is a disappointment that infections and psychiatric conditions are given such a little place in the Johannesburg declaration since they are the diseases which make it harder to achieve sustainable development.

4. Disease and population changes

Epidemiology describes the patterns of disease and their causes in a community. The epidemiological transition is the change in disease that accompanies the change in wealth as a result of development. It follows a recognisable pattern (Omran, 1971).

First, there is an increase in road accidents and unintentional injuries from the improved transport that wealth brings to a community. Following that, a few years later, non-insulin dependent diabetes becomes common. Thereafter heart and circulatory disease increases, and, finally, some years further on still, cancers become common in the community. While all this is going on the infectious diseases become less important as causes of death and loss of years of life (Figure 4).

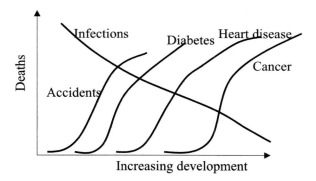

Figure 4. The changes in disease with increasing development: the epidemiological transition (After Omran, 1971).

These changes in disease patterns parallel changes in the structure of a community, its size, growth pattern and age distribution – the demography of a community. This change in the population is known as the demographic transition, in parallel with the epidemiological transition in the disease patterns.

Before a country enters the demographic transition it has a large number of births each year, but an equally high number of deaths, resulting in stable population numbers. When the demographic transition starts the number of deaths begins to fall. Since the number of births remains high at this point the population begins to increase rapidly. Eventually the numbers of births also begin to fall.

When the numbers of births each year equal the number of deaths again the population is expected to become stable again (Figure 5a) (JD VI 47j).

The pattern of the growth of the population also changes through the time of transition. At first, growth is non-existent, with deaths equalling births. As the numbers of births outstrips the numbers of deaths there is usually an ever-increasing growth (exponential), with the total population increasing faster all the time. This is particularly seen in the world population numbers (Figure 5b), which has not yet stabilised.

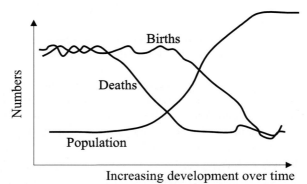

Figure 5a. Increasing development over time brings about increasing population.

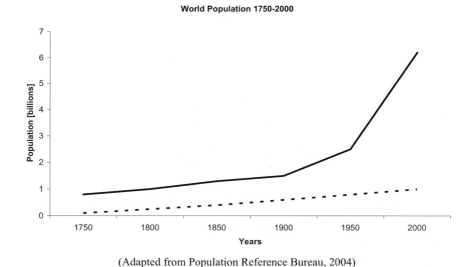

(Adapted from Population Reference Bureau, 2004)

Figure 5b. World population growth over the last 250 years.

Alongside the increase in population numbers there is a change in the age distribution of the population (Figure 6). During the demographic transition the population is dominated by the young: typically, over 50 per cent of the community is under the age of fifteen years (JD VI47f). When a country or community has most of the population under the age of fifteen the problems of development include providing enough food and education at this point and enough jobs in another five years. As the community moves through the transition, the population changes to show a more even spread of ages. The problems also change, to become those of caring for an enlarged elderly, and often frail, community.

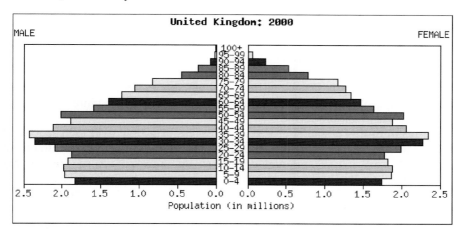

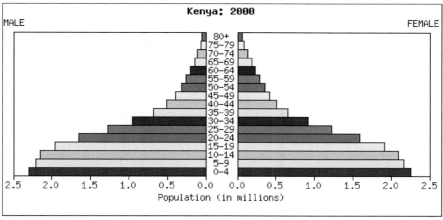

Figure 6. Population pyramids showing the different shape of the distribution of the population in a developed country (UK) and a developing country (Kenya).

Source: US Census bureau "International data base" generated by Ewan Wilkinson <http://www.census.gov/ipc/www/idbpyr.html> (5th July 2003)

Despite the fact that the developing countries are undergoing both the epidemiological and the demographic transition, at different rates in different places, the diseases of poverty are still very important causes of death and loss of years from disability.

The demographic transition arises because there is a drop in the death rate several years before a similar drop in the birth rate. As a country becomes richer the infectious diseases kill fewer people but families are still big (JD VI 47j). So more children live to become adults than previously. The birth rate usually falls, but only some years, perhaps even a generation or more, after the death rate. These extra adults also have children, increasing the speed of growth of the population until there may be too many people for the land to sustain. Eventually, the population is expected to stabilise, but the total population of the country is much greater than before.

The population in the developed countries has stabilised, but over a 100 year period. In some of the developing nations, such as many African states, the growth in population is occurring over a much shorter time span of around 20 years, because of the dramatic impact of immunisation programmes and the widespread use of antibiotics.

The result of this dramatic change is an excess of people over the capacity of the local area to provide for them. This has been called demographic entrapment (http://www.leeds.ac.uk/demographic_entrapment/ Accessed April 2003; King and Elliott, 1997; King 1999a and 1999b). A few voices recognise that Africa is in the middle of such a crisis at the moment (Loeffler, 2003), although many appear to deny it by ignoring it.

But it is not just Africa that has this problem. It is becoming increasingly clear that the whole world is in a similar situation as over-consumption in the developed nations outstrips global resources.

It is said that a baby in the United States will use 150 times the amount of the world's resources of an Indian baby. The USA is by far the largest consumer society using a greater proportion of the world's resources than any other. If this level of consumption were replicated by the rest of the world's population it would rapidly be seen to be totally unsustainable.

There seems to be relatively little discussion as to what is a desirable level of development or consumption. This is likely to be due to the fact that no country is willing to reduce its level of consumption. However, from a health perspective, it must be asked how far the spoken attitude of "health for all" (WHO, 1978) parallels the unspoken attitude of western development and consumption for all (see below).

On top of entrapment and the continuing high death rates from infections in the poor, the rich in the developing nations begin to suffer from the diseases of the "developed" nations. Transition, whether epidemiological or demographic, is not without its difficulties.

Western medical care is only one minor factor in changing the disease pattern in the community. Immunisation programmes in a developing country undoubtedly save lives, but they add to the growing population, as already noted.

Access to health care workers (JDVI 47, 47b), such as nurses in Nicaragua in the 1970s, has also been shown to be important in reducing infant mortality (Sandiford *et al.*, 1991). Hospitals and expensive medical equipment contribute only a little to any reduction in disease rates in a community (Figure 1) (JD VI 47b).

The crippling international debt of many developing countries reduces the amount of money and resources available to develop the community and improve the health of the people, particularly the poorest and most vulnerable (JD VI 48d).

5. Changing behaviour

It is not enough, however, to understand health problems that impair the ability to achieve sustainable development. Information about such situations is unlikely to bring about any change that will be either worthwhile or lasting.

There are several steps that have been identified in a widely accepted understanding of the way change and development takes place (Figure 7).

Figure 7. Behavioural change model [see text for details].
Modified from Puska *et al.*, 1995.

Suppose a village has a well that most people use for their drinking and washing water. Suppose further that the well water is dirty and contaminated, giving rise to diarrhoeal diseases.

Information on its own will not change the way any of the villagers act: knowing that dirty water causes diarrhoeal diseases will not stop most people drinking from the well. The villagers need persuading that a behavioural change is worthwhile. But persuasion itself will only lead to frustration if there is nothing else. The well water is still dirty.

What is needed is a new skill in the community, perhaps the ability to build a well that will remain clean, or the ability to remove the dirt and contamination from the water. However, such skills will not change the health of the community if there is no communal and social support. If the village wants the old, dirty well, then the newly acquired skills are useless.

Along with this communal support there may need to be a change in the natural, social or man-made environment in which community and well is situated. Perhaps the well water becomes dirty because the underground water is contaminated by local sewage. Work needs to be done so that the sewage is contained somehow before the water will remain clean (JD VI 47k, 47l).

These steps (Figure 7) describe the process of change and something of the organisation that needs to be undertaken to alter life and health. But there are two other factors that need consideration at the same time as these steps (JD VI 47e).

First, there is the source of information, as shown by the two boxes on the side of Figure 7, the media and the community. Media messages are useful for spreading information and helping persuade people but they are less good at imparting training or support.

The media can contribute to training and support, but largely through information and advertising or through the ability to influence people towards partaking in training or offering support to leaders and others involved in the processes of change.

Social and community support is crucial. By social we mean the support given to those in the process of change by their closest family, friends, neighbours or work colleagues. Community support is more broadly based throughout the population but, if it is lacking, there will be no lasting change that affects everyone.

The other factor that needs to be noted is that this process cannot be undertaken by one isolated individual. It may start with one or two individuals, but must move outward, as the ripples from a stone thrown into a pond spread across the surface of the whole pond. Without the involvement of groups and communities then no worthwhile change is possible (JD VI 47e).

Change for the better is neither simple nor a single step process, but it is possible. The process of change that lead to improved health and the process of change that leads to sustainable development are the same.

6. Vulnerable groups

In every society there are groups of people who are weaker and therefore more susceptible to a wide variety of problems, including poor health. These groups are less likely to initiate or benefit from any development work in their vicinity. They include women and children, the homeless and ethnic minorities or indigenous peoples (JDVI 47a, 47f, 47I, 48a, 48d).

In most cultures women are the caregivers: they nurse the children and any members of the family who are sick. Even in patriarchal societies the women often run the family.

However it is often the man who is the main wage earner. This puts women in a weaker position with regard to access to education, health care, food and clothes as, if there are costs involved, the women may have to ask for money to reach the service or to pay for the service. What is done with the money the man earns may not help the family at all. It may be spent on alcohol or other inessential items that are of benefit only to the man.

Pregnancy and childbirth can be dangerous. The international variation in maternal mortality reflects the difference that better nutrition and good maternal and obstetric care can make. In some countries the rate is 350 maternal deaths per 100,000 pregnancies while in the developed countries the rate is around 8eight per 100,000 pregnancies.

A maternal death is always a disaster for that family. There are often other young children in the family and their upbringing is likely to suffer. Depending on the culture they will either be distributed round the extended family or a young girl may be brought in to provide basic care. This may reduce the stimulation and development of children. If they are living with another family, it may reduce the likelihood of them receiving much education, as the host family children may get preference.

The traditional approach to education in many of the developing countries is that the boys should take precedence in obtaining education. This occurs as it seen as being a way to increase their earning potential and so the income of the family. However research has shown that improving literacy of women increases child survival (see later).

Research with micro economic projects such as the Gameen Bank in India have found that women are more likely to be successful in their projects than men. Success is assessed as (1) women being more likely to repay the money they have borrowed, and (2) that the financial benefits from the project are likely to benefit the whole family.

The life of families in rural and urban developing countries is often hard. Much of the drudgery falls on the women and children. For example, the collection of firewood and water may take several hours. Technology can make a large difference to life, but it needs to be appropriate to the local situation. Bringing water sources closer to homes may save considerable time, but if it then costs money this means that there has to be a way of earning money with the time saved to pay for the water.

Simple technology can aid women's lives without great expense. An example of this is the way that solar cookers have freed women in parts of Afghanistan and Pakistan. The provision at low cost of solar cookers has cut the need to collect firewood. It has also cut the time spent on cooking. A meal prepared and placed in the solar oven when the woman goes to the fields in the morning will be ready when she returns in the middle of the day. Not only that, but if the woman is delayed in the fields for any reason the meal will not burn since there is no direct heat applied. For a small outlay, simple technology has transformed the lives of many village women.

Transforming women's lives removes some of the factors that lead to ill health. For example, solar ovens reduce the time spent away from the village collecting firewood; this reduces the energy used by these women so that they become less tired. Tiredness reduces immunity to infection.

Transforming women's lives does not just remove unhealthy factors but can strengthen the healthy factors. Giving women greater control over their time they spend cooking through the provision of a solar cooker can enhance the diet of the whole family since the women have more time to spend preparing meals.

Changing women's lives is a necessary adjunct to improving health, but needs to go hand in hand with changing the attitudes of their menfolk. In one instance, the change that solar cookers brought about was so significant that some of the men of the village felt threatened and obtained the removal of the supplier of the solar ovens.

As was noted earlier, change must occur at a number of levels in a community to be fully effective. All change can be threatening and the threats, imagined or real, need addressing and answering if development is to progress. Without this the desired improvement may be lost. The village where the supplier of solar ovens was removed has lost an opportunity to improve the health of its vulnerable women and children.

Children are the future hope of parents, particularly in developing countries. Where there is little prospect of an adequate pension, the support of children is needed in old age.

In subsistence farming children start working in the fields or looking after animals from an early age. They pay their way as part of labour force from around aged ten years or younger, but this can lead to a conflict with the desire for education. In some regards this is a conflict between the short and the long-term benefit. In very poor societies the time for education cannot be afforded, let alone the expense of fees, uniforms and books.

Some of the common infectious diseases, such as polio and meningitis, can result in long-term disability, as indicated earlier, and also birth injuries can result in physical or mental impairment. A disabled child may be a chronic drain on the limited resources in a family or in some case may result in the family breaking up.

Providing services, which reduce the incidence of disability, will reduce the drain on society. Immunisation has done much to reduce the incidence of polio.

In many countries the survival of young children has increased markedly, due to a number of factors including nutrition, improved obstetric care, the management of childhood illness and the benefits of immunisations.

The problem is that if there is no change in the total fertility the number of children rapidly increases. This is seen in many developing countries. Parents who wish support from their children in old age need to see for themselves that the children will survive before they can take the risk of reducing their family size. While the under five mortality is falling in most developing countries, it is still higher than in the developed world.

Also, there is a higher risk of death amongst young adults from infections and accidents and injuries in developing countries and countries in transition than in the developed world, which makes parents reluctant to reduce family size.

The pattern of disproportionate and preventable mortality amongst girl infants and children is a reflection of the poor underlying value given to girls and women. Due to the social structures in some countries boys are seen as more financially beneficial, both in terms of the potential earning power and also not costing much to get married. On the other hand, girls may be seen as having little benefit and being costly to get married. It is only relatively recently in the developed countries that women have had equal opportunities in the job market. Now girls are even starting to out perform boys in academic achievement.

There are other vulnerable groups in society, such as elderly persons and indigenous people and those with disabilities, both physical and mental. To take one example, the disabled are often poorly provided for. For many, simple aids and adaptations may allow them to make a useful contribution to the community but there is often not the knowledge and resources to provide these (JD VI 47a).

For those who are particularly vulnerable, the local culture will often provide help where there is community support, but this will only provide a borderline existence as there is often little to spare for such people. In slum areas there may be no functioning community and little or no help may be offered to those who cannot look after themselves. Marginalised indigenous people often live in slums and have the worst health in a country.

Of those with mental illness, people suffering from schizophrenia form a vulnerable group. Their wild behaviour will not always be tolerated, but their illness can often be controlled with cheap medication. Such treatment requires little investment to enable it to be provided to those in great need and yet mental health often comes low on the list of the priorities in health development.

Depression is often hidden as sadness, or appears in another guise, such as pains throughout the whole body. Much work needs to be done to train communities and their health care workers to recognise and respond to such illnesses, which, as already noted, are a major cause of loss to the patient and the community.

The highest risk areas in developing countries are the slum areas and shanty towns that arise around the cities. Water and sanitation are usually poor. People live in low quality housing. There is little identifiable community life. There is usually poverty with high levels of crime and drunkenness. All this gives rise to high levels of illness and death.

The growth in shanty towns is almost inevitable with the rapidly rising population. Land can only be divided amongst the family so many times before each parcel of land becomes too small to sustain the next generation. There is then little choice but to join the urban drift.

Living in the shanty town gives rise to the higher acceptance of risk taking behaviour in obtaining more dangerous jobs. These areas are usually poorly served by health services and education and therefore children growing up in this area are disadvantaged in terms of being able to break out of the area.

The shanty towns or slums are generally seen by the ruling classes to be of little importance because they have little power and little economic leverage, therefore often decisions are taken without the residents' needs being considered.

7. Economics

Economics is the study of the use of money and resources in a country or system. If anything is to be sustainable it must make economic sense. Health comes into this as much as anything.

Aid (JD VI 47c) can appease the consciences of those with great wealth but can create dependence as groups can become reliant on income from abroad and see it as a right. There is a problem with some organisations, such as famine relief agencies, since the jobs of the employees of the organisation are dependant upon the organisation finding another "famine" or other disaster to give aid to. This can distort the provision of help as well as the local food supply by the inflation of small problems into large ones.

Some aid (JD VI 47c) is given for defined projects of limited length. However, unless the local economy is able to afford to take over funding when the aid project is withdrawn there is a risk of a series of well meaning but short-term projects, which may do little to tackle some of the underlying causes of poor health.

Expectations and hope are raised by the aid projects then dashed by the short time scale or inappropriate schemes since the local people have not been given enough time or involvement to help adapt the aid to the local needs, resources, culture or society (JDVI 47, 47c).

It is not unusual for health related aid (JDVI 47) to donate nice, new hospitals filled with the latest technology. But once the building is finished much aid also finishes. There is no maintenance of building or equipment, there may be no trained staff to run the equipment, or the trained staff move on to better paid jobs elsewhere (JD VI 47d).

How does this deterioration affect health? Disappointment and dashed expectations are one obvious effect on the local population and staff, while lack of improvement in health and health care affects the sick and vulnerable.

HEALTH: A NECESSITY FOR SUSTAINABLE DEVELOPMENT

But a deeper problem is that the money and materials, manpower, imagina and effort expended in the project could have been better spent on other projects which had more suitable and sustainable effects.

An opportunity to improve health that has been lost is an opportunity that has cost someone somewhere health gain. What is known to economists as the opportunity cost of a project is based on the simple idea that to put money (or effort or time or whatever) into something is to be unable to put that same money (or effort or time or whatever) into something else.

Whenever a project to improve health is discussed, the opportunity cost must be considered. One of the problems of the Johannesburg declaration is that it views the health of the developing world from a Western perspective. Heart disease and other chronic diseases are increasing, but infections and psychiatric diseases are still the major causes of ill health in the developing nations.

Putting more effort and money in the developing countries into the chronic diseases, especially if it is at the expense of effort into infections, will inevitably shift the focus away from the commoner causes of death and disability, to the disadvantage and harm of these communities. The richer nations are in danger of forgetting how serious the infectious diseases are for the poorest in the world (Figure 8).

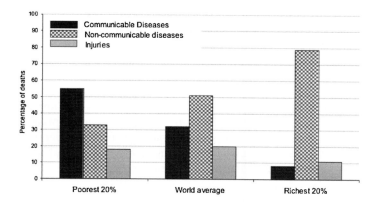

Figure 8. Main causes of death in different communities.

However, economics are not just about costs. An important aspect of any economy is the source of the income used to pay the costs incurred by a country, by development of by investments in health.

There is an ongoing debate about the relative benefits of aid (JD VI 47c) versus trade (JD VI 50) as the source of such investment.

Ideally, trade is more sustainable and better for all concerned. It rewards those working and creates its own self-sustaining organisations, but it also relies on a reasonable degree of honesty in handling money. The reward from corruption is often proportionally larger in a poor society and the pressure to succumb to corrupt practice is possibly greater than in a richer society. The move into or away from corruption is also greatly affected by the underlying values of the culture or subculture as to whether it is accepted behaviour.

However, corruption, by its nature, distorts the people and society it infiltrates. As so often happens, in such a situation the people that suffer are those who are poor and disadvantaged. And when the poor and disadvantaged suffer, for whatever reason, their health suffers, as has been seen already in this chapter.

To work properly trade assumes open markets with equal access for all vendors and purchasers. This does not hold true for many of the more profitable markets. Some markets are subsidised, such as the agricultural subsidises in the European Union and USA. Others suffer from dumping of, for example, foodstuffs, which can wreck markets. This can occur where there is a surplus in the subsidised market in the North. The surplus is then dumped on the South at prices below the local cost price. Other markets have tariffs imposed for imports from outside given areas.

Once again, this distortion of the situation disadvantages people, by not allowing them to develop fair trade. This has severe health effects by keeping the poor poor.

Technical and financial assistance, which can be valuable, is recommended in the declaration, but is mentioned in relation to IT (information technology). Technology, of whatever kind, must be appropriate. Information technology is pushed as the answer to many problems, but even in Europe it has failed to generate many of the promised solutions.

Specialised x-ray machines such as CAT scanners (computerised axial tomography) are very useful, but need upkeep and highly trained staff. The collection of data on computers will not add anything beyond the collection of data on paper if it is not properly organised, funded, maintained and analysed.

But at times the higher technical solution can be more appropriate: e.g. mobile phones are easier to provide and manage than a network of landlines in some countries. This can make the transfer of medical information and advice, whether between a central specialist and a rural practitioner, or a doctor and patient, easier.

One of the basic needs in economics is for suitable jobs. These need to be available at a level corresponding to the levels of training of the local work force. There also has to be a balance between the level of risk regarded as tolerable in the job and the tolerance of the workforce for the job – does the pay justify the risk, is the job satisfaction worth the risk, are the outcomes of the job worthwhile? Imposing Western safety standards may be unrealistic as, if people are unable to gain employment, they may be at much greater risk than in an unsafe job.

Physical labour may be the only source of employment locally. Safety may be too expensive for the local economy. Consideration also needs to be given to what are appropriate local methods of production. For example, importing a stone-crushing machine may be seen to be a faster and safer way of producing aggregate for building, but local people with hammers may produce the same quantity of crushed stone while at the same time creating an income for themselves, and not require any expensive foreign spare parts. Nor will they suffer the injuries associated with machinery; hammers are usually safer than machines with engines. It is also possible that the stone crushing machines will produce more dust than hammers, exposing the fewer workers needed to operate the machinery to a greater risk of lung problems (such as silicosis) than the larger number of hammer wielding workers are exposed to.

Employment and job satisfaction are both related positively to health, while their opposites, unemployment and job dissatisfaction are known to be associated with ill health of many causes, not least increased alcoholism.

Financial assistance can mean grants or loans. Loans, in particular, but also grants, should be clearly seen to be the best value solution to the problem before being approved. The risk is the offer of loan is seen as a free gift, which can lead to later problems. Politicians may see loans as ways of achieving changes in their period in office with the problems accruing to the next generation.

Many countries in Africa spend more per year in debt repayment than they do on health. How then can they increase their spending on health or the factors which affect it? Debt which cripples is an "opportunity-cost" spent against health, not for it.

Loans taken with the best of intentions can lead to later problems when the prices of commodities such as coffee or copper, fall to much lower prices that predicted.

But if aid money is not use to develop services such as water and sewerage either private or public finance will have to be used (JD VI49a). Water is a key resource: only a quarter of Africa and under 50 per cent of Asia has piped water in the household. Those without water in the house are likely to obtain water from a standpipe, well, stream or lake.

Clean water and proper sanitation are major needs in countries where diarrhoeal diseases are important causes of loss of life. Traditional water sources are not sustainable in large urban areas and there is a great need for adequate sanitation.

Developing the infrastructure with private finance makes some financial return on it essential. But creating a market for water can lead to problems: riots have arisen due to the introduction of a water tariff. In some countries private companies are wary of becoming involved, as profits may be low. But without investment of some sort the prospects for health improvement are negligible.

8. Education

Improving the economic situation is all very well, but on its own this will not achieve everything. There is a parallel need for education, as has already been stressed.

When the women in a community are educated, even to a basic level where they only learn to read, the health of their children improves. In Nicaragua in the

1970s, as has already been noted, the improved ease of access to nurses brought about an improvement in infant mortality. Then, in the 1980s, two further things were found, also in Nicaragua, about the way health in children can be improved. The first was that educating women to read and write improved the likelihood that their children would survive an illness. This was despite no corresponding health education having been given to the women. The second finding was that this effect of education on the health of the children was significantly greater among those women with poor access to health services (Sandiford *et al.*, 1995) (JD VI 47b).

The same work also suggested that the effect of education in reducing the risk of malnutrition acts independently of its effect on mortality. Not only that, but both are independent of wealth and of the parents' decision to educate their daughters (Sandiford *et al.,* 1995). In other words, education has an important action of its own in improving health. And education not only improves health, but also leads to the possibilities of sustaining development in the community.

Schooling is thought to impart skills and foster other individual changes that alter women's patterns of social participation, whether in the home or when dealing with bureaucracy, which can include health facilities as well as government departments (LeVine *et al.,* 2001). It would seem likely that the same can be true for men, but one problem with educating men and boys is that there is not always enough employment for them. This leads to frustration, violence, alcohol and drug abuse and ultimately to ill health, accidents, possibly death.

In some countries education has been restructured to try and provide relevant courses at different stages. Education needs to be relevant to what work the students are likely to end up doing, while, at the same time, allowing and even encouraging those who are able to continue into higher education.

Pupils often see education as a way of escape from the situation of the parents, whether they live in rural or urban poverty. However, the aspirations of the pupils do not always match the reality of the prospects. Their basic education does not necessarily give them the skills they require to improve current their life situation. This can rebound on their health, leading to depression, drug taking, violence or stress related heart disease, for example.

For those in rural areas, learning to be better farmers who are able to benefit from relevant research and improved management is, in many ways, a more relevant and beneficial outcome than aspiring to be a clerk in an office.

The effects of better farming far outweigh benefits gained from office work. There is good evidence to show that lower tier office workers suffer poorer health than their bosses; few students from poor rural homes become bosses. Against this, better farming can lead to a wide variety of benefits. For example: sustainable development, more food in the markets, which improves diet and cuts infection, community development with improved social and personal benefits and the maintenance of the family structure with the benefits this brings to women and children.

9. Beliefs and values

For something to be sustainable in a culture it has to be congruent with the beliefs and values of at least a significant proportion of the society (JDVI 47). If it is not, it is unlikely that any change will be continued in the longer term. In the short term it may be sustained while there is other apparent benefit for adopting it, such as additional finance coming into the system. This demonstrates how finance can distort values in a culture. Sometimes this is done deliberately and other times mistakenly by external agencies.

There can be an intellectual arrogance amongst those from the developed world working in developing countries. There is an assumption that because something works in their culture and has some evidence based to it that this is the correct answer for every other culture (JD VI 46-50). For example, in some cultures it is unacceptable to give an individual bad news.

However, the WHO AIDS programme insists that people who are HIV positive must be told that they are HIV positive and given "appropriate" counselling. This may be completely alien to the culture. AIDS is one of the major challenges facing sustainable development in some part of the world, particularly in sub Saharan Africa. For AIDS to be controlled, some cultural practices and beliefs may need to be challenged, but this must be done in an appropriate way.

The development of effective health care systems (JD VI 47) is also affected by the beliefs and values of those in position to make decisions. The majority of deaths and disability occur due to common widespread illnesses such as malaria, respiratory infections and diarrhoea (above). However, in many developing countries there is a desire to have prestigious high-tech hospitals able to offer intensive care and renal dialysis which, while providing essential care to a small number, consume a disproportionate amount of the budget and hence deprive those in need of basic care of simple treatment.

The provision of appropriate and affordable care for the main causes of morbidity and mortality are at the heart of the importance of health for sustainable development. The World Bank (Abbasi, 1999) has asserted the importance of health of the population as an important component in achieving sustainable development. This can be achieved by providing appropriate and accessible healthcare, but this should also supplement other health developments and not be provided in isolation from such development.

Often aid and development are undertaken with little reflection on underlying values (JD VI 47, 47c). What is often seen by the providers as an obvious development and improvement may not be taken up with the expected enthusiasm by the local population.

In one country with a large dairy industry, milking machines were introduced. It was expected that there would be a large demand for them to maximise the production. However it was found that this was not the main priority for farmers; they were more concerned to have a well functioning farm to pass on to the next generation. They did not seek to maximise their income and were cautious of going into debt to buy the machines. The hard work of hand milking was accepted as part of the way of life.

Western medicine is often effective when correctly used, but may not answer questions that are important to local people. It can say what is wrong, such as "you have broken your leg", but has no answer to the question "why did I fall out of the tree today?" The scientific approach does not satisfy this need in local belief systems to know why or who caused an illness or accident.

Much of the world depends on traditional medicine, and some very effective western medicines have been developed from traditional medicines. For example digitalis, a heart drug was developed from a preparation of the Foxglove plant, and Quinghaosu (artemisin), a treatment for malaria, is derived from a traditional Chinese herbal remedy.

However, the role of traditional sources has been largely forgotten or ignored in modern western medicine. Do the developing countries want to import this western approach along with the advances that scientific medicine can bring?

There is scope for fuller integration of the different systems. This is being advocated in some countries but it is not as easy as it might appear. This is because the different systems have different philosophies. One is traditional, depends on the knowledge being handed down through the generations, uses what is found in nature and respects tradition. The other is imported, depends on a scientific theory that is alien to many people, depends on imported and costly treatments and teaches that many traditional beliefs may be harmful to health. There are some projects looking at the scientific basis of traditional medicine, but it is proving difficult to bridge the gap between the systems.

Provision of western medical treatment may include the weaknesses of the market system and the resulting abuse of power, which may not be so common with traditional medicine. Under the imported system patients may be persuaded to have ineffective treatment they can ill afford. These include injections, but even operations are not unknown.

Even if they are given the correct treatment according to the best scientific evidence, the patients may receive counterfeit drugs which have no beneficial effect. The health organisation that is based on western ideas is outside the normal local culture. It is, therefore, more open to abuse than the traditional medical system, which is regulated by the traditional values of the community and culture.

10. Westernisation

Defining development, whether or not it is sustainable, is often difficult to do. Part of development is enabling individuals to achieve their potential. This is usually seen in a western context, with more emphasis on individual than collective achievement. Success is also defined in western terms of achievement, money or power.

The Johannesburg declaration sees development and health from a western perspective; this perspective is not always a good thing. As has been noted already, western beliefs and values may conflict with local ones, leading to a loss to the community, of health amongst other things.

The development of a country is seen to be the achieving of an increase in the gross national product, by increased consumption and production of goods. This is normally achieved by increased education of the population and increased industrialisation.

Development is assumed to be of benefit, but this is not always the case. The increase of traffic leading to traffic jams in Singapore, for example, is seen as a sign of development or "progress". It could also be seen as reflecting a problem with individual greed and acquisitiveness. This shows how things can be interpreted in different ways. Either way, an increase in traffic leads to an increase in accidents and death. Could there be a better way?

It appears that most countries in the world desire the western form of development or increased consumerism. Many countries see the American lifestyle, as portrayed on TV and in films as one to be envied and emulated. This may change as a reaction to events such as the recent war in Iraq.

Westernisation, which is largely what development appears to be, seems to be based on some of Freud's theory of the gratification of individual desires. This approach has been used with great effect within advertising and the consumer society, which has developed in many northern countries. It is striking that even in China, with the recent liberalisation, that smoking and car ownership are rapidly increasing as they become available. Neither of these contributes to the health of the population or the sustainable use of resources. Rather, they clearly pave the way for the epidemiological transition, moving from infections to accidents and cancer.

There is a widely held assumption that the western is the best way. For example, aid (JD VI 47e, 47l, 48b) to some countries is dependent on them adopting western-style multiparty democracy and a free market, yet the US market is subsidised. Strong pressure is put on countries to adopt "democracy" but it has taken hundreds of years for the UK and the USA to develop the form of democracy they have now. It is, therefore, unlikely that any country can rapidly introduce it in the way that seems to be expected. Certainly, importing a culturally inappropriate system is not the answer.

Improving health is not dependent on westernisation. It is helped by increasing the income of the poor and so improving nutrition and access to education and basic health care (JD VI 47b). This can enable a country to allow its population to choose how it develops.

The Chinese have made good progress in rural health improvement without a westernised approach. Unfortunately, as they have begun to move towards greater integration into the world community they have lost their focus on their indigenous medical practice. It is of interest that, at the same time, the health of their rural poor is once more falling.

11. Traffic, air quality and chemicals

Air quality is an important issue in sustainable development (JD VI 49), bringing together many of the themes of this chapter, such as death and disease (JD VI 47f, 47i, 47m, 47o, 48a, 48b, 48c, 48d), employment (JD VI 47d, 47m, 48c), economics (JD VI 47c, 48b, 48d, 49a) and trade (JD VI 50) and the health of vulnerable groups (JD VI 47a, 47f, 47g, 47i, 48a, 48d).

Two of the problems of a western-style development are the increases in traffic, as already noted, and industry. Both of these reduce the quality of the air as they multiply the number of particles and chemicals released into the atmosphere. Particles and many chemicals have serious, adverse effects on health.

But the developing world is not without its own problems of air quality: in rural areas, in particular, the wood smoke from cooking fires is a potent source of particles that can seriously affect health.

There is ample evidence that poor air quality is causing increases in both disease and deaths across the world. As might be expected, vulnerable people are more at risk. For example, children breathe more air relative to their size than adults. They have longer to live with any disability or adverse effects and are, therefore, more susceptible to the problems of pollution.

The poor live nearer to roads, railways and industry than the rich, who like to avoid the noise and smell generated by transport and factories. Such poor are, like children, a vulnerable group. In such situations, they are exposed to high concentrations of pollutants and their health suffers accordingly. The Bhopal disaster in India, where leaking chemicals from a factory killed an unknown number of people and left countless others disabled and diseased, is a prime example of where the poor have borne the burden imposed by the rest of society, indeed, by the rest of the world.

The latest research attributes between 3 per cent and 6 per cent of all deaths to the inhalation of particles (Dockery *et al.*, 1993; Kunzli *et al.*, 2000, Pope *et al.*, 2002). There is also evidence that chemicals give rise to death and disease. For example, an increased incidence of kidney disease has been found in communities living near factory chimneys, which release a variety of chemicals into the local atmosphere. Increased rates of various cancers have been related to working in, or living near, industries using heavy metals or organic compounds.

Although exposure to particles will vary from one part of the world to another, wood smoke from indoor cooking is a major source of particles in the developing world, particularly in rural areas.

It is becoming apparent that it is the smallest particles, called ultrafine particles, that cause lung damage (Maynard and Howard, 1999; Donaldson *et al.*, 2000). These particles are produced by many combustion processes, such as cooking, traffic (the internal combustion engine drives most vehicles), open burning of waste, incineration and power production, and are the hardest to control. High levels of particles in air are known to exacerbate asthma, a condition which affects children more commonly than adults because children have narrower airways.

Traffic fumes (JD VI 49c) are now the most important source of particles in towns and cities. These are mainly from the volume of traffic or from poorly maintained engines.

Factories emit an unknown number of chemicals into the environment , mainly into the air, although releases to water and soil are also possible (JD VI 49c). These chemicals can spread through local communities and produce high levels of disease. With the move of many industries and factories from the developed world to the developing countries, where labour costs are lower and regulations less strict, the diseases associated with western industrial sites can be expected to increase in number and importance in the developing world.

Improvements in air quality can, in principle, be brought about quite rapidly, once governments take appropriate policy decisions. For example, there is an urgent need to reduce the impact of traffic on health by reducing the levels of traffic in cities. Little action, so far, has been taken in the developing world where, as in the richer nations, owning a car has social status that is seen to outweigh any negative effect of the car on the environment or on health.

Road-use charges have been introduced recently in the developed world in an attempt to reduce traffic volume in busy cities. Such charges need to be matched by improvements in public transport. Unfortunately, it is not likely that such pricing systems and improved public transport will be feasible in the developing world for some time, leading to a continued increase in pollution and disease.

In addition to improved public transport, cycling can be beneficial. Cycling helps counter an important negative health effect associated with the car, the reduction in exercise that comes with driving everywhere. Exercise, whether from cycling or walking or some other source, is effective in keeping heart disease, diabetes and some cancers at bay. Cycling is common in certain developing countries and needs to be retained; indeed, it needs to be actively encouraged.

However, cycling is not a simple answer to traffic problems. For example, cycling in a polluted town may cause deeper inhalation of toxins from the air! Cycle taxis and rickshaws are often seen as a sign of poverty and are replaced with motor taxis and rickshaws as soon as possible, reducing exercise and increasing pollution.

Factors that need to be discussed in any move to increase or retain cycling include the relative status of the cyclist and the motorist: should the motorist be seen as richer and more advanced, or the cyclist as wiser and healthier? The monetary costs and the ease and speed of different forms of transportation also affect whether cycling is seen as a realistic option by the community.

Ideally, health economists and others involved in public health should be able to predict the benefit to the community from the implementation of policies. However, it can be difficult to foresee what a policy will affect most. For example, the introduction, for safety reasons, of cycle helmets in one community (in an attempt to reduce head injuries to cyclists) has reduced the total number of cyclists. More people now travel by car. So a measure designed to improve health has actually reduced health.

Perhaps more important, even, than any prediction of the impact of any policy is the possibility that some developing countries could avoid making the mistakes that have already been made in the developed world. Instead, these countries should aim to by-pass these problems and move to more sustainable and healthy development.

However, air quality is not limited to the effects of traffic. Chemical releases from factories and industry are an important matter as well. By choosing to address lead as the only chemical that requires action, to some extent the Johannesburg declaration took the easy option, ignoring a wide variety of other chemical toxins (JD VI 49b, 50).

Lead is a very important issue for child health and, since it is one of the toxic substances that we know most about, must be controlled. Current WHO guidelines recommend a level that is one sixth that which was common in the 1960s in the air of developed countries. In other words, it is widely recognised that progressive lead reduction is achievable. Removing lead from petrol is seen as the single most important step in this.

This may go some way to explain the inclusion of lead to the exclusion of any other chemical. However, there are indications that the current WHO recommended levels may not be fully protective; indeed, there may be no safe, lower limit to lead exposure. Newborn babies with cord blood levels just above the level recommended by WHO have been shown to have impaired thinking skills at the age of two (Bellinger *et al.,* 1987). Further action on lead is necessary and the setting of targets is a good start.

There are some much more intractable problems than lead which have not been mentioned at all in the report. One example is mercury. This is another metal that, like lead, is toxic to the developing brain and nervous system. Programmes to control and reduce the exposure to mercury are not addressed at all in the declaration.

However, an omission from the chemical section of the health chapter of potentially great importance from a public health perspective is the problem of persistent organic pollutants (POPs). These are chemicals which are based on carbon but which are not easily broken down, which is unsurprising, given their chemistry. They survive a long time in the environment and continue to be toxic. The most problematic of these are those that are made from combinations of carbon with halogen atoms (e.g. chlorine and bromine).

The global chemical industries have been producing many millions of tonnes of chlorine each year for several decades. Chlorine gas is a highly toxic by-product of caustic soda production and does not occur naturally, except in minute amounts in volcanoes. The chlorine industry started because this gas could not be disposed of by release into the atmosphere, because it is so poisonous. Chlorine is now mostly combined with carbon to make organochlorine products such as PVC, solvents, paints, pesticides, medicines, floor coverings and furnishings. The problem is that these products are often disposed of by burning. Incomplete combustion of these products gives rise to some of the most persistent and toxic substances known, the dioxins.

Many persistent organic pollutants, including many organochlorine compounds, accumulate in the body. They also accumulate in the animals we eat, resulting in even higher levels in our bodies. Because of their long-lived toxic effects, these chemicals need to be evaluated for the size of their impact on the health of the public on a global scale. There are no unexposed communities. Even those who live far from industry have raised levels of persistent organic pollutants in their body.

Current levels of the persistent organic pollutants are high enough in some women's bodies to be associated with problems in the development of their children's immunity, brain, nervous and reproductive systems. Mothers with the highest levels of dioxins or the similarly toxic polychlorinated biphenyls (PCBs) had the most affected children in recent large studies in Holland (ten Tusscher, 2002; Koopman-Esseboom *et al.*, 1996; Patandin *et al.,* 1999). Some of these effects have been shown to last from birth to beyond seven years of age.

The solution to these problems is not difficult: reduce the production of the chemicals that give rise to dioxins or the other persistent organic pollutants. That is, reduce global chlorine and bromine production and eventually the levels of the associated persistent organic pollutants will fall. Herein lies the problem: politically and economically this is a difficult problem. The chemical industry contributes 7 per cent of the global economy and up to 20 per cent of the chemical production companies are concerned with halogen products.

Reducing these, or other, chemicals that pollute and cause disease and disability will not be easy. The Johannesburg declaration recognises that children represent one of a number of vulnerable groups (JD VI 47f). The recognition of such vulnerable groups is important and suggests that an approach based on caution would be a possible way forward.

The Precautionary Principle, as it is termed, is a well-recognised method for environmental protection, dating from the United Nations Conference on Environment and Development in Rio de Janeiro, Brazil in 1992.

Its use to protect people and health should be at least as vital as its use to protect the environment. We would suggest the following as a guiding principle concerning the adverse effects of developments that affect health:

"Where there is a significant threat to health, lack of full scientific certainty should not be used as a reason for postponing measures to avoid or minimise such a threat."

However, we need to be more than just minimising threats, we need to be actively promoting health, as this chapter has shown.

12. Conclusions

The major health problems that need tackling are deaths and disability arising from infection. At the same time, increasing westernisation needs to be examined carefully and questions asked about what it offers.

The poor seem to suffer everywhere. With the failure to improve everyone's health by the year 2000, WHO has stopped using the slogan "Health for All" (Yamey, 2002). While this can be seen as a retreat from a global commitment it could be the opportunity to review the relevance of a western approach to health and development and start to rethink what we are doing. We need to identify and examine every assumption, both in the ways that we set out to improve health and in the achieving of sustainable development.

Politicians usually work to short time scales, which are closely related to the next election. Dealing with complex problems of public health often needs a time scale and a vision over decades.

To ask a politician to take steps that will (1) reduce income to the exchequer and (2) potentially be of benefit in something like 30 years is to ask for a lot. However, it is to meetings such as the global environment and health meeting in Johannesburg that society looks. It seems that we are not yet at a stage where we can rely on such meetings to take the long-term, wise choices that need to be taken.

Perhaps local action in and by the developing countries needs to show the way forward, indicating to politicians, economists, aid and development workers, health care staff and industrial leaders, local communities and developed countries that there are workable answers. After all, a recent article on the World Health Report for 2002 was entitled "*Simple measures could increase life expectancy by 5-10 years*" (Dyer, 2002).

Development can lead to problems with disease, with population, with westernisation, with assumptions and beliefs, with air quality, traffic and chemicals. Sustainable development aims to provide the advantages of development without the weaknesses and shortcomings that the current approach produces.

Health and sustainable development are interdependent. Just as we need two legs to be able to stand comfortably for long periods, so we need the two legs of health and sustainable development to allow our children and their children to live their lives to the full with hope and good prospects.

References

Abbasi, K.: 1999, 'The World Bank and world health – Healthcare strategy', *British Medical Journal,* **318**, 933–936.

Bellinger, D., Leviton, A., Waternaux, C., Needleman, H. and Rabinowitz, M.: 1987, 'Longitudinal analyses of prenatal and postnatal lead exposure and early cognitive development', *New England Journal of Medicine,* **316**(17), 1037–1043.

Dahlgren, G. and Whitehead, M.: 1991, *Policies and Strategies to Promote Social Equality in Health*, Institute of Future studies, Stockholm.

Dockery, D.W., Pope, C.A., Xu, X.P., Spengler, J.D., Ware, J.H., Fay, M.E., Ferris, B.G. and Speizer, F.E.: 1993, 'An association between air-pollution and mortality in 6 United-States cities', *New England Journal of Medicine,* **329**, 1753–1759.

Donaldson, K., Stone, V., Gilmour, P.S., Brown, D.M. and MacNee, W.: 2000, 'Ultrafine particles: mechanisms of lung injury', *Philosophical Transactions of the Royal Society of London*, **358**, 2741–2749.

Dyer, O.: 2000, 'Simple measures could increase life expectancy by 5–10 years', *British Medical Journal*, **325**, 985.

Gwatkin, D.R. and Heuveline, P.: 1997, 'Improving the health of the world's poor. Communicable diseases among young people remain central', *British Medical Journal*, **315**, 497–498.

King, M.H. and Elliott, C.M.: 1997, 'To the point of farce: A Martian view of the Hardinian Taboo – the silence that surrounds population control', *British Medical Journal*, **315**, 1441–1443.

King, M.H.: 1999a, 'Commentary: Bread for the world – another view', *British Medical Journal*, **319**, 991.

King, M.H.: 1999b, 'The US Department of State is policing the population policy lockstep', *British Medical Journal*, **319**, 998–1001.

Koopman-Esseboom, C,. Weisglas-Kuperus, N., de Ridder, M.A.J., van der Paauw, C.G., Th. Tuinstra, L.G.M. and Sauer, P.J.J.: 1996, 'Effects of polychlorinated biphenyl/dioxin exposure and feeding type on infants' mental and psychomotor development', *Pediatrics*, **97**(5), 700–706.

Kunzli, N., Kaiser, R., Medina, S., Studnicka, M., Chanel, O., Filliger, P., Herry, M., Horak, F., Jr., Puybonnieux-Texier, V., Quenel, P., Schneider, J., Seethaler, R., Vergnaud, J.C. and Sommer, H.: 2000, 'Public health impact of outdoor and traffic-related air pollution: a European assessment', *Lancet*, **356**(9232), 795–801.

Levine, R.A., Levine, S.E. and Schnell, B.: 2001, '"Improve the women": mass schooling, female literacy, and worldwide social change', *Harvard Educational Review*, **71**(1), 1–50.

Loeffler, I.: 2003, 'The population trap', *British Medical Journal*, **326**, 507.

Maynard, R.L. and Howard, C.V. (Eds): 1999, *Particulate Matter: Properties and Effects upon Health*, Bios Scientific Publishers, Oxford, ISBN 1 85996 172X.

Omran, A.: 1971, 'The epidemiological transition: a theory of the epidemiology of population change', *Millbank Quarterly*, **49**, 509–538.

Patandin, S., Lanting, C.I., Mulder, P.G., Boersma, E.R., Sauer, P.J. and Weisglas-Kuperus, N.: 1999, 'Effects of environmental exposure to polychlorinated biphenyls and dioxins on cognitive abilities in Dutch children at 42 months of age', *Journal of Pediatrics*, **134**(1), 33–41.

Pope, C.A., 3rd, Burnett, R.T., Thun, M.J., Calle, E.E., Krewski, D., Ito, K. and Thurston, G.D.: 2002, 'Lung cancer, cardiopulmonary mortality, and long-term exposure to fine particulate air pollution', *Journal of the American Medical Association*, **287**(9), 1132–1141.

Population Reference Bureau.: 2004, http://www.prb.org.

Puska, P., Tuomilehto, J., Nissinen, A. and Vertiainin, E.: 1995, *The North Karelia project. 20 Years Results and Experiences*, The National Public Health Institute (KTL) Finland, Helsinki.

Sandiford, P., Cassel, J., Montenegro, M. and Sanchez, G.: 1995, 'The impact of women's literacy on child health and its interaction with access to health services', *Population Studies – A Journal Of Demography*, **49**(1), 5–17.

Sandiford, P., Morales, P., Gorter, A., Coyle, E. and Smith, G.D.: 1991, 'Why do child mortality rates fall – an analysis of the Nicaraguan experience', *American Journal of Public Health*, **81**, 30–37.

ten Tusscher, G.: 2002, *Later Childhood Effects of Perinatal Exposure to Background Levels of Dioxins in The Netherlands*, Thesis (PhD), University of Amsterdam, ISBN: 90-9016271-2.

WHO – World Health Organisation.: 1978, *Declaration of Alma Ata, 1978*, World Health Organisation, Geneva, http://www.who.dk/AboutWHO/Policy/20010827_1, Accessed April 2003.

WHO – World Health Organisation.: 2000, *Global Burden of Disease*, World Health Organisation, Geneva, http://www.who.int/evidence/bod, Accessed April 2003.

Yamey, G.: 2002, 'WHO's management: struggling to transform a "fossilised bureaucracy"', *British Medical Journal*, **325**, 1170–1173.

CHAPTER 8

SUSTAINABLE DEVELOPMENT IN SMALL ISLAND DEVELOPING STATES
The case of the Maldvies

FATHIMATH GHINA

International Coral Reef Action Network (ICRAN), c/o United Nations Environment Programme-World Conservation Monitoring Centre (UNEP-WCMC), 219 Huntingdon Road, Cambridge, CB3 0DL, UK (e-mail: FGhina@icran.org; fax: C44 1223 277136; tel.: C44 1223 277314)

Abstract. This paper explores the status of sustainable development in small island developing states (SIDS), through the presentation of a case study on the Maldives, which is a typical small island developing state in the Central Indian Ocean. At the outset, a brief history of sustainable development as related to SIDS on the international agenda is outlined, starting from Rio to Barbados to Johannesburg. SIDS are expected to face many challenges and constraints in pursuing sustainable development due to their ecological fragility and economic vulnerability. It is the position of this paper that issues related to environmental vulnerability are of the greatest concern. A healthy environment is the basis of all life-support systems, including that of human well-being and socio-economic development. Priority environmental problems are: climate change and sea-level rise, threats to biodiversity, threats to freshwater resources, degradation of coastal environments, pollution, energy and tourism. Among these, climate change and its associated impacts are expected to pose the greatest threat to the environment and therefore to sustainable development. For small islands dependent on fragile marine ecosystems, in particular on coral reefs, for their livelihoods and living space, adverse effects of climate change such as increased frequency of extreme weather events and sea-level rise will exacerbate the challenges they already face. It is concluded that the 'paper' path from Rio to Barbados to Johannesburg has made significant progress. However, much remains to be done at the practical level, particularly by the developed countries in terms of new and additional efforts at financial and technical assistance, to make sustainable development a reality for SIDS.

Key words: biodiversity, climate change, coastal ecosystems, coral reefs, energy, freshwater, Maldives, pollution, sea-level rise, Small Island Developing States, tourism, vulnerability.

Abbreviations: AOSIS – Alliance of Small Island States; BPOA – Barbados Programme of Action; CSD – Commission on Sustainable Development; CVI – CommonWealth Vulnerability Index; ENSO – El Niño Southern Oscillation; EVI – Environmental Vulnerability Index; GEO – Global Environment Outlook; GHG – Green house gases; IPCC – Intergovernmental Panel on Climate Change; IS92a/e – GHG Emission Scenarios of IPCC; MHAHE – Ministry of Home Affairs, Housing and Environment, Maldives; MoT – Ministry of Tourism, Maldives; MPND – Ministry of National Planning and Development, Maldives; MRf – Maldivian Rufiya; SIDS – Small Island Developing States; SOPAC – South Pacific Applied Geoscience Commission; SST – Sea Surface Temperature; UN – United Nations; UNDP – United Nations Development Programme; UNEP – United Nations Environment Programme; UNFCCC – United Nations Framework Convention on Climate Change; UWICED – University of West Indies Centre for Environment and Development; WSSD – World Summit on Sustainable Development.

Readers should send their comments on this paper to: BhaskarNath@aol.com within 3 months of publication of this issue.

L. Hens and B. Nath (eds.), The World Summit on Sustainable Development, 183–209.
© 2005 *Springer. Printed in the Netherlands.*

1. Introduction

The challenges and needs faced by small island developing states (SIDS) (see Box 1) in pursuing sustainable development are widely recognised. It has been on the international agenda since the early 1990s, beginning with the United Nations Conference on Environment and Development or the Earth Summit at Rio de Janeiro in 1992. SIDS are thought to be both ecologically and economically fragile (Agenda 21; UNEP, 1999a,b,c). Ecologically, most of them are coastal entities with small and dispersed (in the case of archipelagic states) land areas. They generally possess a rich diversity of highly endemic flora and fauna but relatively few natural resources. Their geographical isolation, small size of the economy and dependence on a narrow range of products often leads them to be highly dependent on international trade and therefore are vulnerable to external shocks. Agenda 21 clearly recognises that there are special challenges to planning for and implementing sustainable development in these islands.

In 1994 the UN Global Conference on the Sustainable Development of SIDS was held in Barbados. This was the first global conference on sustainable

---

Box 1. List of small island developing states.

State/country	Population in thousands (year)	State/country	Population in thousands (year)
• *Africa*		• *Europe*	
Cape Verde	434.8 (2000)	Cyprus	689.5 (2001)
Comoros	578.4 (2000)*	Malta	391.7 (2000)*
Sao Tome and Principe	159.9 (2000)*	• *Latin American & the Caribbean*	
Mauritius	1178.8 (2000)	Antigua and Barbuda	59.4 (1999)
Seychelles	79.3 (2000)*	Aruba	69.5 (2000)*
• *Asia and the Pacific*		The Bahamas	294.9 (2000)*
Bahrain	650.6 (2001)	Barbados	274.5 (2000)*
Cook Islands	20.4 (2000)*	Cuba	11217.1 (2000)
Fiji	832.5 (2000)*	Dominica	71.5 (2000)*
Kiribati	91.9 (2000)*	Dominican Republic	8442.5 (2000)*
Maldives	270.1 (2000)	Grenada	89.0 (2000)*
Marshall Islands	43.4 (2001)	Haiti	6867.9 (2000)*
Micronesia	133.1 (2000)*	Jamaica	2573.7 (1998)
Nauru	11.8 (2000)*	Netherlands Antilles	210.1 (2000)*
Niue	2.1 (2000)*	St. Kitts and Nevis	38.8 (2000)*
Palau	18.7 (2000)*	St. Lucia	156.3 (2000)*
Papua New Guinea	4926.9 (2000)*	St. Vincent & the Grenadines	115.5 (2000)*
Samoa	170.3 (2002)	Trinidad & Tobago	1175.5 (2000)*
Singapore	4151.3 (2000)*	U.S. Virgin Islands	120.9 (2000)*
Solomon Islands	466.2 (2000)*		
Tokelau	1.4 (2000)*		
Tonga	102.3 (2000)*		
Tuvalu	10.8 (2000)*		
Vanuatu	189.6 (2000)*		

Note: There is no single definitive list of SIDS, but for the purposes of this paper, those included by the UN-Department of Economic and Social Affairs is used.
*Estimated figures; Source: CIA Factbook. The rest of the population figures are from national censuses.

development and the implementation of Agenda 21. The main outcome of the conference was a framework for planning and implementing sustainable development in SIDS, the Barbados Programme of Action (BPOA), taking into consideration the special characteristics and constraints faced by these islands. This framework provides a synopsis of recommended actions and policies to be implemented over the short-, medium- and long-terms in 14 priority areas. These are: climate change and sea-level rise; natural and environmental disasters; management of wastes; coastal and marine resources; freshwater resources; land resources; energy resources; tourism resources; biodiversity resources; national institutions and administrative capacity; regional institutions and technical co-operation; transport and communication; science and technology; human resource development and implementation, monitoring and review.

A reaffirmation of the principles and commitments to sustainable development as incorporated in Agenda 21, the Barbados Declaration and the BPOA was declared during the 22nd special session of the UN General Assembly for the review and appraisal of the implementation of the program of action for the sustainable development of SIDS. World Leaders attending the Millennium Summit also expressed their commitment to continue to address the special needs of SIDS accordingly.

Most recently, at the World Summit on Sustainable Development (WSSD), the special circumstances of SIDS in relation to their environment and development as underlined in Agenda 21, the decisions adopted at the 22nd special session of the General Assembly and the BPOA were revitalised and given new impetus. The WSSD Plan of Implementation calls for action in the following areas:

- accelerating national and regional implementation of the BPOA for sustainable development for SIDS with adequate financial resources, including through Global Environmental Facility (GEF) focal areas;
- transferring environmentally sound technology and assistance for capacity building;
- implementing sustainable fisheries management and strengthening regional fisheries management initiatives;
- assisting SIDS in the sustainable management of their coastal areas and exclusive economic zones (EEZ);
- developing and implementing SIDS-specific components within programmes on marine and coastal biological diversity;
- assisting in the implementation of sustainable freshwater programmes;
- addressing waste and pollution and their health-related impacts by undertaking by 2004 initiatives aimed at implementing the Global Programme of Action for the Protection of the Marine Environment from Land-based Activities;
- taking account of SIDS in World Trade Organisation (WTO) work programme on trade in small economies;
- developing community-based initiatives on sustainable tourism by 2004;

- extending assistance towards hazard and risk management, disaster prevention, mitigation and preparedness and relief from the consequences of disasters, extreme weather events and other emergencies;
- supporting operationalisation of economic, social and environmental vulnerability indices and related indicators;
- mobilising adequate resources and partnerships to address adaptation to the adverse effects of climate change, sea-level rise and climate variability;
- capacity building and institutional arrangements to implement intellectual property regimes;
- supporting the availability of adequate, affordable and environmentally sound energy services and new efforts on energy supply and energy services by 2004;
- capacity building for and strengthening health-care services with emphasis on HIV/AIDS, tuberculosis, diabetes, malaria and dengue fever;
- capacity building for maintaining and managing water and sanitation services in rural and urban areas;
- implementing initiatives aimed at poverty eradication;
- undertaking a comprehensive review and appraisal of the implementation of the BPOA in 2004 and request the General Assembly to convene an international meeting for the sustainable development of SIDS.

This paper explores the constraints and challenges faced by SIDS in their path to sustainable development, through the presentation of a case study on the Maldives, a SIDS located in the Central Indian Ocean (Figure 1). These characteristics are explored in relation to environmental and socio-economic vulnerability. Environmental vulnerability is of utmost importance as it is concerned with the risk of damage to the island's natural ecosystems or environment which underpin much of the socio-economic processes on SIDS. The paper also demonstrates that by their very ubiquitous nature, climate change and its associated impacts through sea-level rise and increased frequency of extreme weather events will greatly enhance these challenges and vulnerability. As emphasised by H.E. Ambassador Tuiloma Neroni Slade (2002) in his statement on behalf of Alliance of Small Island States (AOSIS), at the second session of UN Commission on Sustainable Development (CSD 10) in preparation for WSSD: 'climate change is an additional and exacerbating problem that goes directly to the roots of their sustainability. This problem for small island communities is understated and seriously underestimated by the international community.' Finally, progress on the implementation of sustainable development of SIDS according to the BPOA and the implications of WSSD Plan of Implementation is assessed.

2. Vulnerability and small islands

Small islands are by no means homogenous in their geographical distribution, physical characteristics, and social, cultural, economical and political contexts. However, they all share similar characteristics which constrain them in their path to implement sustainable development. Over the last decade, these characteristics have been increasingly associated with the concept of 'vulnerability' of SIDS (UNEP, 1999a,b,c; Kaly et al., 2002).

The BPOA (Paragraphs 113 and 114) called for the development of vulnerability indices and other indicators that reflect the status of SIDS integrating ecological fragility and economic vulnerability. It is now widely accepted that indices such as the per capita income or GDP (which is often used as an indicator of the status of development of a country) are not adequate indicators of status of development in small island states. They neither reflect the complex interactions between the environmental resources and economic and social issues, nor the structural and institutional weaknesses facing SIDS.

2.1. MEASURING VULNERABILITY

Vulnerability can be defined as the extent to which the environment, economy or social system is prone to damage or degradation by external factors. *Economic vulnerability* is taken to refer to the risks faced by these economies from exogenous shocks to the systems of production, distribution and consumption. *Environmental vulnerability* is concerned with the *risk* of damage to the island's natural ecosystems or environment (e.g. coral reefs, mangroves, freshwater, coastal areas, forests), including physical and biological processes, energy flows, diversity, genes, ecological resilience and ecological redundancy (UWICED, 2002; Kaly et al., 2002). *Social vulnerability* reflects the degree in which societies or socio-economic groups are affected negatively by stresses and hazards. In all the cases, the causal factors can be natural or anthropogenic or both, and can vary with time and place. All three types of vulnerability are a function of (i) the risk of hazards occurring, (ii) the intrinsic resilience (referred as the innate characteristics of a country that would tend to make it more or less able to cope with natural and anthropogenic hazards) and (iii) the extrinsic resilience (described as the ecological integrity or level of degradation of the ecosystems) (Kaly et al., 2002). Box 2 presents the natural and anthropogenic characters that are thought to contribute to their enhanced vulnerability.

Among the three components of vulnerability, issues relating to environmental vulnerability are of the greatest significance. A healthy environment is the basis of all life-support systems, including that of human well-being and development. A team of experts from the South Pacific Applied Geoscience Commission (SOPAC) has been working on the development of a global environmental vulnerability index (EVI) to characterise the vulnerability of natural systems. The EVI will allow comparisons among countries and through time. Priority environmental problems in SIDS have been identified to be climate change and sea-level rise, threats to biodiversity, threats to freshwater resources, degradation of coastal environments, pollution, energy and tourism (BPOA, UNEP, 1999a,b,c; United Nations, 1999b).

Vulnerability to climate change and sea-level rise is an area that deserves extra attention. This is noted by the fact it is on the top of the list of priority areas in

---

Box 2. Characteristics of SIDS leading to their vulnerability.

- Geographical isolation
- Small physical size
- Ecological uniqueness and fragility
- Rapid human population growth and high densities
- Limited natural resources
- High dependence on marine resources
- Sensitivity and exposure to extremely damaging natural disasters
- Susceptibility to climate change and sea-level rise
- Small domestic market and high dependence of exports
- Limited terrestrial natural resource endowments and high import content
- Small economies with limited diversification possibilities
- Inability to influence international prices
- Peripherality (related to remoteness and isolation): high per unit transport costs, marginalisation, uncertainties of supply, need to keep large quantities of stocks)
- Trade vulnerability: High dependence on trade taxes, vulnerability of domestic industries, dependence on trade preferences, inability to utilise the TRIPS agreement (Agreement on trade-related aspects of intellectual property rights), dispute settlement mechanism or accession
- Limited ability to exploit economies of scale
- Limitations on domestic competition
- Difficulties in absorbing FDI (foreign direct investment)
- Limited investment opportunities, including in communication services
- Problems of public administration
- Dependence on external finance
- Remittances

(List compiled from various sources: Kaly et al. (2002), UWICED (2002), Barbados Program of Action (1994), Witter et al. (2002))

---

the BPOA, the special session of UN General Assembly to review the status of implementation of BPOA and as recently pointed out by Ambassador Slade at the CSD 10. The IPCC defines vulnerability to climate change as the degree to which a system is susceptible to or unable to cope with the adverse effects of climate change, including climate variability and climate extremes. It is a function of the character, magnitude and rate of climate change and variation to which a system is exposed, its sensitivity and adaptive capacity (IPCC, 1998). It has been established that SIDS, as a consequence of their inherent characteristics, are particularly vulnerable to the effects of climate change (Hoegh-Guldberg et al., 2000; Burns, 2001; Nurse et al., 2001). For example, Gommes et al. (1998) calculated an index of vulnerability to sea-level rise, which was composed from the product of an insularity index (ratio between the length of the coastline and the total land area that it encloses) and population density. Out of all the AOSIS members Maldives was shown to be the most vulnerable and protection cost is estimated to constitute 34% of GDP (Gommes et al., 1998). Hoegh-Guldberg et al. (2000) developed a simplified vulnerability index for Pacific small islands based on the following factors: physical exposure, political stability, population density/pressure, foreign aid per head and subsistence activities. The outcome showed that all the small states were highly vulnerable (Tuvalu, Kiribati, Cook Islands, Palau), while the lowest scores were obtained by the larger states of Vanuatu, Solomon Islands, Fiji and American Samoa.

In response to the BPOA, various groups of island experts have attempted to develop vulnerability indices that reflect the relative economic and ecological susceptibility to exogenous shocks. One example is the Commonwealth Vulnerability Index (CVI). The CVI is based on two principles: (i) the impact of external shocks over which the country affected has little or no control and (ii) the resilience of a country to withstand and recover from such shocks. In their analysis using a sample of 111 developing countries (37 small and 74 large), it was shown that small states were more vulnerable than larger countries, irrespective of income. Furthermore, of the 25 most vulnerable countries, 24 were small states out of which 17 were small islands. According to the study, the most vulnerable country in the world is Vanuatu. A review of many of these indices by an *ad hoc* expert group of the United Nations Department of Economic and Social Affairs in 1997 concluded that, 'as a group, small island developing states are more vulnerable than other groups of developing countries' (United Nations, 1998). However, at present, it seems there is no single satisfactory quantitative measure of vulnerability or vulnerability index (United Nations, 1998).

3. Maldives – a typical small island developing state

In the following section, the environmental, economic and social contexts of Maldives, a small island state, is analysed in order to explore the validity of the 'special' status given to SIDS, particularly in relation to environmental vulnerability and the priority environmental problems identified in the previous section. Maldives shares many similarities in terms of environment and socio-economic features with other SIDS, particularly the more vulnerable, smaller and low-lying islands of the SIDS group such as Seychelles in the Indian Ocean, Tuvalu, Kiribati and Nauru in the Pacific, as well as islands such as St. Lucia, Barbados and the Bahamas of the Caribbean.

The Republic of Maldives consist of a double chain of coral atolls, 820 km in length and 130 km at its widest point within a total area of 90 000 km^2, situated in the Central Indian Ocean (MPND/UNDP, 1999) (Figure 1). Geologically, the atoll chain is formed on the crest of a volcanic mountain range which extends 2000 km from the Lakshadweep islands near India to the Chagos islands south of the equator. There are an estimated 1192 coralline islands and found within the

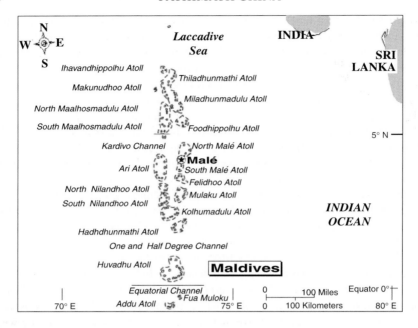

Figure 1. Location and map of Maldives (Source: MapQuest).

lagoon are microatolls, faros, patch reefs and knolls. The abundance and variety of these reef forms, particularly ring shaped faros and microatolls are unique to the atolls of the Maldives (Naseer, 2000; MoT, 2000). It is estimated that the atolls of the Maldives are over 10 000 years old and have been known to be inhabited for up to 2500 years (MHAHE, 2001).

The islands are distributed among 26 natural atolls, which are grouped into 20 units for administrative purposes (MHAHE, 2001). The maximum height above sea level recorded is around 3 m. They vary in shape and size from small sandbanks with sparse vegetation to elongated strip islands. Although the total land area is estimated to be only 300 km^2, the maritime area of the country's EEZ under the jurisdiction of the state amounts to more than 859 000 km^2 (MHAHE, 2001). Many of them are highly dynamic, being moved around the reef surface, or totally destroyed or formed during storms. Most of the islands are quite small (average 0.7 km^2) with only nine being larger than 2 km^2. The largest island is about 5.2 km^2. Over 80% are low-lying with an average elevation above mean sea level of 1–1.6 m (MHAHE, 2001).

The climate is tropical, oceanic with little diurnal or seasonal temperature variation. Annual mean temperature range from 28°C to 32°C and on average receives 8 h of sunshine per day, throughout the year. The climate is governed by the southwest monsoon from April to August, characterised by strong winds and the northeast monsoon with gentler winds. Average annual rainfall is around 1855 mm per year (MHAHE, 2001).

The inherent nature of the islands predispose them to frequent damage from storm and wave surges. This vulnerability has direct implications for a number of activities related to the economy and indeed livelihoods.

3.1. DEMOGRAPHY

In 2000, the total population of the Maldives was 270 101 (MPND, 2002). The average population growth rate was 2.8% in the period 1990–1995 and 1.9% in 1995–2000. If the current growth rate remains unchecked, the population can be expected to double in a space of 50 years. The population is spread over 199 islands and approximately 79% are below the age of 35 years. The most populous of the inhabited islands and the centre of commerce is the capital Malé. The population of Malé in 2000 was 74 069, which constitutes 27.4% of the total population (MPND, 2002). Further, at any one time there may be around 16 000 expatriates (MoT, 2000). With a total area of around 1.8 km^2, Malé is one of the most densely populated cities in the world (about 50 000 persons/km).

The population density on some other islands is also very high. Nearly half the inhabited islands have population densities over 2000 persons/km^2. Of these inhabited islands, 90 have fewer than 500 people, 72 have between 500 and 1000, 38 have between 1000 and 5000 and only 3 have more than 5000 (MPND/UNDP, 1999; MNPD, 1999). Hence the population is extremely fragmented and dispersed.

For hundreds of years, the Maldives people lived more or less in harmony with their environment. This was made possible due to the low level of populations in the islands which fluctuated around 60 000–70 000 people from the early 1900s (Pernetta, 1989). Improved health care, for example the eradication of malaria and treatment of childhood diseases such as dysentery led to an explosive population growth from the late 1960s to a population of 200 000 in the 1980s, in just 20 years.

The rapid growth of population and high densities in some islands has put great pressure on the natural and economic resources. For example, the high densities of population in Malé have led to depletion of the freshwater aquifer and quality of living space. An average house in some of the overcrowded islands can have more than 12 adults, in addition to a number of children. Often, the houses are poor in quality of construction and in accessibility to essential services like water and sanitation. The level of overcrowding and lack of privacy are believed to be an increasing source of emotional problems and stress, which have been linked to such problems as juvenile delinquency, drug abuse, crime and child abuse, all of which appear to be on the increase. Other problems include increased number of school leavers competing for limited number of jobs and potentially high number of dependants with increased life expectancy. It is important to address these demographic patterns in order not to exceed the carrying capacity of the islands.

3.2. Socio-economic status

The Maldives relies totally on coastal and marine resources for subsistence and its economic development. Tourism and fishing are the major economic activities, contributing 18.5% and 6.9% to the GDP respectively. Historically, fishing was the main traditional occupation and economic activity of Maldivian people and was also the major source of foreign exchange earnings. Fishery is focused on tuna and reef fish such as 'baitfish', grouper, emperor shark and a variety of aquarium fish. However, in the last 15 years or so, tourism has become more important and the contribution of fisheries to the GDP has shown a declining trend. In 1978, fisheries contribution to the GDP was estimated at about 22% but by 1999 this figure has decreased to 6.5% (MHAHE, 2001). Trade in marine products is the next most important sector. Analysis of the fishing sector also shows that the number of persons employed in the sector is in decline, manpower base is ageing, fish catch is flat, some marine resources are being overexploited and tuna prices show marked fluctuations (MPND, 1999). Agriculture is carried out mainly at subsistence level, as land is scarce and the soils of the Maldives are poor, composed of parental coral rock and sand. They form a layer, from only few centimetres to about 20 cm thick, are alkaline and deficient in nitrogen. Agriculture is focused on mainly coconut, chillis, watermelon and a variety of root crops. The contribution of agricultural sector to GDP is around 7.7% (MHAHE, 2001).

The rate of tourism growth has been impressive, from a mere 1097 tourists (2 resorts) in 1972 to over 467 154 tourists (86 resorts) in 2000 (MoT, 1999). Tourism now generates approximately 40% of government tax and non-tax revenues. These earnings play a major role in the socio-economic development of Maldivian society, providing a source of funds for investment in essential social infrastructure such as educational, health institutions, transport and power generation. The contribution of tourism to total foreign exchange earnings and to the Balance of Payments is highly significant, even after taking into account the 'leakages'. Leakages arise mainly due to imports of supplies, materials and equipment, remittances by overseas nationals, commissions, interest and profit share payments to overseas investors and lenders including tour operators, and can be as much as one third of total earnings (MoT, 2000). In 1998, net foreign exchange earnings from tourism contributed about two thirds of gross foreign exchange earnings and also covered 65% of imports other than tourism imports (MoT, 2000). Maldives economy is highly dependent on imports for commodities.

Tourism also generates a considerable share of employment and contributes directly to growth in household incomes generated from remittances associated with resorts and tourists. At atoll or island level, income generation opportunities are significant and many report an increase in their income and associated living standards. This is mainly through increased fishing, particularly reef fishing, carpentry and masonry in resort construction and maintenance, by supplying safari boats, local handicrafts, selling souvenirs, mat weaving and thatch for roofing material in resorts. In 1998, about 50% of the workforce were employed in the tourism sector. However, it is important to note that over 40% of jobs in the sector are occupied by expatriates.

The lack of Maldivian employees in the tourism industry manifests a general shortage of skills in all sectors of modern economy and government in the Maldives. In 1995, only 32% of the population had completed primary schooling, while less than 6% had completed secondary, pre-university and university education (MPND, 1999). In the same period, it was estimated that one in five of all jobs in Maldives were occupied by expatriates (MPND, 1999). Rapid growth in tourism and other sectors mean an ever-increasing expatriate workforce and therefore greater leakage of foreign exchange earnings.

Real GDP per capita is currently around US$ 700 per person per annum with a growth rate of 6.2% in the period 1996–97. However, great disparities and inequalities in income and access to essential infrastructure exist between islands (MNPD/UNDP, 1999). This is most pronounced when Malé is compared with other islands. Income disparities between Malé and the atolls are reported to be in the order of 2 : 1 and access to social and physical infrastructure and services at 4 : 1 (MNPD/UNDP, 1999). The inequality in access to social and physical infrastructure is partly related to the high cost of providing these services to the small and dispersed populations. For example, the average construction cost of a primary school or a health clinic on an atoll in Maldives can be five to six times higher than that would cost in a non-SIDS developing country like Sri Lanka, due to the high cost of maritime transport, need to import building materials and failure to achieve economies of scale (MPND, 1999). The average income for households on Malé is around Maldivian Rufiya (MRf) 35 (~US$ 3) per person per day and MRf 20 (less than US$ 2) per person per day in the atolls. Approximately 1 in 4 of all Maldivians live on incomes of less than US$ 1 per day, or below the World Bank's definition of poverty (MNPD/UNDP, 1999). Presently, a population and development consolidation strategy is being pursued in order to develop islands with the greatest potential for growth and expansion, with a view to attain equitable and sustainable development for the entire population (MHAHE, 2002).

3.3. FRESHWATER

Freshwater is confined to an underground 'freshwater lens' which comprises a freshwater zone separated by a transition zone of a few metres' thickness between the freshwater and underlying seawater. This lens is found 1.5–2.0 m below the land surface and changes continuously with the tide. The water is alkaline and availability depends upon the rate of abstraction and recharge by rain. During the dry season, up to 25% of household in all atolls report a shortage of water (MHAHE, 2002). In the island of Malé (the capital) this lens has been severely degraded from overexploitation. The thickness of the freshwater zone has decreased from 20 m in 1973 to 6–8 m in 1983, and less than 3 m in 1993 (Pernetta, 1993). Additionally, due to the porous nature of the soil and poor waste disposal methods, the water is susceptible to pollution and contamination. Currently, the main source of freshwater in Malé is from desalination – which is extremely costly, requires dependence on imported fuels and contributes to increased green house gas (GHG) emissions. Hence, it does not represent a very sustainable source. In view of the declining quality of water in most inhabited islands, high priority is now given to increasing rainwater harvesting by construction of rainwater storage tanks at both the community and individual levels (MHAHE, 2002).

3.4. ENERGY

The Maldives has no reserves of coal, oil or gas. Wood fuel is the main indigenous and most important source of energy. It is primarily used in the resident sector for cooking and a small percentage in fish processing, palm sugar making and lime making industries. In 1994, wood fuel made up almost 55% of total energy consumption (FAO-REWDP, 2000). The total energy consumption per capita was estimated at 4570 kWh for 1998 (IEA, 2000).

To a lesser extent, other biomass sources such as dried coconut husks, shells and leaves, various types of dried grasses and waste paper are also used. The use of wood for energy is a major cause of deforestation and at the same time presents serious health implications from indoor-air pollution, particularly to women and children.

Imported petroleum products is the chief source of commercial energy, of which the major part is used to produce electricity. Each island has to have its own power generation system and infrastructure, with the capital island Malé having the highest power generation and consumption. Electricity is provided by the government owned State Electric Company. Over 85% of the total electricity consumption is by the domestic sector, the rest being attributed to commercial and government consumption (Idris, 2000). The percentage of population with access to electricity has grown over the past decades. Now, more than 60 of the inhabited islands have electricity 24 h a day, accounting for 55% of the population. However, 21% of the population have less than six hours or no access to electricity (MPND/UNDP, 1999).

Solar energy is the only renewable form of energy utilised in the Maldives. For example, to power navigational lights, communication transceivers on fishing boats and for power supply at the remote installations in the national telecommunications network. The telecommunications network is the single biggest producer and user of renewable energy, in the form of solar power and solar–diesel hybrid systems. The largest site has a capacity of 3.5 kW and the aggregated capacity of the total of 177 sites is approximately 130 kW. The installations are not connected to the grid and are privately owned (Idris, 2000). There is a need to proliferate such renewable, sustainable forms of energy throughout the country. Access to energy is important, as it plays a major role in raising the living standards and quality of life of communities.

3.5. POLLUTION

The disposal of solid waste is a particularly critical problem. Limited land area makes the option of landfill unsustainable in the long term and other options of collection and disposal, such as incineration have so far proved to be economically unfeasible. Solid waste is produced in a much larger volume by tourist resorts than local inhabited islands. Estimated waste production is within the range 40–204 tons per year per resort, depending on the size of the resort, with up to 16.5 kg waste per visitor per week (Brown et al., 1997). Compared to other atolls, solid waste generated per capita in Malé is much higher, with an average of 2.48 kg per capita per day in Malé, as opposed to 0.66 kg of waste per person per day in the atolls (MHAHE, 2002). Waste from Malé, Hulhulé International Airport and many resort islands are transported to 'garbage island' to be disposed in a landfill. In many inhabited islands solid waste is just dumped near the beach or buried in unlined pits. Due to the high permeability of the coral limestone bedrock, the aquifer is susceptible to pollution from such activity with the risk of spreading diseases. Often wetland areas such as swamps and mangroves are also used as waste disposal areas, thus destroying these fragile habitats. Moreover, considerable amount of the waste is discarded at sea and in close vicinity of reefs, particularly in the case of outer atolls and heavily used tourist islands.

The only public sewerage system in the country was established in Malé in 1988 and more recently in the island of Villingili, where untreated sewage is discharged directly into the sea on both the lagoon and ocean sides of the island. Effluents from septic tanks and raw sewage are discharged directly into the sea from tourist islands and on more isolated islands open beaches are frequently used. Nutrient rich waste water affect the growth of hard corals which favour nutrient-poor conditions. The impact is manifested in enhanced growth of seagrasses and algae, which although localised at first, may become more extensive depending on the site conditions and level of discharges.

Another significant source of pollution comprise fish processing factories and other islands where fishing is an important economic activity. Fish are cleaned at the beach and the waste is dumped directly into the lagoon. Such activity not only poses risk to the reefs, but to human health as well via spreading of diseases. Studies in fishing villages of Laamu Atoll, where such pollution and anthropogenic enrichment of lagoon systems by fish wastes occur, an increase in sea grass has been observed (Miller and Sluka, 1999). The sea grass has been observed to encroach upon corals of lagoonal patch reefs and although the impact of sea grass competition on these reefs have not been investigated, it is suggested that it could pose a threat to these reefs.

With the increased frequency of sea transport, pollution by diesel and oil is another potential threat. Increased shipping traffic with the associated risk of oil spills and dumping, and oil pollution from the increasing mechanisation of fishing boats especially in and around fishing ports and harbour areas are of particular concern. Moreover the used oil is often just disposed of in the sewerage system or dumped with other solid waste in containers.

Presently, since Maldives does not practice large-scale agriculture or farming, pollution from pesticides or fertilisers is a minor threat. The disposal of hazardous waste is another major issue, though fortunately, it is believed that the generation of hazardous waste, particularly from other atolls is minimal (MHAHE, 2002).

3.6. BIODIVERSITY

The extent of biological diversity including flora and fauna present in the islands of the Maldives has not been adequately documented, but as is common for atolls/islands highly endemic flora and fauna are found. Vegetation consists mainly of coconut palms, banyans, bamboo, pandanus, banana, mango, and breadfruit trees. Based on published data, 583 species of plants have been recorded and of these 55% are cultivated species. Occasionally, mangroves may be found fringing the ocean sides of the islands, but they occur more commonly associated with inland brackish water bodies. Terrestrial animal species are rather limited, however, about 165 species of seabirds, shorebirds and landbirds are recorded (Zuhair, 1997).

In contrast to the terrestrial biodiversity, the marine biodiversity is amongst the richest in the region. The coral reef area of Maldives is one of the largest and support the greatest diversity of corals and associated organisms, along with the Chagos Archipelago, in South Asia (Rajasuriya et al., 2000). This is also because Maldivian atolls form part of the so-called 'Chagos stricture' representing an important link or stepping stone between the reefs of the Eastern Indian Ocean and those of the Eastern African region; and as such the fauna combines elements of both eastern and western assemblages (Spalding et al., 2001). The total coral reef area is estimated at 8920 km^2 and contributes 5% of the world's reef area (Spalding et al., 2001). In the Maldives, over 250 species of scleractinian corals (representing over 60 genera) and over 1200 species of reef fishes are recorded (Pernetta, 1993). 400 reef fish species have been identified and catalogued out of which 7 species are endemic (Naseer, 2000). A great diversity of other marine species including 36 species of sponges, 400 species of molluscs and 350 species of crustaceans are also found. Five species of turtles are found in the Maldivian waters, all of which are endangered, including the loggerhead, green, Hawksbill, Olive Ridley and leatherback turtles (MHAHE, 2002). Seven species of dolphins and nine species of whales are also recorded. Though not well documented, mangroves and seagrass systems are also present associated with coral reefs of Maldives. Seagrass beds are often found in shallow lagoons behind the coral reefs. It provides a habitat to various crustaceans, molluscs and fish. More importantly, it is also believed that these habitats provide a breeding ground for many coral reef and other marine fishes.

Coral reefs possess great ecological, social and economic value wherever they occur in the world. Members of practically all phyla and classes are believed to be present in coral reef ecosystems. They provide vast number of people all over the world with food, recreational possibilities, coastal protection, as well as aesthetic and cultural benefits and are described to have tremendous value as life-support systems to society (Moberg and Folke, 1999). One estimate suggests that reef habitats provide living resources (e.g. seafood) and services (e.g. tourism, coastal protection) worth US$ 375 billion annually (Bryant et al., 1998). It is estimated that outside of the Western Pacific, the Maldives is the nation that is most dependent on coral reefs for the maintenance of land area, food, export earnings and foreign currency from tourism revenues (Spalding et al., 2001).

The high densities of people, lack of environmental awareness and poor management and developmental activities have placed great stress on the fragile coral reef ecosystems. Though, on the whole, Maldives coral reefs are in good condition, localised degradation has been experienced around those islands with high level of population and development. Causes of reef degradation include coral mining, dredging, land reclamation activities, pollution, badly engineered coastal constructions, channel clearance and tourist activities. About 11% of reefs are estimated to be at risk (Spalding et al., 2001) and about two to five percent are estimated as irreparably damaged prior to the 1998 bleaching event (Rajasuriya et al., 2000).

3.7. CLIMATE CHANGE AND SEA-LEVEL RISE

One of the greatest environmental threats to Maldives and other similar island states is climate change and sea-level rise. In the regions where small island states are located including the Indian Ocean, review of past and present trends indicate that temperatures have risen by as much as 0.1°C per decade and sea levels by 2 mm year^{-1}. It is reported that the 1990s was the warmest decade and 1998 the warmest year in instrumental record since 1861 (IPCC, 2001). For example, in the Maldives analysis of surface air temperature data available for Malé (central atolls) for the 30 year period (1969–1999) indicate that annual maximum temperatures have increased by 0.17°C per decade and the annual minimum temperatures by 0.07°C per decade (MHAHE, 2001).

Global ocean temperatures have increased significantly since the late 1950s and more than half of the increase in heat content has occurred in the upper 300 m of the ocean at a rate of about 0.04°C per decade. In the Maldives, for the period 1950–2000, a significant increasing trend of 0.16°C per decade is observed (Edwards et al., 2001). Evidence also shows that the ocean thermohaline circulation is weakening with consequences to global ocean heat distribution (IPCC, 2001).

Warm episodes of the El Niño-Southern Oscillation (ENSO) phenomenon (which consistently affects regional variations of precipitation and temperature over much of the tropics, sub-tropics and some mid-latitude areas) have been more frequent, persistent and intense since the mid-1970s, compared with the previous 100 years (IPCC, 2001).

Globally, average temperatures are predicted to rise by 1.4–5.8°C for the period 1990–2100 (IPCC, 2001). Mean rainfall intensity is also projected to increase by about 20–30%. The frequency of extreme temperatures during the dry season is likely to increase with the implication of increased thermal stress conditions during the 2050s and more so during the 2080s. Such predictions imply more frequent episodes of droughts as well as floods for the region (Nurse et al., 2001).

Some studies also predict an increase of approximately 10–20% in the intensity of tropical cyclones under enhanced CO_2 conditions (Nurse et al., 2001). Another cause of concern is the projected increase, though small, in amplitude of ENSO events. The ENSO phenomenon is the strongest natural fluctuation in climate on interannual timescales (IPCC, 2001). Though it has its core activity in the tropical Pacific, changes associated with it can have far reaching consequences in other regions, as manifested by the mass coral bleaching events around the globe due to sustained high sea surface temperatures (SSTs) linked to the ENSO events. The Central Indian Ocean was the hardest hit during the most recent and strongest ENSO episode in 1998. Box 3 provides some examples of extreme weather events and their implications in Maldives.

Sea level is projected to rise at a rate of $5 \, \mathrm{mm \, year^{-1}}$ (with a range 2–$9 \, \mathrm{mm \, year^{-1}}$) and may rise in the range 0.09–0.88 m by 2100 (Nurse et al., 2001). For low-lying SIDS this represents perhaps the greatest threat. Already in the island

---

Box 3. Extreme weather events.

April 1987: Severe swell waves caused widespread damage to Malé, the international airport at Hulhule and other surrounding inhabited islands and resorts. On Malé the swells either washed away or inundated a large part of the $600\,000 \, \mathrm{m^2}$ of land reclaimed from the shallow lagoons along the southern and western coasts. The area had been reclaimed between 1979 and 1986 at a cost of Rf 50 million (US$ 4.2 m) (MPND, 1999). A large part of the retaining wall on the southern seafront was also destroyed and the cost of breakwater repairs and rebuilding was estimated at US$ 5 million (Edwards, 1989; MPND, 1999). In addition, a waste disposal compound was badly affected, spreading refuse to surrounding areas leading to outbreaks of diarrhoeal diseases (MPND, 1999). To prevent a recurrence of such events and to set up an improved system of coastal defence around Malé, an estimated US$ 51 million was utilised, acquired through foreign aid (MPND, 1999).

June–July 1988: Flooding and wave damage to mainly the western side of the archipelago due to severe southwest monsoon. Thulhadoo was flooded for up to 30 m inland and 32 houses had to be evacuated (Edwards, 1989). About 3–5 m of beach area was also eroded on the south side of the island.

May 1991: Severe storms swept over the archipelago, with worst effects on the southern islands. The storms damaged or uprooted about 60 000 banana trees and thousands of mango and breadfruit trees in Addu Atoll alone. Around 2000 buildings were also reported to be damaged (MPND, 1999)

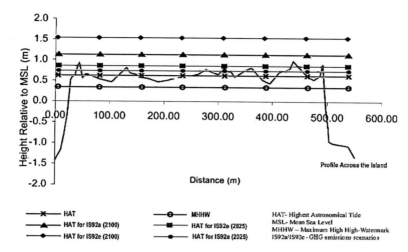

Figure 2. Traverse across Hithadhoo in Baa atoll showing potential impacts of sea-level rise based on projected climate change scenarios (Source: MHAHE, 2001).

of Tuvalu and some islands of Papua New Guinea in the Pacific are experiencing storm over wash and shrinkage of their land area by 20 cm per year (Boyd, 2001). For a typical island of the Maldives, the potential effects of sea-level rise under different climate change scenarios projected for Maldives is presented in Figure 2. Based on the IS92a and IS92e GHG emission scenarios of the IPCC, climate models predict that by the end of this century the temperature may have increased by 2.0–3.8°C and sea level may rise by 49–95 cm.

Figure 2 illustrates that with a maximum height above mean sea level of less than 1 m, potential sea-level rise of even few centimetres will have tremendous impact on these low-lying islands, culminating in total inundation. Moreover, this is expected to occur in the very near future (2025).

Another critical concern to the Maldives would be impact on the groundwater availability. Rising sea levels would decrease the thickness of the freshwater lens and therefore the availability of freshwater. Moreover, storm over-wash of the islands by increased frequency and intensity of storms will lead to increased incidences of contamination of freshwater by saltwater.

The scientific community agrees that though there is a large degree of uncertainty about the mechanisms involved and about the likelihood or timescales of such transitions, the possibility for rapid and irreversible changes in the climate system exists (IPCC, 2001). Furthermore, they establish that there already is a global commitment to climate change and sea-level rise as a result of greenhouse forcing arising from historic emissions. Even with a fully implemented Kyoto Protocol, by 2050 warming would be only about 1/20th of a degree less than what is projected and therefore, climate change impacts are inevitable (Nurse et al., 2001).

3.8. Implications of Climate Change to Sustainable Development

The United Nations Framework Convention on Climate Change (UNFCCC) and its Kyoto protocol stated its main objective to achieve stabilisation of GHG concentrations in the atmosphere at a level that would prevent dangerous anthropogenic interference with the climate system (UNEP/IUC, 1999a,b). Simultaneously, it also proposed that such a level should be achieved within a time frame sufficient to allow ecosystems to adapt naturally to climate change, to ensure that food production is not threatened, and to enable economic development to proceed in a sustainable manner (UNEP/IUC, 1999a,b). Unfortunately, progress in meeting these goals has been minimal.

For countries like Maldives and other SIDS that depend to a large extent on fragile coastal and coral reef ecosystems for livelihoods and income, the current situation is distressing. They are very sensitive to changes in their environment, and therefore would be one of the first ecosystems to be affected by global climate change. The potential severe degradation of coral reefs and other associated coastal ecosystems will greatly hinder the ability of SIDS to promote sustainable development. Ironically, SIDS have contributed least to GHG emissions (Table I) but will be subjected to some of the most significant adverse effects of climate change.

When physiologically stressed, corals may lose much of their symbiotic algae (zooxanthellae), which supply nutrients and colour. In this state corals are referred to as 'bleached'. Corals can recover from short-term bleaching by regenerating the symbiotic relationship with the zooxanthellae. However, during this period there would be slowed growth and reproduction, increased susceptibility to diseases, lowered ability to compete and withstand other stress factors. For instance, increased sedimentation or reduction in salinity due to heavy rains will impede recovery and lead to death. Prolonged bleaching can cause irreversible damage and subsequent mortality (Pomerance, 1999). Factors that cause bleaching can be both natural and human-induced. For example, prolonged high or low temperatures, high or low levels of visible light and UV radiation, low tides (long periods of exposure to air), low or high salinity, pollution and diseases. Increases of SSTs above the normal warmest period maximum is believed to be the major cause of mass bleaching events over the past two decades (Strong et al., 1998; Hoegh-Guldberg 1999; Wilkinson et al., 1999; Edwards et al., 2001). During the

TABLE I. Relative CO_2 emissions between regions.

Area	Population (millions)	CO_2 emissions per capita (t)	Total CO_2 emissions (Mt)
Maldives	0.24	0.54	0.13
Pacific Islands	7.1	0.96	6.82
OECD	1092.3	11.09	12117.05
World	5624.4	4.02	22620.46

Source: MHAHE (2001).

coral bleaching event of 1998, an estimated 90% of the reefs of the Maldives were bleached either partially or totally (Edwards et al., 2001). Fortunately, recovery is recorded in many areas. However, if the increasing trend in SSTs of Maldives at 0.16°C per decade continues, and the April–May high temperature anomalies as experienced in 1998 (which led to the mass bleaching) would become a regular event in only 30 years' time (Edwards et al., 2001).

The health of reefs are directly correlated to the economic mainstays of the nation: fishery and most importantly, reef-based tourism which is the driving force of the Maldives economy. Reef fishery presently comprises only a small percentage of total fishing activity, which is based on tuna fishery. However its importance is increasing, and moreover, the baitfish which are dependent on coral reefs is of great importance to the pole and line tuna fishery. In this way, reef fishery and tuna fishery are inextricably linked. Though some changes in the diversity and abundance in reef fish have been noted following the mass coral bleaching event, no major changes in the fishery have been reported. It is believed that effects on fishery would take a longer time manifest.

Around 45% of all tourists visiting Maldives are divers, with 69% of the divers making more than 5 dives during their stay (Westmacott et al., 2000). One estimate suggests that annual total number of dives made by visiting tourists is around more than half a million, at a cost of US$ 35 per dive (MHAHE, 2001). The financial losses to tourism in Maldives due to the April–May 1998 mass coral bleaching event was estimated at US$ 3 million. A further survey focusing on tourists' willingness to pay (WTP) for better reef quality identified that for 47% high quality reef was of utmost importance. Global welfare loss to the Maldives due to the bleaching event was estimated at US$ 19 million (Westmacott et al., 2000).

It is estimated that close to 70% of tourists also visit Maldives for its white sandy beaches (MoT, 2000). Hence, loss of beach area through increased erosion related to increased frequency of storms and other extreme weather episodes or general degradation of reefs which supply the sand to islands and sea-level rise would represent an additional considerable loss to tourism.

Another area of major concern is the impact to valuable infrastructure. For example, currently, an average investment for a resort with 200 beds is estimated to be over US$ 9 million and for a modern 700 bed resort US$ 43 million (MoT, 2000). Notwithstanding the fact that islands are so small that the entire island has to be considered a coastal entity, tourist demand encourages building rooms as close to the beach as possible and even over the lagoon itself on stilts. An estimate from one resort indicate that US$ 60 000 per year is needed to maintain coastal protection of the island for example in the form of groynes. (MHAHE, 2001). Impact to other important investments such as the airport on the island of Hulhulé is also of major concern. Hulhulé is between 1.0 and 1.7 m above sea level. The runway itself is only 1.2 m above mean sea level and has only 0.5 m clearance at highest high water (MHAHE, 2001). Damage caused to Hulhulé due to the 1987 tidal wave episode was considerable and cost of repairs were estimated at US$ 4.5 million (Edwards, 1989) (see also Box 3). The total investment in the airport to date is estimated at around US$ 57 million, excluding the cost of recently opened Hulhulé Island Hotel and other investments by local businesses (MHAHE, 2001).

The coral reefs provide natural protection to the islands from waves, storm surges and flooding. At the same time, the shapes and size of the small islands are determined by the tidal and current patterns. The beach systems on them are highly dynamic and have directional shifts within the shoreline in accordance with the prevailing seasonal conditions (MHAHE, 2001). These features make them extremely vulnerable to erosion. It is estimated that 50% of all inhabited islands and 45% of all tourist resort islands suffer from varying degrees of beach or coastal erosion (MHAHE, 2001). The cause is not clear but changes in the intensity of wind and the resulting changes in currents and waves could be one reason as well as reef degradation and erosion. In a recent study of the reefs of Chagos following the 1998 coral bleaching and mass mortality event, 1.5 m of reef surfaces have been shown to be eroded, together with reduction in the three-dimensional structure (Sheppard et al., 2002). Another reason could be badly engineered coastal structures such as groynes, jetties, and causeways which alter currents and sedimentation patterns. Coral and sand mining, dredging, land reclamation and the consequent destruction of coral reefs can be another major cause.

Already a 1.5 km long breakwater at a cost of US$ 14 million with has been constructed on the southern side of Malé following the 1987 flooding by tidal waves. More recently, protective seawalls have been constructed on the western, eastern and southern perimeter of Malé at a cost of US$ 100 million. The estimated cost for the seawall that is presently under construction on the northern side is US$ 20 million, bringing the total cost of protection of Malé alone to US$ 134 million (MHAHE, 2001). These defence structures were built only to protect Malé from waves of height 2 m above present mean sea level (as experienced in extreme weather episodes) and at the time, the potential impacts of climate change were not taken into consideration at all.

Maldives has limited options to respond to coastal erosion and inundation. Three possible coastal responses were proposed by Biljsma et al. (1996) in the IPCC Second Assessment Report (cited in McLean et al., 2001):

1. *Protect*: which aims to protect the land from the sea so that existing land uses can continue, by constructing hard structures (e.g. sea walls) as well as using soft measures (e.g. beach nourishment).
2. *Accommodate*: which implies that people continue to occupy the land but make some adjustments to, for instance, elevating buildings on piles.
3. *Retreat*: which involves no attempt to protect the land from the sea; in an extreme case the coastal area has to be abandoned.

Given the natural features of the islands – the small size and overall low elevation of entire islands, adaptation measures such as accommodation and retreat may not be viable options. Protection appears to be the most feasible adaptation option. However, even under protection, soft (also less expensive) measures such as beach nourishment are also difficult to implement, as beach material is limited in supply. Additionally, their extraction from lagoon, as has been practised, has been observed to further exacerbate erosion problems. In the light of these limitations, hard engineered coastal protection structures such as seawalls seem to be the appropriate option. But as noted, this type of protection measure is extremely expensive and presents a major financial drain to poor countries such as Maldives. Looking at the extremely high cost of protection by such methods, it seems obvious that it cannot be the solution for all the 199 inhabited islands. One estimate of the cost of protecting only 50 of these islands was projected to be over US$ 1.5 billion (MHAHE, 2001). With a GDP of US$ 161 million (in 1999), the cost of protection of only 50 of the 200 inhabited islands would be 9 times more than the country's GDP. Therefore, without external aid, it would be an impossible task to achieve.

Coral reefs have survived major global climatic changes in geological history showing that they are potentially resilient and robust. Research has shown that corals and reef communities possess numerous mechanisms for acclimatisation and adaptation through diverse reproductive strategies, flexible symbiotic relationships, physiological acclimatisation, habitat tolerance and community interactions (NOAA, 1998). However, it should be noted that, periods of coral reef growth have been interspersed *with major periods of no reefs*; and fossil record and climatic history suggest that 'coral reefs' as we know them are geologically rare features (Buddemeier and Smith, 1999). Presently, the increased stress of human activities have already placed many reefs at risk. At the same time, coral reefs are expected to face a multitude of changing environmental factors such as rising SST, rise in sea level and increased intensity of extreme weather events at unprecedented rates. In the past, it may be that the absence of other stressors helped them to cope, for instance, with rising sea levels of about 20 cm per decade (Burns, 2001). Therefore, if environmental changes exceed the adaptive and acclimative capacities established under previous rates and ranges of disturbance, modern coral reefs will probably lose their resilience and robustness (NOAA, 1998).

4. Progress in implementation and future prospects

Although the paper has focused mainly on the problems faced by Maldives in relation to sustainable development, and in particular due to climate change, many of them are shared by most of the SIDS to varying degrees. It is worth noting that the United Nations Environment Programme (UNEP) has recently published state of the environment reports on the islands of the Pacific, Caribbean and Western Indian Ocean regions which provide comprehensive and valuable information on the environmental problems and policy priorities in the context of development, as part of their Global Environment Outlook (GEO) series. These reports provide an ideal source of information to compare with the situation of Maldives.

In summary, the importance of the relative vulnerability of the natural environment and economy of SIDS is now well established. Even though similar problems are present in most or all developing countries, because of the small size they are felt more acutely in SIDS. As clearly illustrated by the case study on Maldives, the interdependency between environment and socio-economy is tremendous. For example, the extreme paucity of land-based resources, limited mineral resources, scarcity of arable land, and a lack of durable, sustainable building materials leads to over reliance on marine resources, destruction of coral reefs and mangroves for building and other purposes. The high population densities in some islands and coastal areas combined with high rates of population growth put increasing pressure on the natural and economic resources. Moreover, the population being scattered over numerous islands constrains equitable distribution of goods and services. The low elevation above sea level of much of the land area increases the susceptibility to coastal erosion, inundation and flooding by high waves associated with storms and other extreme weather episodes. The narrow economic base focused on fishery and tourism leads to economic vulnerability as these are sectors that are largely dependent on international markets, not to mention leading to the overexploitation of fish resources and degradation of the natural environment by tourism development. Additionally the high dependence on imports makes the economy very dependent on foreign exchange earnings, which in turn places heavy reliance on exports. Low income and purchasing power of the people, diseconomies of scale in production of goods and services, infrastructure and transport, lack of financial and human capital, and lack of skilled human resources are other factors contributing to its economic vulnerability. While SIDS face all these problems already, climate change will greatly enhance them and put tremendous burden on the SIDS governments to address the issue.

In terms of economic development, outlook for many of the SIDS is worrying. According to the UNCTAD (2002), out of the nine least developed countries (LDC) – SIDS, two are slow-growth economies (Sao Tome and Principe and Haiti), four are regressing economies (Kiribati, Vanuatu, Comoros and Solomon Islands) and only 3 are high growth economies (Maldives, Samoa and Cape Verde).

Annual review and appraisal of progress in the implementation of the BPOA for the Sustainable Development of SIDS is carried out by the UN-DESA. It was seen that, in general, *overseas development assistance to SIDS has substantially declined* (United Nations, 1999c). It was also seen that the pattern of bilateral Official Development Assistance (ODA) and Foreign Direct Investment (FDI) flow was largely determined by historical and geographical ties disadvantaging some SIDS. Analysis of trends in the external development support to SIDS between 1992 and 1997 showed that a considerable number of programme areas have not received adequate attention in terms of ODA. Looking at total bilateral and multilateral commitments, the larger shares were in human resource development, transport and communication, freshwater resources, land resources, coastal and marine resources and energy resources. Climate change and sea-level rise, biodiversity resources and management of wastes, all considered as high priority areas requiring urgent action, received the lowest level of assistance (United Nations, 1999c). In the light of this, the UN called for an intensification of efforts to provide external development assistance through new and additional commitments and disbursement of resources. A mere shift in sectoral allocation of ODA resources will not be sufficient.

Similarly, at a meeting of representatives of donors and SIDS in 1999, under the auspices of UN–DESA and UNDP on the mobilisation of resources to assist SIDS in effectively implementing the BPOA, SIDS representatives pointed out that notwithstanding their efforts at national and regional levels, progress in the implementation of the Programme of Action has been impeded by, among other problems, lack of financial support from the international community, inadequate human resources with appropriate training, inadequate institutional capacity, inadequate capacity for the enforcement of environmental legislation and regulations, and inadequate investment resources (United Nations, 1999a).

The BPOA is far from being fully implemented and at the same time, it is estimated that 70% of the tasks and efforts outlined in the programme has been carried out by SIDS themselves (UWICED, 2002). One notable and valuable achievement of the BPOA is the establishment of SIDSnet, an internet site hosted under UN-DESA that provides a platform for communication and information exchange within SIDS and between SIDS and the rest of the world. Most SIDS also have in place National Environment Action Plans and National Development Plans which incorporate sustainability.

With regard to the WSSD Plan of Implementation, there are just seven time-bound targets in the plan: to halve the number of people without access to proper sanitation by 2015; to restore depleted fish stocks by 2015; and to significantly reduce the extinction rate of he world's plants and animals by 2010. Time-bound targets directly related to sustainable development and SIDS were only established in relation to addressing waste and pollution, sustainable tourism, energy sector and for the review and appraisal of the BPOA by 2004. Sadly, there was no target date for the universal ratification of the Kyoto Protocol on Climate Change. Indeed, considerably more attention needs to be given to the major principles of sustainable development as they relate to SIDS, particularly, to Principles six and seven of the Rio Declaration and Article 3 of the UNFCCC.

> *Principle 6 of Rio Declaration:* The special situation and needs of developing countries, particularly the least developed and those most environmentally vulnerable, shall be given special priority.

> *Principle 7 of Rio Declaration:* In view of the different contributions to global environmental degradation, States have common but differentiated responsibilities. The developed countries acknowledge the responsibility that they bear in the international pursuit to sustainable development in view of the pressures their societies place on the global environment and of the technologies and financial resources they command.

> *Article 3 of the UNFCCC:* The Parties should protect the climate system for the benefit of present and future generations of humankind, on the basis of equity and in accordance with their common but differentiated responsibilities and respective capabilities. Accordingly, the developed country Parties should take the lead in combating climate change and the adverse effects thereof.

Although there were calls and pledges at the WSSD to increase ODA, no specific figures or dates were set. Presently, only five countries: Norway, Denmark, the Netherlands, Sweden and Luxembourg have so far met the 0.7% of GDP on development aid. The USA, the world's richest country spends only 0.1% of its GDP on ODA (OECD, 2002).

It is clear that until the gap between commitments made and translation into action is bridged, the future remains bleak for sustainable development of SIDS.

5. Conclusion

SIDS appear to be the most vulnerable group of countries in the world, in terms of their ecology and economy. They face a multiplicity of challenges and constraints, related to their ecological fragility and environmental vulnerability in their path to achieve sustainable development. SIDS have a most intricate and sensitive relationship between their environment, socio-economy and culture and environmental vulnerability is a critical issue. The greatest environmental threat facing most of the SIDS is climate change and its associated impacts through sea-level rise and increased frequency of extreme weather events. Through the solidarity and hard work of SIDS, they have been able to establish international recognition of their special status and need for co-operation from the international community to assist them in their pursuit of sustainable development. This is manifested by the references to SIDS as a special case in the major international agreements and agenda related to sustainable development since the Earth Summit at Rio, culminating in the inclusion of a specific chapter devoted to SIDS and sustainable development in the WSSD Plan of Implementation. Hence the 'paper' path from Rio to Barbados to Johannesberg has made significant progress. However, in practical terms, much remains to be done. SIDS by themselves are working hard to pursue sustainable development, but they urgently need the assistance and cooperation from developed countries in terms of *further and additional* financial and technical assistance to persist in their path to sustainable development. Capacity building at all levels and sectors of development is of the highest priority – a prerequisite for self-reliance and lasting sustainability of any state. Otherwise, for most of the SIDS sustainable development will remain just a distant dream.

References

Boyd, A.: 2001, *Oceania, Pacific Beat, Beware the Rising Tide*, Asia Times, www.atimes.com (viewed 26.07.2001).

Brown, K., Turner, R.K., Hameed, H. and Bateman, I.: 1997, 'Environmental carrying capacity and tourism development in the Maldives and Nepal', *Environmental Conservation* 24(4): 316–325.

Bryant, D., Burke, L., Mcmanus, J. and Spalding, M.: 1998, *Reefs at Risk: A Map-Based Indicator of Threats to the World's Coral Reefs*, WRI/ICLARM/WCMC/UNEP, World Resources Institute, website: http://www.wri.org (last update: 24 July 2002).

Buddemeir, R.W. and Smith, S.V.: 1999, 'Coral adaptation and acclimatisation: a most ingenious paradox', *American Zoologist* **39**, 1–9.

Burns, W.C.G.: 2001, 'The possible impacts of climate change on Pacific Island state ecosystems', *Int. J. Global Environmental Issues* **1**(1), 56.

Edwards, A.J.: 1989, *The Implications of Sea-Level Rise for the Republic of Maldives* – Report to the Commonwealth Expert Group on Climate Change and Sea level Rise, Commonwealth Secretariat.

Edwards, A.J., Clark, S., Zahir, H., Rajasuriya, A., Naseer, A. and Rubens, J.: 2001, 'Coral bleaching and mortality on artificial and natural reefs in Maldives in 1998, sea surface temperature anomalies and initial recovery', *Marine Pollution Bulletin* **42**(1), 7–15.

FAO-REWDP,: 1998–2000, *Regional Wood Energy Development Program In Asia (1997)* available online at http://www.rwedp.org/c_mld.html.

Gommes, R., du Guerny, J., Nachtergaele, F. and Brinkman, R.: 1998, *Potential Impacts of Sea-Level Rise on Populations and Agriculture*, Food and Agriculture Organisation of the United Nations (FAO), http://www.fao.org (posted: March 1998, viewed 30.09.2002).

Hoegh-Guldberg, O.: 1999, 'Climate change, coral bleaching and the future of the world's coral reefs', *Marine Freshwater Resources* **50**, 839–866.

Hoegh-Guldberg, O., Hoegh-Guldberg, H., Stout, D.K. and Cesar, H.: 2000, *Pacific in Peril – Biological, Economic and Social Impacts of Climate Change on Pacific Coral Reefs*, GreenPeace.

Idris, A.R.: 2000, *Renewable Energy In The Maldives*, Ministry of Communication, Science and Technology, Maldives.

Intergovernmental Panel on Climate Change (IPCC): 1998, *The Regional Impacts of Climate Change, An Assessment of Vulnerability, A Special Report of IPCC Working Group II*, R.T. Watson, M.C. Zinyowera and R.H. Moss (eds.), Cambridge University Press.

Intergovernmental Panel on Climate Change (IPCC): 2001, *Third Assessment Report, Technical Summary of the Working Group I:* available online at http://www.ipcc.ch.

International Energy Agency (IEA): 2000, *World Energy Outlook 2000*, IEA Publications.

Kaly, U.L., Pratt, C.R. and Howorth, R.: 2002, *Towards Managing Environmnetal Vulnerability in Small Island developing States (SIDS)*. SOPAC Miscellaneous Report 461.

Mclean, R.F., Tsyban, A., Burkett, V., Codignotto, J.O., Forbes, D.L., Mimura, N., Beamish, R.J. and Ittekkot, V.: 2001, 'Coastal Zones and Marine Ecosystems', in J.J. McCarthy, O.F. Canziani, N.A. Leary, D.J. Dokken and K.S. White (eds.), *Climate Change 2001: Impacts, Adaptation and Vulnerability. Contribution of Working Group II to the Third Assessment Report of the Intergovernmental Panel on Climate Change (IPCC)*, UK, Cambridge University Press, pp. 344–379.

Miller, M.W. and Sluka, R.: 1999, 'Coral-seagrass interaction in an anthropogenically enriched lagoon', *Coral Reefs* **18**, 368.

Ministry of Home Affairs, Housing and Environment (MHAHE): 2001, *First National Communications to the United Nations Framework Convention on Climate Change (UNFCCC)*, Malé, Maldives, MHAHE.

Ministry of Home Affairs, Housing and Environment (MHAHE): 2002, *National Assessment Report: Progress towards Sustainable Development, from Rio 1992 to Johannesburg 2002*, Malé, Maldives.

Ministry of Planning and National Development (MPND) and United Nations Development Program (UNDP): 1999, *Maldives: Vulnerability and Poverty Assessment 1998*, Malé, Maldives, MPND and UNDP.

Ministry of Planning and National Development (MPND): 1999, *Report to the 6th Round Table Meeting with UNDP*, Malé, Maldives, MPND.

Ministry of Planning and National Development (MPND): 2002, *Statistical Year Book of Maldives 2002*, Maldives, MPND.

Ministry of Tourism (MoT): 1999, *Tourism Statistics*, Ministry of Tourism, Malé, Maldives.

Ministry of Tourism (MoT): 2000, *Social, Economic & Environmental Impact Study of Tourism*, Ministry of Tourism, Malé, Maldives.

Moberg, F. and Folke, C.: 1999, 'Ecological goods and services from coral reefs', *Ecological Economics* **29**, 215–233.

Naseer, A.: 2000, *Reefs of Maldives*, http://www.reefsofmaldives.com (viewed: 26.07. 2000).

National Oceanic and Atmospheric Association (NOAA): 1998, *Coral Reefs and Global Change: Adaptation, Acclimation or Extinction?* Initial Report of a Symposium and Workshop, Jan 3–11, Boston, USA, available online at http://coral.aoml.noaa.gov/themes/coral_cg.html.

Nurse, L.A., Sem, G., Hay, J.E., Suarez, A.G., Poh, P.W., Briguglio, L. and Ragoonaden, S.: 2001, 'Small island states', in J.J. McCarthy, O.F. Canziani, N.A. Leary, D.J. Dokken and K.S. White (eds.), *Climate Change 2001: Impacts, Adaptation and Vulnerability. Contribution of Working Group II to the Third Assessment Report of the Intergovernmental Panel on Climate Change (IPCC)*, UK, Cambridge University Press, pp. 75–103.

OECD: 2002, *A Mixed Picture of Official Development Assistance in 2001: The US Becomes the World's Largest Donor Again; Most EU Members' Aid Also Rises* http://www.oecd.org/EN/document/0,,EN-document-590–17-no-12–29438–590,00.html (viewed: 10. 10.2002).

Pernetta, J.C.: 1989, *Cities on Oceanic Islands: A case study of Malé the Capital of the Republic of Maldives* in R. Frasetto (ed., 1991), Impacts of Sea Level Rise on Cities and Regions: Proceedings of the First International Meeting – 'Cities on Water', Venice, 11–13 December 1989, Marsilio Editori.

Pernetta, J.C.: 1993, *Marine Protected Area Needs in the South Asian Seas region*, Vol. 3, IUCN, Maldives.

Pomerance, R.: 1999, *Coral Bleaching, Coral Mortality and Global Climate Change*, Report presented by Deputy Assistant Secretary of State for the Environment and Development to the US Coral Reef Task Force, 5–6 March 1999, Mauii, Hawaii.

Rajasuriya, A., Zahir, H., Muley, E.V., Subramanian, B.R., Venkataraman, K., Wafar, S.M., Khan, M.H. and Whittingham, E.: 2000, *Status of Coral Reefs in South Asia: Bangladesh, India, Maldives and SriLanka* in Status of Coral Reefs of the World 2000, available online at http://www.cordio.org.

Sheppard, C.R.C., Spalding, M., Bradshaw, C. and Wilson, S.: 2002, 'Erosion vs recovery of coral reefs after 1998 El Niño: Chagos Reefs, Indian Ocean', *Ambio* **31**(1), 40–48.

Slade, T.N.: 2002, *Statement by H.E. Ambassador Tuiloma Neroni Slade Permanent Representative of Samoa to the United Nations on Behalf of the Alliance of Small Island States (AOSIS) at the Second Session of CSD 10 Acting as the Preparatory Committee for the World Summit on Sustainable Development, General Debate, New York, 1 February 2002*, available online at http://www.sidsnet.org/docshare/other/AOSIS_WSSD_final_Statement.pdf.

Spalding, M.D., Ravilious, C. and Green, E.P.: 2001, *World Atlas of Coral Reefs*, Berkeley, USA, University of California Press.

Strong, A.E., Goreau, T.J. and Hayes, R.L.: 1998, *Ocean Hotspots and Coral Reef Bleaching: January–July 1998*, Reef Encounters, pp. 20–21.

UNCTAD: 2002, *The Least Developed Countries, 2002 Report – Escaping the Poverty Trap*, United Nations, available online at http://www.unctad.org/en/docs/ldc02.en.pdf.

UNEP: 1999a, *Pacific Islands Environmental Outlook-1999*, Nairobi, UNEP.

UNEP: 1999b, *Western Indian Ocean Environment Outlook-1999*, Nairobi, UNEP.

UNEP: 1999c, *Caribbean Environment Outlook-1999*, Nairobi, UNEP.

UNEP/IUC: 1999a, *United Nations Framework Convention on Climate Change (UNFCCC)*, Geneva, UNEP/IUC.

UNEP/IUC: 1999b, *Kyoto protocol to the Convention on Climate Change*, Geneva, UNEP/IUC.

United Nations: 1998, Report of the Secretary-General on *Development of a Vulnerability Index for Small Island Developing States* (E/1998/100).

United Nations: 1999a, Report of the Secretary-General on *Meeting of Representatives of Donors and Small Island Developing States*, 24–26 February 1999 (E/CN.17/1999/18), available online at http://www.un.org/esa/sustdev/sids/sids22–4.pdf.

United Nations: 1999b, Report of the Secretary-General on *Progress in the Implementation of the Programme of Action for the Sustainable Development of Small Island Developing States*, Commission of the Sustainable Development, Seventh session, 19–30 April 1999 (E/CN.17/1999/6), available online at http://www.un.org/esa/sustdev/sg6–99.pdf.

United Nations: 1999c, Report of the Secretary-General on *Progress in the Implementation of the Programme of Action for the Sustainable Development of Small Island Developing States, Current Donor Activities*, Commission of the Sustainable Development, Seventh session, 19–30 April 1999, E/CN.17/1999/7, available online at http://www.un.org/esa/sustdev/sids/sids22–4.pdf.

University of the West Indies Centre for Environment and Development (UWICED): 2002, *Vulnerability and Small Island Developing States*, prepared for the UNDP/CAPACITY 21 Programme of Action for the Sustainable development of Small Island States, available online at http://www.sopac.org.fj/Projects/Evi/Files/Env%20vuln%20of%20SIDS.pdf.

Westmacott, S., Cesar, H., Soede, L.P. and Linden, O.: 2000, *Assessing the Socioeconomic Impacts of the Coral Bleaching Event in the Indian Ocean, Report prepared for the Coral Reef Degradation Indian Ocean (CORDIO) program*, Resource Analysis (Netherlands), Institute for Environmental Studies, Free University Amsterdam.

Wilkinson, C., Linden, O., Cesar, H., Hodgson, G., Rubens, J. and Strong, A.E.: 1999, 'Ecological and scoioeconomic impacts of 1998 coral mortality in the Indian Ocean: an ENSO impact and a warning of future change?', *Ambio* **28**(2), 188–196.

Witter, M., Briguglio, L. and Bhuglah, A.: 2002, *Measuring and managing Economic Vulnerability of Small Island Developing States*, Paper prepared for the Global Roundtable – Vulnerability and Small Island Developing States: Exploring Mechanisms for Partnerships, Montego Bay, Jamaica, 9–10 May, 2002.

Zuhair, M.: 1997, *Biodiversity Conservation in Maldives – Interim Report to the Convention on Biological Diversity*, Ministry of Planning, Human Resources and Environment, Male, Maldives.

SUSTAINABLE DEVELOPMENT –A NEW CHALLENGE FOR THE COUNTRIES IN CENTRAL AND EASTERN EUROPE

ISTVÁN LÁNG

Hungarian Academy of Sciences, Roosevelt tér 9, Budapest, Hungary
(e-mail: ilang@office.mta.nl.hu; tel.: C36-1-269-2656; fax: C36-1-269-2655)

Abstract. The paper gives an overview on the transformation process of 10 Central and Eastern European (CEE) countries leading to a change of political structure and the emergence of market mechanisms. These countries intend to get admission into the European Union (EU) in the near future. Therefore, the legal system of the EU is the standard for them and, in particular, they follow the respective environmental protection measures.

The implementation of sustainable development is a new challenge for the CEE countries. Beside environmental protection, economic and social dimensions are also to be considered, and these three pillars are in mutual interaction. In the CEE countries, the preparations to implement elements of sustainable development began in the last few years. Thus,the documents of the Johannesburg Summit on Sustainable Development stimulate this recent process. The successful implementation requires the close cooperation of the governments, the various stakeholders and the civil society.

Key words: CEE countries, integration of the policies, Johannesburg Summit, sustainable development.

Abbreviations: CEE – Central and Eastern Europe; COMECON – Council for Mutual Economic Assistance; ECO – Environmental Citizens Organization; EU – European Union; GDP – Gross Domestic Product; NGO – Non-Governmental Organization; REAP – Regional Environmental Accession Programme; REC – Regional Environmental Center (Szentendre, Hungary); UN – United Nations Organization; USD – United States Dollar.

1. Historical overview

After the Second World War, most of the Central and Eastern European (CEE) countries came under the Soviet sphere of influence. Within a few years, the communist parties wiped out the opposition forces and assumed total power. Single-party rule and central planning were introduced. The Soviet Union and its European allies established the Council for Mutual Economic Assistance (COMECON) for the coordination of economic activities on a regional level, and formed the Warsaw Pact for close military cooperation. The fundamental aim of building a socialist social order was attained uniformly in the countries in question, with some allowance for national character.

As is known, the UN Conference on Human Environment was held in Stockholm in 1972. Originally, the Soviet Union and its allies planned to send government

Readers should send their comments on this paper to: BhaskarNath@aol.com within 3 months of publication of this issue.

L. Hens and B. Nath (eds.), The World Summit on Sustainable Development, 211–222.
© 2005 *Springer. Printed in the Netherlands.*

delegations. However, due to a diplomatic complication, they decided to boycott the conference – with the exception of Romania. The reason was that the German Democratic Republic was not invited as a participant with full rights, since it was not yet a member of the United Nations Organization (UNO) at the time. According to contemporary statements, these absentee countries too concurred with the recommendations set forth in the documents of the conference. Nevertheless, the fact that the environmental policy and environmental protection experts of these countries were not present at the conference and, therefore, could not establish personal contact with their counterparts in the western world put them at a disadvantage for years.

In the 1980s, the Soviet Union and the CEE countries began to gradually open the doors to economic, cultural and scientific cooperation with Western Europe and the United States. Bilateral (East–West) relations were established in a number of areas including, among others, environmental protection.

In 1989–1991, fundamental political changes took place in the region. There were important events preceding and leading up to the changes: the national fight for freedom in Hungary in 1956, the 'Prague Spring' in 1968 and the social movement of the Solidarity Worker's Union in the early 1980s in Poland. The Soviet Union and its regimes disintegrated. The democratization process started, the parliamentary framework of a multi-party system was established, enforcement of human rights was reinstituted, and the transition to a market economy began. A new phrase was coined to describe the region: 'countries in transition'.

However, the transformation demanded great sacrifices. It was accompanied by economic depression that led to a significant fall in Gross Domestic Product (GDP) and declining living standards in the 1990s, particularly in the first half of the decade. Structural changes were carried out in the industrial sector. Major decisions were made primarily in heavy industry, as a result of which a number of companies were dissolved or partially closed. By the turn of the millennium, the economy stabilized, inflation decreased, new industrial mechanisms and relations developed.

Environmental protection movements played an active role in the transformation processes, and in many countries they became one of the forces underlying political change. Social sensitivity toward environmental problems developed gradually. The protection of the environment was also embraced by civil organizations, as a result of which a forum was created between government organs and non-governmental organizations (NGO), ending sometimes in cooperation and sometimes in confrontation.

In 1992, the UN Conference on Environment and Development was organized in Rio de Janeiro. Every CEE government sent a delegation, whose members also included the representatives of national NGOs. The documents of the conference (Rio Declaration, AGENDA-21, Convention on Climate Change, and the Convention on Biodiversity) exerted a positive influence on the national environmental policy of the CEE countries. New state organizations (ministries, advisory commissions) were set up, short- and medium-term environmental protection action plans were framed, environmental education intensified.

Ten years later, the World Summit on Sustainable Development was held in Johannesburg. The central theme was no longer environmental protection, but sustainable development which comprised environmental protection and related economic activities, as well as their human dimension.

In Johannesburg, the countries awaiting admission into the European Union (EU) (the so-called associate members) attended the reconciliatory discussions already as an independent group, known as the Central Group. It was comprised of 13 countries: Poland, Czech Republic, Hungary, Slovenia, Slovakia, Estonia, Latvia, Lithuania, Romania, Bulgaria, Cyprus, Malta, and Turkey. They all signed an association agreement with the EU and, with the exception of Cyprus, Malta, and Turkey, once belonged to the Communist Bloc. Fundamentally, the members of the Central Group supported the EU in order to attain the maximum results possible under the circumstances at the Johannesburg Conference (WSSD, 2002).

Two documents were adopted in the World Summit:

- The Johannesburg Declaration on Sustainable Development;
- Plan of Implementation.

Articles 79 and 80 of the Plan of Implementation reiterate that the European regional ministerial conference preparing the World Summit on Sustainable Development acknowledged the important role and great responsibility of this region in carrying out global actions to further sustainable development. Since the countries in the region represent various levels of development, the fulfillment of the recommendations of AGENDA-21 requires a different approach and method. The document makes reference to the efforts concerning the realization of sustainable development as stated in EU resolutions and by the European environment ministers.

2. Conferences of the European Ministers for environment

The ministers recognized the necessity of the Pan-European environmental cooperation when they decided to organize regular ministerial conferences. The Czechoslovak Minister for Environment, J. Vavrousek (1944–1995), who died at an early age, played an important part in the initiative. The program was named Environment for Europe.

2.1. THE FIRST MINISTERIAL CONFERENCE, DOBRIS 1991

The cornerstone of the Pan-European environmental protection strategy was laid down at this conference. Special attention was paid to resolving the environmental problems of (former socialist) countries whose economy was undergoing transformation. A decision was made to prepare a publication summarizing the state of the environment in Europe. The book, titled *The European Environment: Dobris Assessment*, was published in 1995 (Stanners and Bourdeau, 1995).

2.2. THE SECOND MINISTERIAL CONFERENCE, LUZERN 1993

CEE countries were again in the focus of attention and action programs for cooperation were worked out for them. The following were defined as the important elements of the long-term Pan-European environmental program: technical cooperation, integrated pollution prevention control, economic incentives, evaluation of national environmental performance, information on the environment, participation in the implementation of EU environmental policy, and the working out and application of international legal regulations.

2.3. THE THIRD MINISTERIAL CONFERENCE, SOFIA 1995

The participants reviewed the progress of the implementation of the earlier environmental programs. The need for the integration of environmental policy into other sectoral policies, for propagating the concept of sustainable development appeared as new elements. Among the working documents, the Pan-European program for the conservation of biological and landscape diversity may be considered to be of outstanding importance.

2.4. THE FOURTH MINISTERIAL CONFERENCE, AARHUS 1998

At this occasion the second assessment of Europe's environment (EEA, 1998) first was published. The necessity for cooperation with the business sphere in environmental matters came to the fore. The representatives of 32 countries signed a declaration on the use of unleaded gasoline. The so-called Aarhus Convention, or, to be exact, the Convention on Access to Information, Public Participation in Decision Making and Access to Justice in Environmental Matters, is the most important document of the Conference.

The essence of the Convention is that environmental matters must be transparent in order to enable local communities to participate in environmental policy decisions and in supervising the effects of the implemented measures. As a result, relations between the government sphere and civil society would grow stronger and acquire new meaning. Access to justice also promotes social support for environmental policies.

In case of an emergency, the authorities are obligated to make public all information that may prevent or reduce the danger arising from a threatening situation. The Convention pays special attention to the inventory of contamination sources. The public must be informed early on in the decision-making process concerning the environment about the proposed action and its expected impact on the environment.

2.5. THE FIFTH MINISTERIAL CONFERENCE, KIEV 2003

The action plans based on the documents of the Johannesburg Summit have been worked out at this conference.

3. The state of the environment in the region

The following are the more important environmental problems on the European continent and, in particular, in the CEE region:

- climate change;
- damage to the ozone layer in the stratosphere;
- traffic and industrial accidents;
- acidification;
- harmful effects of tropospheric ozone and other photochemical oxidants;
- pollution of freshwater resources;
- devastation of forests;
- endangered seashore zones;
- reduction of the amount of waste and the treatment of waste;
- stress caused by urbanization;
- risks of using chemicals.

Life expectancy at birth in CEE countries has been much lower during the last 20 years than the European average. In some CEE countries life expectancy has decreased in the last 5 years, e.g., in Russia from 62 to 58 years for men (in Western Europe this figure is 76.5 years).

Economic development has been very diverse in the Western and CEE Region. In the last 30 years, growth was about 30% in Europe as a whole, masking large differences: in 1999, GDP per capita was 25 441 United States Dollar (USD) in Western, 3139 USD in Central and 1771 USD in Eastern Europe. GDP decline over the same period was more than 50% in the Eastern European Countries.

3.1. ATMOSPHERE

According to CEE experts, safeguarding the atmosphere against pollution constitutes the first priority in the region. The main sources of pollutants are the combustion of fossil fuels, urban traffic and heating systems, and industrial activity which pays little attention to the protection of the quality of air. Electric power plants, oil, chemicals, paper, cement, steel, and non-ferrous metal industries are primarily responsible for the emission of pollutants, such as carbon dioxide, sulfur dioxide, heavy metals, and particulates (dust, ash, soot, etc.).

During the preceding period, there were big investments to develop heavy industry in the region. In the utilization of energy resources, low-grade coal predominated, and the technological standard in use was well below that in Western European countries. As a result, the emission of sulfur dioxide was considerable, which contributed to the formation of acid rains and to forest degradation.

In the 1990s, as a result of the transformation of the industrial structure in the region, carbon dioxide emission declined in a number of CEE countries. Table I shows core data on the state of the environment.

TABLE I. Environmental indicators in the CEE countries (World Resources, 2000).

Country	Per capita CO_2 emission (kg)	Annual fertilizer use (kg/hectare cropland)	Average annual internal renewable water resources (m^3/capita)
Bulgaria	6543	43	2188
Czech Republic	12282	107	1464
Estonia	11180	29	9105
Hungary	5834	83	598
Latvia	3714	30	7104
Lithuania	3728	41	4239
Poland	9228	122	1419
Romania	5270	41	1657
Slovakia	7389	77	2413
Slovenia	6537	258	9317
Europe (average)	8414	89	3981

3.2. SOIL

The most severe forms of decline in the quality of the soil cover in the region are water and wind erosion, acidification and, in some areas, heavy metal pollution. Further degradation is caused by the packing of soils due to the use of heavy agricultural machines and by decreasing organic-matter content, by the harmful effect of wind erosion due to overgrazing and salinization due to inadequate irrigation methods.

The main pollutants are persistent organic compounds, heavy metals, nitrates, and phosphates released into surface or subsurface waters. The situation is aggravated by waste dumping and sewage sludge that release considerable amounts of pollutants into the waters. Oil pollution is frequently found in the vicinity of military and industrial establishments.

As a result of the nuclear accident in Chernobyl in 1986, a considerable amount of radionuclides was released in the atmosphere and from there into the soil of the neighboring countries.

The use of artificial fertilizers greatly declined in the CEE countries during the last decade of the 20th century, primarily for economic reasons. The transition to market economy is gradually ending earlier subsidies for the use of artificial fertilizers. Consequently, their use has declined. This is considered a favorable process from the point of view of environmental protection (see Table I).

During the period 1995–1997, the European average per hectare was with 89 kg relatively low. However, it must be noted that the data for the Russian Federation are also included in the European average, but since that figure was only 17 kg, the average remained low.

Sustainable forestry management is promoted by the Pan-European Forest Certification in Central Europe to decrease wood production from the natural resources and enhance biodiversity. Economic instruments, such as timber extraction charges were introduced in Croatia, Czech Republic, Hungary, Lithuania, and Poland for forest protection.

The impact of agriculture, urban development, and other human activities on biodiversity are key issues, because CEE countries still have quite well-preserved landscapes and a relatively high number of species compared to Western Europe.

The introduction of genetically modified organisms into agriculture is intensively debated in the CEE region. In the interest of public awareness and openness regarding this issue, NGOs promote a regional bio-safety process and an open dialogue since 1995.

3.3. WATER

In a large part of the region the quantity and quality of fresh water is a big problem. On the whole, there is no water shortage in the CEE Region, but there is a considerable difference in available water in the various areas. In the second half of the 20th century, high amounts of phosphate and nitrogen were released into the rivers and lakes, which made these waters predisposed to eutrophication and toxicity. The sewage of settlements contributed significantly to the rising phosphate concentrations, while in some areas agriculture is responsible for 80% of the nitrogen content and 20–40% of the phosphate content of surface waters. The nitrate concentration of ground water often exceeds the limit considered safe for human consumption. Therefore, the rural population is especially at risk.

Disease caused by contaminated water has been registered in a number of countries. For instance, bathing is not recommended at certain reaches of the Danube river, especially downstream of urban centres. Most of the sewage of cities along big rivers is released in the water without biological purification. As a result, biological decomposition takes place in the rivers. During the past 10 years there have been significant investments in the development of water purification capacities in the region's cities, thus, noticeable improvement is expected to take place soon.

The Johannesburg World Summit recommendation to reduce the number of those who have no access to safe drinking water be reduced by half by 2015 means in CEEs case that every inhabitant should be guaranteed of drinking water the quality of which is in compliance with EU directives.

Table I contains data on freshwater resources per capita. In four of the 10 countries, per capita water withdrawal is higher than the European average; in six countries it is less. Hungary has conspicuously limited renewable freshwater resources (Kereszty, 1998).

3.4. BIODIVERSITY

Compared to other parts of the world, the number of species in the region is relatively low, while the number of endangered species is high. At the same time, there are more habitat types and the number of indigenous species is also higher in CEE than in Western Europe.

TABLE II. Nature conservation (World Resources, 2000).

Countries	Protected areas (percent of land area)	Number of biosphere reserves	Number of wetlands of international importance
Bulgaria	4.5	17	5
Czech Republic	15.8	6	10
Estonia	11.1	1	10
Hungary	7.0	5	19
Latvia	12.5	1	3
Lithuania	9.9	0	5
Poland	9.1	8	8
Romania	4.6	3	1
Slovakia	22.1	4	11
Slovenia	5.9	0	2
Europe	4.7	139	632

Biological diversity in the region is influenced by mass tourism (especially at the seaside and in the mountains), intensive agriculture (pastures and wetlands), deteriorating water quality, industrial forestry, transport and energy-related activities.

Nature conservation is well organised in the region. The admission of the candidate countries will, no doubt, significantly enrich the EU.

The data in Table II show that the values expressed in terms of protected areas as a percent of land area are significantly higher than the European average, except in Romania where this value corresponds to the European average.

One third of the designated European biosphere reserves are found in the region.

At the same time, there is a considerable disproportion with respect to the number of wetland habitats of international importance. Only 12% of the Ramsar areas in Europe are found in the region.

4. General problems

Natural disasters, such as floods, forest fire, and man-caused disasters, such as industrial accidents, oil spills and nuclear reactor problems raise issues of safety, preparedness, and emergency planning. During the last two years, floods were the most important natural disasters in the whole of Europe, urging transboundary river basin cooperation between East and West. The mining accident at Baia Mare in Romania, which caused an ecological disaster in the Tisza river in 2000, calls for policy responses, such as the joint protocol of the liability of the Helsinki Convention and the Trans-boundary Effects of Industrial Accidents.

A high quality of life standard has always been a demanding goal. The dramatic change of human conditions urges the growing concern for sustainable development. Equity between societies, present and future generations, social well-being, the carrying capacity of the ecosystem raise questions for societies: how well we are, how well is the ecosystem, how are people and the ecosystem affecting each other?

The goal-setting and the monitoring of progress call for a general, realistic, accessible and reliable indicator set. The wider application of Prescott-Allen's (2001) well-being index, combining the human well-being with the ecosystem well-being would give the region a tool that would allow easy overview, good understanding, and a reliable basis for political decision making.

5. Regional Environmental Center

The Regional Environmental Center (REC) for CEE is a non-partisan, non-advocacy, not-for-profit organization with a mission to assist in solving environmental problems in CEE. The Center fulfils this mission by encouraging cooperation among NGOs, governments, businesses, and other environmental stakeholders, by supporting the free exchange of information and by promoting public participation in environmental decision-making.

The REC was established in 1990 by the US, the European Commission and Hungary. Today, the REC is legally based on a Charter signed by the governments of 28 countries and the European Commission, and on an International agreement with the Government of Hungary. The REC has its headquarters in Szentendre, Hungary, and Country Offices in each of its 15 beneficiary CEE countries, these are: Albania, Bosnia and Herzegovina, Bulgaria, Croatia, Czech Republic, Estonia, Hungary, Latvia, Lithuania, FYR Macedonia, Poland, Romania, Slovakia, Slovenia, and Yugoslavia.

Recent donors include the European Commission and the governments of the US, Japan, Austria, Canada, Czech Republic, Croatia, Denmark, Finland, France, Germany, Hungary, Italy, the Netherlands, Norway, Slovakia, Switzerland, and the UK, as well as other inter-governmental and private institutions.

The REC structure is as follows: In the headquarters, an international staff of 60 people work, most from the CEE region. Also the centre has offices in the capitals of its beneficiary countries. In these offices another 70 persons are employed (all are nationals of that country). Each office has a registration in the country it is sited.

The Program Areas of the REC are:

- environmental capacity building;
- information;
- support to NGOs/ECOs;
- local initiatives;
- business and environment;
- environmental policy;
- public participation;
- climate change;
- environmental law.

These program areas vary in size depending on the projects being managed within them. Almost all of these departments and programs will be involved in implementing REAP activities.

The Country Offices are managed by Country Office Directors, they usually have a small staff made up of Financial and Grants Officers, Project Managers or Officers and Information specialists.

The overall budget of the centre is around 10 million Euro per year. The majority comes from donations or project funds from donor countries, though a small percentage comes from the beneficiary countries (REC, 2000).

6. The EU and the CEE countries

Eight of the 10 CEE countries under study will join the EU in 2004. The probable date of admission of two countries (Bulgaria and Romania) is 2007. In the light of this it is understandable that these countries make great efforts to render the rules of the EU to their national legal system, and to fulfil the assumed obligations.

The three most significant environmental problems in all countries are:

- air pollution and climate change;
- water pollution and water resource management;
- waste minimization and waste treatment.

The 6th Environmental Action Plan of the EU is the standard for the environmental protection programmes of all CEE countries. These programmes were worked out in all CEE countries and are in different phases of implementation. Scientific research and technological development pays great attention to environmental protection and to the rational use of natural resources. Ever growing intellectual and material reserves are being mobilized in these areas. Interdisciplinary approaches gain fortunately more and more ground in scientific research.

A new challenge for the CEE countries is the acceptance of the concept of sustainable development and, on its basis, the elaboration of complex (environmental, economic, social) development projects. At the Gothenburg Summit in June 2001, the EU accepted the document entitled *A Sustainable Europe for a Better World: A EU Strategy for Sustainable Development* (European Commission, 2001). The main priorities are as follows:

- improve policy coherence;
- getting prices right to give signals to individuals and businesses;
- invest in science and technology for the future;
- improve communication and mobilize citizens and business;
- take enlargement and the global dimensions into account;
- limit climate change and increase the use of clean energy;
- address threats in public health;
- manage natural resources more responsibly;
- improve the transport system and land-use management;
- combat poverty and social exclusion;
- deal with the economic and social implications of an ageing society.

In several CEE countries the elaboration of the national strategy of sustainable development has started. In that framework attempts are made to integrate the environmental, economic and social policies. This task means new challenges to all CEE countries. The political will of the governments may give large impetus to this process; its lack may hinder the progress.

The document entitled *Estonian National Report on Sustainable Development 2002* (Vetik, 2002) was presented at the Johannesburg Summit. According to this report, to ensure better integration of different fields of the society for supporting continuous sustainable development process, it is essential to take account of the following aspects:

- implementation of development strategies;
- political consistency and involvement of stakeholders;
- knowledge-based policy formulation.

The Estonian document may serve as a model for the other countries to compile similar reports.

7. Conclusions

The Stockholm Conference (1972) concentrated mainly to the protection of the environment. The Rio Conference (1992), beside the environment, highlighted the concept of development, as well, thereby urged the integration of environmental and economic policies. The Johannesburg Conference (2002) emphasized the significance of the human dimension. As a result, environmental protection, economic growth, and social cohesion must go hand in hand. This concept meets the interest of the CEE countries, and after Johannesburg, it is the common interest and responsibility of national governments, local business spheres, and civil societies to gradually implement the sustainable development. Science and technology, as well as education have to contribute to this process.

Acknowledgements

The author of the article thanks Dr. János Zlinszky and Dr. Éva Csobod of the REC for CEE for their contribution in providing information and references.

References

European Environment Agency (EEA): 1998, *Europe's Environment: The Second Assessment,* Copenhagen, Denmark, 293 pp.

European Commission: 2001, *A Sustainable Europe for a Better World: A European Union Strategy for Sustainable Development,* Brussels, Belgium, 16 pp.

Kereszty, A: 1998, *Tények könyve. Zöld [Book of Facts. Green]* (in Hungarian), Budapest, Hungary, Greger-Delacroix, 464 pp.

Prescott-Allen, R.: 2001, The *Wellbeing of Nations: A Country-by-Country Index of Quality of Life and the Environment*, Washington, D.C., Island Press, 219 pp.

Regional Environmental Centre (REC): 2000, *'Europe Agreeing': 2000 Reporting the Status of the Implementation of Multilateral Environmental Agreements in the European Region*, Szentendre, Hungary.

Stanners, D. and Bourdeau, P.: 1995, *Europe's Environment, The Dobris Assessment*, Copenhagen, Denmark, European Environment Agency, 676 pp.

Vetik, R.: 2002, *Estonian National Report on Sustainable Development*, Tallinn, Estonia, Estonian Commission on Sustainable Development, 56 pp.

World Summit on Sustainable Development, Plan of Implementation (WSSD): 2002, http://www.johannesburg.org.

World Resources: 2000, *World Resources 2000–2001. People and Ecosystems. The Fraying Web of Life*, Washington D.C., World Resources Institute, 389 pp.

CHAPTER 10

WSSD 2002, LATIN AMERICA AND BRAZIL: BIODIVERSITY AND INDIGENOUS PEOPLE

ALPINA BEGOSSI[1] and FERNANDO DIAS DE ÁVILA-PIRES[2]

[1]*Núcleo de Estudos e Pesquisas Ambientais (NEPAM), Universidade Estadual de Campinas, Campinas, SP, 13081-970, Brazil;*
[2]*Instituto Oswaldo Cruz, Rio de Janeiro, Brazil*
(e-mail: alpina@reitoria.unicamp.br, favila@matrix.com.br)

Abstract. Latin America comprehends notable variations in terms of natural environment, availability of natural resources, living standards, and demographic patterns. Latin America is a mosaic of cultures, post- and pre-Columbian. The rich variety of life forms discovered and described by chroniclers and traveling naturalists in the Neotropics contributed to the proposal, in mid-XVIIIth century, of a new system of classification and a scientific code of nomenclature for all organisms. Biodiversity was, for many centuries, a source of resources to be exploited *in natura*. In scientific circles, its inventory became the domain of taxonomists. But modern technology showed how important the miriad of life forms really are as sources of chemical molecules to be engineered as drugs and reassembled as novel manufactured products. We are on the brink of a new agricultural and medical revolution, thanks to the techniques of genetic engineering, which will lead eventually to the elimination of hunger and malnutrition.

In this essay, the Brazilian environmental and social heterogeneity will serve as an example to illustrate some key points, which have influenced sustainability policies. The Amazon deforestation and indigenous knowledge (IK), subjects often associated with areas of high biodiversity, are usually the focus of environmental debates. The importance of IK in integrating development, reducing poverty and sustainability are considered together with the intellectual property rights of native populations.

In the World Summit on Sustainable Development (WSSD) Implementation Plan, a few paragraphs were dedicated to Latin America, because of the pre-existing Action Platform on the Road to Johannesburg 2002, approved in Rio de Janeiro in October 2001. This paper calls attention to the need to draw up specific environmental policies for a region which shows an extremely high cultural and biological diversity, associated with a high availability of forests and water, among other resources.

Key words: Amazon, biodiversity, biotechnology, Brazil, Central America, clone, culture, deforestation, demography, epidemiology, ethics, genetic engineering, health, healthcare, indigenous peoples, Latin America, medical ecology, microorganisms, nature conservation, neotropical region, parasites, South America, sustainable development, systematics, taxonomy.

Abbreviations: CAN –Andean Community of Nations; CBD – Convention on Biological Diversity; CDM – Clean Development Mechanism; EU – European Union; GEF – Global Environmental Facility; GTA – Amazon Working Group (Brazil); FAO – Food and Agriculture Organization; IPAM – Amazon Institute of Environmental research (Brazil); IK – Indigenous Knowledge; ISA – Socioenvironment Institute (Brazil); IUCN – International Union for the Conservation of Nature and Natural Resources; MEA – Multilateral Environment Agreements; NGO – Non-Governmental Organization; NUFFIC – Netherlands Organization for International Cooperation in Higher Education; OEA – Organization of the American States; TKDL – Traditional Knowledge Libraries; TRIPs – Trade Related Aspects of Intellectual Property Rights; UNCTAD – UN Commission on Trade and Development; UNDP – United Nations Development

Readers should send their comments on this paper to: BhaskarNath@aol.com within 3 months of publication of this issue.

L. Hens and B. Nath (eds.), The World Summit on Sustainable Development, 223–239.
© 2005 *Springer. Printed in the Netherlands.*

Programme; UNEC – United Nations Economic Commission; UNEP – United Nations Environmental
Programme; UNESCO – United Nations Educational Scientific and Cultural Organization; WIPO – World
Intellectual Property Organization; WHO – World Health Organization; WHRC – Woods Hole Research
Center (USA); WSSD – World Summit on Sustainable Development.

1. Introduction

In the sixties, at least two books had a special role in announcing that Humankind
was endangered by pollution, population trends, food and resource distribution.
These books were, *Silent Spring*, by Carson (1962), and *The Population Bomb*, by
Ehrlich (1968). Concerns related to the natural environment began to grow in the
XIXth century, with conservation groups being created, such as The Commons,
Open Spaces and Footpaths Preservation Society in Britain in 1865, some British
ornithological societies, national parks, such as Yellowstone in 1878, New Zealand
parks at the end of the century, and the *Sierra Club*, in 1892 in USA, which is
considered the first Non-Governmental Organization (NGO) for political action to
conserve nature (Hatcher, 1996: 65).

In the seventies and eighties of the XXth century, the Stockholm Conference in
1972, among other meetings, and books such as The Limits of Growth (Meadows
et al., 1972), and The Global 2000 (Barney, 1980), analyzed the Earth's limitations
on the growth of human population and the use of natural resources.

Besides opening a public debate on the need for a sustainable use of resources,
the Rio Conference in 1992 and the Agenda 21, also decided on conservation
principles for public policies. Even when a final balance on the effectiveness of
the concrete results of Rio has not been published yet, there is few doubt about
the benefits brought for a global dialogue (Leis, 1996: 74). For example, the
Rio Declaration on Environment and Development entails a set of principles that
should be the basis for a sustainable development policy. Some of these are the
precautionary principles, the equity principles; the polluter pays principle and the
user pay principle (Hens, 1996: 86). Agenda 21, includes actions for SD targeted
policies. It includes about 900 pages addressing international problems on environment and development (Hens, 1996: 91–92). Along with these, international
financing mechanisms, such as the GEF (Global Environmental Facility) provided
financial support for projects on climate change, biodiversity, and ozone depletion.
(Hens, 1996: 100).

1.1. Scientific Networks

A positive development, which occurred in between the Rio Summit 1992 and the
World Summit on Sustainable Development (WSSD) 2002 at Johannesburg, was
the increasing participation of the academy. In fact, since the 1990s, the international academy, including ecological scientific groups, established combined
programmes aimed at environmental questions. Examples of such approaches
were the Sustainable Biosphere Initiative (in the *Bulletin of the Ecological
Society of America,* 1992) and subjects on the meanings and indicators of

sustainable development (Nath et al., 1996). According to Clarke (2002), although governmental attempts to address sustainability are still confused, researchers are slowly building up a picture of what science can contribute. The examples given by Clarke (2002), for the WSSD, are networks of scientists, such as the Initiative on Science and Technology for Sustainability (Harvard University) and the Third World Academy of Sciences (Trieste, Italy). In Latin America, there are also contributing networks, such as the Instituto de Pesquisa Ambiental da Amazonia (Brazil) (IPAM), working together with the Woods Hole Research Center (USA) (WHRC), which played an important role in Johannesburg.

Documents by International Council of Science (ICSU), such as background papers prepared for the WSSD, illustrate how groups of scientists took an active part in an attempt to integrate science and public policies. One example is the document addressing new ways of integrating information from the natural and social sciences as well as on economic research, where resilience is considered a management tool and three general policy recommendations are drawn. The first being, the importance of policies that highlight interrelationships between the biosphere and a prosperous development of society. The second concerns the need of policies to create space for flexible and innovative collaboration towards sustainability. The third is about how to establish operational sustainability in the context of social–ecological resilience (Folke et al., 2002: 8). The use of resilience as an ecological concept is useful to analyze the interaction between native people, the environment, and their capacities to respond to changes. In Brazil, examples are provided for the native caboclos of the Amazon, and for caiçaras of the Atlantic Forest coast, which live in areas of high biodiversity (Begossi, 1998: 129–133, 2000). For the indigenous populations of Latin America, the dilemma is how to access development, improve livelihoods and guarantee a sustainable way of living. Considering indigenous knowledge (IK), a challenge is how to link already existing knowledge to environmental policies.

Four Latin American sub regional preparatory meetings were held in 2001 (Cone Sul, Andina, Mesoamericana, and Caribe), in an attempt to negotiate common regional interests and to influence the Johannesburg agenda (Guimarães, 2001). In 1990, Latin American forests accounted for 56% of tropical forests (Togeiro de Almeida, 2002: 145), and between 1950 and 2000 the population of Latin America and the Caribbean increased from 167 to 519 million (Guimarães, 2001).

This article focuses on how the WSSD Implementation Plan is applied to a diverse continent such as is Latin America. Latin America includes various ecosystems and cultures. There is a continental heterogeneity in terms of economic, environmental, health, demographic and social aspects, which might be, at least partially, represented through information on a diverse country like Brazil. Latin America holds 8% of the planet's population, 23% of cultivable lands, 27% of mammal species, and 43% of bird species (Gitli and Muriló, 2002: 78), and challenges include integrating the positive aspects of development with the necessary ecological needs.

2. The diversity of life forms

The primeval need to know the animals and plants surrounding their dwellings and sometimes existing only in their imagination, leads our early ancestors to establish primitive systems of utilitarian classifications and rudimentary, descriptive, nomenclature. The system of scientific classification and international nomenclature was formally proposed by the Swedish botanist Carolus Linnaeus. The discipline of classification became known as Taxonomy or Systematics. Nowadays, over 1.5 million species of animals and 300 thousand plants have been described and named but recent estimates of the total number of existing organisms varies from a conservative 10–50 millions. The term biodiversity refers not only to species but also to infraspecific genetic diversity and to ecosystems.

Tropical regions offer a more diversified choice of ecological niches and provides for increased biological activity; as a result, ecological webs are more complex and biodiversity greater than in other latitudes. South and Central America lies in the Neotropical Region, with the greater portion of Mexico in the Neartic. The largest continuous rain forest of the globe is found in South America, although other biomes are important for sustainable exploitation. Until the end of the Triassic period South America and Africa, already separated from Antarctica, Australasia, and India, were part of the supercontinent of Pangea. By the end of the Cretaceous, some 70 million years ago, South America had broken free and had drifted away from Africa. At the beginning of the Cenozoic, a continental bridge, the isthmus of Panama connected the Americas. These two continents became separated again in the early Paleocene to be reunited by the uplift of the isthmus of Panama in the Pliocene. During the Tertiary, South America was isolated and developed a remarkable endemic fauna (Fittkau et al., 1968) which became extinct after the Pliocene–Pleistocene transition. The modern native biota is a mixture of forms that evolved from ancestors dating from times of the Pangea, and from others that migrated from the Northern continent, arriving through the Panama land bridge or by island hopping through the Caribbean sea.

Natural productions from the American continent began to be introduced in Europe after the first voyage of Columbus. A choice of spices, maize, potatoes, manihot, cuys, monkeys, birds, as well as medicinal plants like the cinchona, ipeca, tobacco, vanilla, salsaparilha, copaiba, and coca would change nutritional habits and add to the repertoire of pets and *Materia Medica*. From Europe came sugar cane, wheat, coffee, cattle, horses, sheep, goats, dogs, cats, poultry, together with herbs, weeds, roaches and rats, viruses and bacteria. The first results of ecological disruptions soon began to appear. Blood-sucking bats increased in numbers when large supplies of blood from cattle, horses, and chickens became available. The introduction of the rabies virus would prove to be a serious problem, and the first suspicions were vented by the Dutch physician Piso (1658) in Brazil, who questioned if the poison inoculated by vampire bats would not be the same as that of rabid dogs, a notion that would be scientifically proved in the XXth century. Parasites of the Old World found new hosts and native complacent reservoirs, while autochtonous pathogens infected the newly arrived colonizers, as the syphilis bacillus and the trypanosome of Chagas disease. Thanks to the lack of specific immunity virgin soil epidemics were rampant across the American continent, eliminating whole tribes and nations. Over exploitation of native plant and animal resources, habitat changes, and deforestation, which continues nowadays, although somewhat controlled.

Although some of the larger parasitic worms, as *Ascaris*, *Enterobius*, and *Taenia* were known from antiquity, their role was disputed until late in the XIXth century (Ávila-Pires, 1998a). Some physicians considered them beneficial for 'cleansing' the intestines of their hosts from poisonous byproducts of the process of digestion. Microscopic organisms, though, were seen for the first time in the XVIIth century, after the invention of the microscope. The diversity of microorganisms had to wait for Pasteur to be fully recognized, and their biological, medical, and ecological roles understood. Parasites did not attract the attention of amateur zoologists and demanded special techniques to be seen, studied and classified. Being so, it was a hopeless task to try and make sure that diagnoses from the vague descriptions of symptoms in old chronicles, before the discovery of techniques nowadays adopted by paleoparasitologists. The inventory of microorganisms is an enormous task which is being too slowly accomplished. Zoological groups which attract medical attention are given priority, and we tend to forget that newly discovered agents of diseases are old unremarked members of the biotic assemblage.

Old conceptions of disease as a divine punishment, or a resultant of the exposure to a miasmic atmosphere were an obstacle for a general theory of disease. Furthermore, the recognition of the importance of zoonoses depended upon a change in the philosophical conception of the position of man in the biosphere. Diseases of man were considered to be exclusively human, with few exceptions. The discovery and description of non-human hosts, vectors, and reservoirs of pathogens by the end of the XIXth century forced a change in that outlook. The work of Laveran in France, Manson and Ross in England, Bassi in Italy, Finlay in Cuba, Reed in Panama, and Lutz in Brazil helped to pave the way to our modern understanding of the epidemiology and ecology of transmissible diseases. The isolation of the yellow fever virus in Latin America at the beginning of the XXth century was another important breakthrough and today we have identified some 150 virus infections of zoonotic nature. Thanks to the work of the laboratories established by the Rockefeller Foundation in Latin America, more than 100 arboviruses were described and studied. That Foundation was instrumental also in the introduction of modern methods for the study of zoology and ecology. Yellow fever was successfully eradicated, only to be reintroduced decades later. The discovery of Chagas disease and the confirmation of its importance as a public health problem of large proportions was instrumental in attracting the interest of medical researchers from the traditional institutes dedicated to the study of tropical medicine (Delaporte, 1999).

It is important to recognize that in many instances, popular lore preceded and guided scientific discovery. Empirical observations lead to the association of yaws with flies in the XVIth century in Brazil, cutaneous leishmaniasis with phlebotomines in Peru, Texas fever with ticks. The correct identification of all animals involved in the biological cycle and in the ecological webs of parasites and viruses is of paramount importance (Hoogstraal, 1956) and the immense biodiversity in Latin America poses a challenge for taxonomists and non-taxonomists alike, as well as for ecologists. The need to make extensive and intensive scientific collections of all forms of life is hampered by the small number of active systematists. Many taxa of plants and animals have not been revised lately or ever, and there is an urgent need for catalogues and field guides to help non-taxonomists the same being true for the organisms involved in the epidemiological chains. Furthermore, epidemiologists tend to restrict their models to those species directly involved in the chains of transmission, leaving their preys, predators, and symbionts, and associates out.

The concern with the conservation of natural renewable resources is an old one. Forests were preserved since medieval times for the building of ships or as the hunting grounds for royalty and nobility. The birth of scientific ecology in the late XIXth century, showed the need to approach the question from an ecosystem point of view, and proved to be of epistemological importance for the health sciences. During 1960–1980 Latin American scientists had an important role in the definition of terms and concepts, and in the active participation in the international movement that forced governments to implement international agreements, treaties and pass national conservation laws. Expressions like 'protection', 'conservation', 'rational use', and concepts as 'endangered' and 'rare' became the subject of long academic arguments and were eventually defined by the International Union for the Conservation of Nature and Natural Resources based in Morges, Switzerland since 1961. The IUCN was established in Fontainebleau in 1948 under the auspices of UNESCO with the name of International Union for the Protection of Nature and seated in Brussels. It was renamed during the General Assembly held in Edinburgh in 1956, under the consensus that the term *conservation*, defined as the rational utilization of natural resources, was a better expression of its philosophy (Bourlière, 1964). This concept was superseded by that of sustainable development proposed in the Rio Conference of 1992.

Mexico was the first Latin American country to establish a national park, El Chico, in 1898, followed by Argentina in 1907 where Francisco Moreno donated the $75 \, \mathrm{km}^2$ of the famous Nahuel Huapi National Park, created by law in 1934. In March, 1968, took place in San Carlos de Bariloche, Argentina the Latin American Regional Conference on the Conservation of Renewable Natural Resources. As a follow-up the first official list of endangered species was drawn in Brazil, based mostly upon the professional experience and personal opinion of individual zoologists and botanists. Latin American natural scientists served on the special commissions of the IUCN, as the noted Survival Service Commission. During last two decades, the conservation of nature a popular subject and began to be addressed by other professionals and to be adopted as political by individuals in general.

The relationships between ecological diversity and health arose from the amalgamation of notions coming from fields as diverse as cultural anthropology, ethnobotany, medical geography, medical ecology, epidemiology, human ecology, and economic development. A pioneer contribution resulted from the International

Biological Program in the form of a book (Dasman et al., 1973), in which animal and human health problems were viewed as a resultant of man's interference with natural ecosystems. It shows a considerable advance over the rather limited treatment of this question found in the UNESCO report on the 1968 Intergovernmental Conference of Experts on the Scientific Basis for the Conservation of Resources of the Biosphere. The sustainable development of Latin American tropics was the subject of a meeting held in Venezuela in 1974, under the auspices of IUCN, UNEP, FAO, UNDP, UNEC, and OEA, but as far as health is concerned, the resulting report (IUCN, 1975) offers a single contribution on the question of the excessive utilization of pesticides. In 1980, the IUCN together with the UNEP and the World Wildlife Fund formulated a policy to integrate economic development with the demands of ecologists, adopting three general principles: maintenance of ecological processes, preservation of genetic diversity, and the sustainable exploitation of species and ecosystems. There is no direct mention to health. Actually, the concern with health arose from the general concern with economic development and the costs of disease. Even so, health is usually disregarded or treated as an accessory or optional item in Environmental Impact Assessments. For that matter, the Report of the European Commission to the Rio Conference (EC, 1992) deals on the subject of biodiversity, but not directly to health. In 1997, though the World Health Organization (WHO) published an important report on the subject of health, environment, and development (WHO, 1997). For a short survey of the international initiatives and accomplishments in the field of conservation see Ávila-Pires (1999, pp. 155–157).

Having begun as a concern of naturalists, the conservation movement directed its attention to the preservation of the natural environment. This emphasis was translated in the proposal or revision of national legislation on the protection of plants and animals, lead by Brazil in 1965 and 1967, and followed by Colombia, Peru, and Bolivia. Sustainable development became a concern for economists, who see biodiversity through the optics of economics as an exploitable resource. It is up to medical ecologists, zoologists, and medical researchers with an interest in the zoonoses to further our present knowledge on the implications of biodiversity conservation on animal and human health.

The development of molecular biology coupled with the new technologies of genetic engineering promises a revolution in medical care, health promotion, disease control, as well as in agriculture and in many other traditional areas which will acquire new dimensions and goals. Biotechnology will be the agent of a profound socioeconomic revolution in the XXIth century, as it was for biology in the last decades of the XXth century. In fact, the current concept of biodiversity crosses over the limits of the biological entities, to include genes and molecules (Sener, 2002), disregarding the frontiers of the levels of complexity (Grant, 1963; Simpson, 1964). Grant and Simpson draw a sharp dichotomy between molecular biology and organismal biology, and warn about the dangers of excessive reductionism. Simpson goes on to question the validity of the very expression 'molecular biology', recognizing that phenomena taking place at the molecular level throws light on what happens at the cellular level but at the same time acknowledging the peculiarities inherent to the organismal level.

In the style of the *Arabian Nights* which in the literary trade is known as "boxes within boxes", where one story unfold into another prior to its end, before we completed the inventory of what nature has to offer we started with artificial selection of varieties by selective breeding of some 20 domesticated organisms; and without reaching the end of *that* story, biochemistry and molecular biology unfolded an endless choice of new pathways and of new ways of modeling organisms according to our needs. Of course, all those new developments poses unforeseen ecological, ethical and legal problems as the safety of transgenic organisms and products and its impact on natural ecosystems and biomes; the utilization of trunk-cells; the cloning of human cells (Nuffield Council on Bioethics, 2002). And they bear heavily upon developing countries with a rich biodiversity and an urge to develop rapidly.

3. The Brazilian heterogeneity: environmental, demographic, and socio-economic diversity

Brazil consists of heterogeneous characteristics, from its varied environmental landscapes to its economic and social features, such as earnings and living standards, sanitary conditions, educational levels, among others. In 2000, 81% of 170 millions of Brazilians lived in urban areas (Carmo, 2002: 172). Brazilian ecosystems also show differences in population growth and development indicators, besides the typical environment features, such as climate and vegetation, among others.

As described by Hogan (2002: 13–18), the Amazon region is still sparsely populated, with low population densities (4 persons per km^2). The cerrado (savannah) is the second largest Brazilian ecosystem, with a population density of 6 persons per km^2 where about 37% of its original vegetation has been converted to pasture, and crops, among others. The caatinga is the Brazilian semi-arid region, located in the Northeast, having a population density of 66 persons per km^2, and an area where outgoing migration has been chronic. The last 5% of remnants of the Atlantic Rainforest are included in five Southeastern Brazilian States. This is also the region where the population–environment balance is the most precarious, and is where most Brazilians live (105 persons per km^2). The Southern Campos do Sul (savannah) is located in a highly urbanized region, including hilly lands, with areas of forest and with a population density of 36 persons per km^2.

The population in Brazil is now represented by fertility declines, which helps to support the view that what is needed for the WSSD is a more complex view of the population–environment question, including the distribution factor (Hogan, 2002: 21). According to Sawyer (2002: 230), the fertility decline in Brazil is now close to the replacement level, with a fertility rate of 2.2 children per woman. The unequal distribution of resources in Latin America, including Brazil, calls for attention in order to carry out an analysis of such distribution not just following a North-South world dichotomy, but also including the differential access to resources within the continent or the country. Sawyer (2002: 227) sustains that in a less simplistic vision, it is recognized that there is also over-consumption among high-income segments in the South. Viola (1996: 31) stresses the social asymmetry in countries like Brazil, Bolivia, Nicaragua, Peru, and Haiti, among others, and the opportunity that the globalization brings to change such inequalities.

Unequal distribution refers to access to resources, employment, sanitary conditions, health, and water, among others. These distribution factors are reinforced by looking at data on some sanitary conditions and water availability in Brazil, moreover these data may well represent the Latin American reality as well. When looking at sanitary conditions, for example, in 2000 about 52% of Brazilian municipalities had some kind of sewer system; when considering toilet facilities, there are Brazilian states, such as Maranhão and Piauí where 40% and 43% of the population, respectively, have no bathroom or toilet facilities. This contrasts to areas such as São Paulo, where one finds a figure of 0.4% of the population without bathrooms (Carmo, 2002: 172–173).

South America has about 23% of the whole planet's fresh water, and 12% of the fresh water is located in Brazil. Water distribution varies greatly, especially when considering demographic density and the Brazilian river basins. For example, in the Eastern Atlantic, availability per capita ranges from 5 to $13\,m^3\,day^{-1}\,inh^{-1}$ whereas in the Amazon it is $1723.1\,m^3\,day^{-1}\,inh^{-1}$ (Carmo, 2002: 169–171). Therefore, one realizes how crucial the importance of the Amazon is, considering its extremely high biodiversity, its availability of water, and its potential contribution to the global climate. Once again, the inequality in Latin America is also observed in terms of the efforts made to protect the Biodiversity, with countries such as Belize and Costa Rica representing high efforts towards keeping to the regulation and protection of the use of their natural resources, compared to Brazil and Mexico, showing high environmental rhetoric, but low effective conservation policies (Viola, 1996: 35).

4. The Amazonian dilemma

As a central part of the challenges towards sustainability, there are the carbon dioxide concentrations and associated greenhouse effects. Even if environmentally friendly approaches were adopted from now on, carbon dioxides would still rise until 2050 (Gewin, 2002). In that respect, Brazil has great responsibilities since it is responsible for 3% of the global emissions due to the deforestation that is occurring in the Amazon (Moutinho et al., 2002). Representing 40% of the world's remaining tropical rainforests (Laurance et al., 2001a), the Brazilian Amazon plays a pivotal role for biodiversity conservation and climate change. Hence, the dilemma is in fact a global dilemma, with deep roots coming from diverse kinds of necessities, from the destruction of habitats that may extinguish many animal species to the survival of the local indigenous peoples. As pointed out by Fraser and Mabee (2002), the Convention on Deforestation will not prevent a poor farmer from cutting down a tree, and declarations for sustainable development will not change North America's appetite for consumption.

Legal Brazilian Amazon comprises $5.1 km^2$ including seven Brazilian states. According to Instituto Nacional de Pesquisas Espaciais (INPE) about 11.5% of Legal Amazon has been deforested (data in Sydenstricker-Neto, 2002: 57). In the Brazilian Amazon there are about 17 million people earning less than US\$ 100 per month (Nepstad et al., 2002). Considering a rural and forested area, such as the Sustainable Reserve of Mamirauá (a state reserve), where sustainable management has been carried out since 1996, data show that the annual mean earnings of the families living in a focal experimental area of the Mamirauá Reserve is of US\$ 900, 53% being spent on the acquisition of food. Fishing is an activity with high revenues, representing 72% of the value of the total earnings in Mamirauá. Earnings of a domestic group are obtained from the sale of fish, timber, and manioc flour (Lima, 1999: 260; Viana et al., in press).

In the eighties, Fearnside (1986) was one of the researchers who showed and modeled the impact of roads in the Amazon, including feedback mechanisms of opening roads, increasing population density, and deforestation. This kind of impact is suggested by Moutinho et al. (2002), who show that more than 70% of deforested areas in the Amazon are located on a side stretch of 50 km off the roads. Their analyses show that governmental programs, such as the 'Avança Brasil' could be responsible for a great increase in carbon dioxide emissions related to road construction, paving and associated deforestation. Paved highways increase accessibility to forests, creating networks of secondary roads, increasing forest exploitation (Laurance et al., 2001b). The side event *Frontier Governance and Sustainable Development in Amazonia* held in Johannesburg, included presentations by the IPAM (Amazon Institute of Environmental Research), by the ISA (Socioenvironment Institute) and GTA (Amazon Working Group) (Issue #1 WSSD p. 6) with criticisms concerning the government's plans to increase infrastructure in the Amazon.

Laurance et al. (2001a) estimated the rate of deforestation in the Brazilian Amazon as of 2 million ha per year. In spite of disagreements on the actual rate of deforestation in the Brazilian Amazon (see debates on *Science* 291, 2001, and Nepstad et al., 2002; Laurance and Fearnside, 2002), there is a general agreement that half of the investments of the 'Avança Brasil' project (over \$ 20 billion) will be used for the construction of major highways and infrastructure projects that might cause negative impacts on the Amazonian forests. Moreover, that many projects are designed to support soybean, logging, and cattle-ranching industries, which offer limited benefits to the poor (Laurance et al., 2001b). IPAM figures indicate between 8 and 18 million ha of deforestation in the following 25–35 years due to building of four major roads (Bonnie et al., 2000).

According to Laurance et al. (2001a) forest deforestation in the Brazilian Amazon has had causal factors such as the increase in non-indigenous population, industrial logging and mining, construction of roads and highways, besides human-ignited wildfires. However, it is worth noting that about 66% of the Amazonian population are concentrated in cities, not in forests (overall density of 4 persons km^{-2}).

Therefore, it is a difficult task to consider that environmental threats to the Amazon are due to population pressure (Hogan, 2002: 14–15). Considering the capacity of forest regeneration, conditions for recovery have been increasingly difficult especially when edge effects are taken into account. Gascon et al. (2000) observed that the ability of forest regeneration depends on the 'harshness' of the matrix, which depends on the variety of habitats surrounding the forest patches. For example, fragments of the Atlantic Forest surrounded by sugar cane are at risk, because the forest is unable to regenerate at the edge.

Carbon sequestering is a debated ecosystem service in the context of the Kyoto Protocol, but Costa Rica forested conservation areas are credited with income for their services (Folke et al., 2002: 14). Taking into account the importance of the carbon-offset funding for avoiding deforestation through the Kyoto Protocol, and the Clean Development Mechanism (CDM) Brazil might obtain credit for avoiding deforestation (Fearnside and Laurance, 2001). The formal adoption of forest carbon markets should increase incentives for developing nations to protect forests (Bonnie et al., 2000). Other environmental agreements that concern Latin American countries are some Multilateral Environmental Agreements (MEA), including the International Timber Agreement among Brasil, Colombia, Ecuador, Peru, and Venezuela in January 1997, and the Acordo-Marco on Environment for the Mercosul, in June 2001 (Togeiro de Almeida, 2002: 124, 132). Other countries, such as Colombia and Guatemala, have been given taxes for river basin water uses (Guimarães, 2001).

The group of megadiverse countries, which comprehends 15 countries (Bolívia, Brazil, China, Colômbia, Costa Rica, Ecuador, India, Indonesia, Kenya, Malaysia, Mexico, Peru, Phillippines, South Africa, and Venezuela) with 70% of the planet biodiversity and 45% of the world population is taking the leadership in associating human health, biodiversity, and related ecosystem services (Joly, 2002; www.megadiverse.com). What is expected to improve after the WSSD at Johannesburg are improvements moving beyond principles and government agreements, in order to reach effective management practices, based on scientific data and on current IK. As suggested by Guimarães (2001), development processes should be sustainable in the form of environmental, social, cultural, and political processes.

5. Indigenous peoples and sustainability

Indigenous knowledge is an intrinsic part of Latin America, since part of its population has a livelihood based upon forest resources, and embodies a local culture associated with this livelihood. The literature approaches a vast array of definitions of what should be considered ecological, indigenous, traditional, or local knowledge. A special issue (volume 2) of *Environment, Development, and Sustainability* includes different approaches and examples on the local knowledge in the tropics and its relevance to conservation and management (Begossi and Hens, 2000). In Brazil for example, excluding native people who are a mix with Portuguese

Colonists, there are about 218 indigenous peoples, and among these 180 keep their original language (Azevedo and Ricardo, 2002: 184). However, the interaction of indigenous peoples with nature has been changing, and communities have been substituting local health practices and healing plants by western medicine. In the Atlantic Forest of Brazil, elder and a few women are key segments for the maintenance of the local knowledge on folk medicine, but younger often prefer health clinics instead of local care based on traditional medicine (Begossi et al., 1993; 2002).

Many periodicals and documents have been providing support for the maintenance of indigenous knowledge, so that there might be a closer relationship and integration with conventional science, and for action towards sustainability. Examples are *IK Worldwide (Nuffic*, The Netherlands), documents from international agencies, such as The World Bank (1998), as well as associated documents concerning indigenous knowledge and WSSD.

According to the World Bank (1998: 19), generally selected features of IK include: agriculture, soil and land classification, animal husbandry and ethnic veterinary medicine, post harvest technology and nutrition, use and management of natural resources, environment protection, handicrafts, primary health care, preventive medicine, psycho-social care, saving and lending, community development, and poverty alleviation. The same document stresses that IK is the social capital for the poor (p. 4). Declarations associated to the WSSD included the development of partnerships between science and traditional knowledge (Lewis, 2002), the reward of indigenous knowledge for the commercial exploitation, such as suggested by South Pacific states in September 17, 2002, at New Caledonia (Schiermeier, 2002), the introduction of traditional knowledge libraries (TKDL) (Gupta, 2002), the Johannesburg Declaration on Biopiracy, Biodiversity, and Community Rights, which reflects views from the Valley of 1000 Hills Declaration (Kwazulu Natal, South Africa, March 2002), the Rio Branco Commitment (Rio Branco, Brazil, May 2002), and the viewpoints of participants at the Second South-South Bio piracy Summit, Johannesburg, August 2002. This declaration is also controversial to WIPO (*World Intellectual Property Organization*), considering its attempts to develop systems for the protection of IK as being inappropriate, since indigenous rights should be defined by communities themselves (Biowatch SA – Johannesburg Declaration, www.SciDev.Net.).

The Indigenous People's Plan of Implementation on Sustainable Development, Johannesburg (www.tebtebba.org) includes principles on Cosmo vision and spirituality, self-determination and territory, children, women, sacred sites, food security, intellectual property rights, biodiversity, protected areas, mining, energy, tourism, fisheries, water, climate change, health, desertification, and education, among other issues. In terms of the Intellectual Property Rights, it demands the annulment of agreements under Trade Related Aspects of Intellectual Property Rights (TRIPs) that takes indigenous knowledge into account, in order to assert the rights of effective participation in decision making arenas on biodiversity and traditional knowledge, such as the Convention on Biological Diversity (CBD), WIPO,

UN Commission on Trade and Development (UNCTAD) and Andean Community of Nations (CAN). Another IK document, the Declaration of Atitlán, April 2002, Guatemala (www.tebtebba.org) called for the right for food, and recommended a World Food Summit five years later to the WSSD. It recommended, among other things, that all States should ratify the Convention on the Elimination of Persistent Organic Pollutants and the Kyoto Protocol on Climate Change. There are of course many other side events and declarations by indigenous peoples, such as on tourism (January 2002, Chiang Rai, Thailand), but such descriptions are outside the scope of this brief essay.

Actually, the preparatory meetings, such as the Regional Preparatory Conference of Latin America and the Caribbean for WSSD, Rio de Janeiro, October 2001 (www.johannesburgsummit.org) called for the ratification of the Convention on Biological Diversity, of the Kyoto Protocol, and for other pertinent issues, but they did not mention direct actions concerning indigenous people and their knowledge, except for a sentence on equitable access to the benefits afforded by the use of genetic resources. However, policies towards international trade were emphasized.

6. Conclusions

The challenge of attaining and maintaining an acceptable level of sustainable development based upon the rational utilization of natural resources and the conservation of biodiversity in Latin America will demand much ingenuity and political will. Biodiversity and health are closed related and interdependent, as we shall see. Biotechnology has opened a box of Pandora, allowing us to see natural resources in a new light, dimly perceived fifty years ago. At the same time the concept of geosystem demands a new ethics in conservation, by making everyone responsible for the preservation of biodiversity at the very source of the products we use and enjoy.

Ecologists have long been aware of the importance of preserving the living and non-living environment in the biosphere. During the 1960 the conservation movement became an international concern and the Rio Conference of 1992 and the recent Johannesburg World Summit on Sustainable Development showed some advances but also some difficulties in providing the necessary conditions and political will to proceed to a rational use of natural resources to maintain a sustainable development. Economy has direct effects upon health, although economic indexes do not reflect the general sate of health of a country. GNP does not take into consideration the distribution of revenues and the profound regional inequalities we find throughout Latin America. At the time that governments began to take notice of environmental problems, medicine began to define health and disease in relation to environmental conditions, but in a manner far more complex than the simple Hippocratic correlations of airs, waters, and soils.

In spite of the WSSD Implementation Plan having taken into account world needs, such as poverty eradication, waste management systems, the protection and management of natural resources, among others, it dedicated about two paragraphs (73 and 74) to Latin America. Such a brief approach was due to the existence of the Platform for Action on the Road to Johannesburg 2002, approved in Rio de Janeiro, October 2001 (Plan of Implementation: 49).

Considering the importance and heterogeneity of natural resources in Latin America, and its key position in terms of water availability and forestry resources, the Implementation Plan is far from being realistic in terms of Latin American becoming more sustainable. Considering that some aspects are key elements for building a sustainable realistic program, we have focused on a few, but high priority elements for sustainability: health, alleviation of poverty and associated sanitary conditions, population growth, the Amazon deforestation, and the participation of indigenous peoples as far as biodiversity is concerned. There is much more to go into detail about on Latin America, especially if the interaction of poverty, environment, and trade is considered. Such a task would involve transgenic soybean in Argentina, ecological conflicts and shrimp exports in Ecuador, Honduras, Colombia, just to mention a few examples (Togeiro de Almeida, 2002: 201).

The importance of the indigenous populations in Latin America represents more than merely conserving local or traditional knowledge; it is also a challenge in terms of how to integrate better living standards with a sustainable use of natural resources through policies that take current local management practices into account. There is a high diversity of indigenous cultures in Latin America. Even in Brazil, besides the Native Indians, the native populations that mixed with Portuguese colonists are the Caboclos of the Amazon and the Caiçaras of the Atlantic Forest coast. Such populations have been participating in the economic cycles since colonization, by doing things such as rubber tapping in the Amazon or making sugar cane rum (aguardente or cachaça) on the Atlantic coast. Their importance for certain current economic sectors, such as artisanal fishing, adds to their cultural value, thereby making the crucial link between indigenous peoples and sustainability in a complex system involving demography, local regional and international trade, living standards, and ecological factors.

Acknowledgements

A. Begossi is grateful to FAPESP for research grants, to CNPq for a research productivity scholarship, and to Paulo Moutinho (IPAM) for the kindness of sending some references.

References

Ávila-Pires, F.D.: 1998, 'Parasites in history', *Rev. Ecol. Lat. Amer.* **5**(1–2), 1–11.
Ávila-Pires, F.D.: 1999, 'Tropical forest ecosystems', in B. Nath, L. Hens, P. Compton and D. Devuyst (eds.), *Environmental Management in Practice,* Vol. 3, London, Routledge, pp. 151–164.

Ávila-Pires, F.D., Mior, L.C., Aguiar,V.P. and Schlemper, S.R.: 2001, 'The concept of sustainable development revisited', *Found. Sci.* **5**, 261–268.

Azevedo, M. and Ricardo, F.: 2002, 'Indigenous lands and peoples: recognition, growth and sustenance', in D.J. Hogan, E. Berquó and H.S.M. Costa (eds.), *Population and Environment in Brazil: Rio +10*, CNPD (National Commission on Populations and Development), ABEP (Brazilian Association of Population Studies), NEPO (Population Studies Center, Unicamp), chapter 8, Campinas, pp. 183–206.

Barney, G.O.: 1983, *The Global 2000 Report to the President of the U.S.* 5th edn., New York, Pergamon Press Inc.

Begossi, A.: 1998, 'Cultural and ecological resilience among caiçaras of the Atlantic Forest and caboclos of the Amazon, Brazil', in C. Folke and F. Berkes (eds.), *Linking Social and Ecological Systems*, chapter 6, Cambridge, Cambridge University Press, pp. 129–157.

Begossi, A.: 2000, 'Livelihood and sustainability in tropical environments: Atlantic and Amazonian cases. *Paper presented at the Congress of the International Society of Ecological Economics Congress -+2000* (ISEE 2000), Canberra, Australia, July 5–8.

Begossi, A. and Hens, L.: (Guest Editors) 2000, Special Issue: Local knowledge in the tropics: relevance to conservation and management. *Environment, Development, and Sustainability* **2**.

Begossi, A., Leitão-Filho, H.F. and Richerson, P.J.: 1993, 'Plant uses at Búzios Island (SE Brazil), *Journal of Ethnobiology* **13**(2), 233–256.

Begossi, A., Hanazaki, N. and Tamashiro, J.: 2002, 'Medicinal plants in the Atlantic Forest (Brazil): knowledge, use, and conservation', *Human Ecology* **30**, 281–299.

Bonnie, R., Schwartzman, S., Openheimer, M. and Bloomfield, J.: 2000, 'Counting the cost of Deforestation', *Science* **288**, 1763–1764.

Bourlière, F.: 1964, 'The evolution of the concept of nature protection', *IUCN Bulletin* 1–2.

Carmo, R.L.: 2002, 'Population and water resources in Brazil, in D.J. Hogan, E. Berquó and H.S.M. Costa (eds.), *Population and Environment in Brazil: Rio +10*, CNPD (National Commission on Populations and Development), ABEP (Brazilian Association of Population Studies), NEPO (Population Studies Center, Unicamp), chapter 7, Campinas, pp. 167–182.

Carson, R.: 1962, *Silent Spring*, New York, Fawcett Crest Books.

Clarke, T.: 2002, 'Sustainable development: wanted: scientists for sustainability', *Nature* **418**, 812–814.

Crichton, M.: 1970 *Five Patients*, New York, Ballantine.

Dasmann, R.F., Milton J.P. and Freeman, P.H.: 1973, *Ecological Principles for Economic Development*, London, John Wiley.

Delaporte, F.: 1999, *La Maladie de Chagas,* Paris, Payot.

Ehrlich, P.R.: 1968, *The Population Bomb,* New York, Ballantine Books, Inc.

EU-European Commission: 1992, *Rapport de la Commission des Communautées europeénnes à la Conférence des Nations unies sur l'environnement et le développement – Rio de Janeiro, june 1992.* Luxembourg, Office des publications officielles des Communautés européennes.

Fearnside, P.: 1986, *Human Carrying Capacity of the Brazilian Rainforest*, New York, Columbia University Press.

Fearnside, P.M. and Laurance, W.F.: 2001, 'Response to P. Frumhoff and B. Stanley on The future of the Brazilian Amazon', *Science* **291**, 438–439.

Fittkau, E., Illies, J., Klinge, H., Schwabe, G. and Sioli, H.: (eds.), 1968, *Biogeography and Ecology in South America*, 2 Vols, Vol. 1, The Hague, Junk.

Folke, C., Carpenter, S., Elmqvist, T., Gunderson, L., Holling, C.S., Walker, B., Bengtsson, Berkes, F., Colding, J., Danell, Falkenmark, M., Gordon, L., Kasperson, R., Kaustky, N., Kinzig, A., Levon, S., Mäler, K., Moberg, F., Ohlson, L., Ohlson, P., Ostrom, E., Reid, W., Rockström, J., Savenije, H. and Svedin, U.: 2002, 'Resilience and Sustainable Development: building adaptive capacity in a world of transformations', *ICSU Series on Science and Development no. 3*, Science Background Paper commissioned by the Environmental Advisory Council of the Swedish Government in preparation for WSSD.

Fraser, E.D.G. and Mabee, W.: 2002, 'Summit: vague answers to well known problems?', *Nature* **418**, 817.

Gascon, C., Williamson, B. and Fonseca, G.A.B.: 2000, 'Receding forest edges and vanishing reserves', *Science* **288**, 1356–1358.

Gewin, V.: 2002, 'UN predicts long way to repair environment', *Nature* **417**, 475.

Gitli, E. and Murilo, C.: 2002, 'O futuro das negociações sobre investimentos e Meio Ambiente', in *Comércio e Meio Ambiente, uma agenda para a América Latina e o Caribe*, chapter 3, Ministério do Meio Ambiente, Brasília, pp. 61–95.

Grant, V.: 1963, *The Origin of Adaptations*, New York, Columbia University Press, p. 20.

Guimarães, R.P.: 2001, 'La sostenibilidad Del desarrollo entre Rio-92 y Johannesburgo 2002: eramos felices y no sabiamos', *Ambiente e Sociedade* **4**, 5–24.

Gupta, A.K.: 2002, 'Intellectual property for traditional knowledge on-line, indigenous knowledge key documents', www.SciDev.Net.

Hatcher, R.L.: 1996, 'The Pre-Brundtland Commission Era, in B. Nath, L. Hens and D. Devuyst (eds.), *Sustainable Development,* chapter 2, Brussels, VUB University Press, pp. 57– 81.

Hens, L.: 1996, 'The Rio Conference and thereafter', in B. Nath, L. Hens, and D. Devuyst (eds.), *Sustainable Development,* chapter 3, Brussels, VUB University Press, pp. 81–109.

Hogan, D.: 2002, 'Population and environment in Brazil: Stockholm + 30', in D.J. Hogan, E. Berquó and H.S.M. Costa (eds.), *Population and Environment in Brazil: Rio + 10,* CNPD (National Commission on Populations and Development), ABEP (Brazilian Association of Population Studies), NEPO (Population Studies Center, Campinas), chapter 1, Campinas, pp. 11–28.

Hoogstraal, H.: 1956, 'Faunal explorations as a basic approach for studying infections common to man and animals', *East African Medical Journal*, November, pp. 1–8.

IUCN: 1975, *The Use of Ecological Guidelines for Development in the American Humid Tropic,* Morges, IUCN.

IUCN/UNEP/WWF: 1980, *World Conservation Strategy: Living Resource Conservation for Sustainable Development,* International Union for Conservation of Nature and Natural Resources, United Nations Environment Programme and World Wildlife Fund, Gland.

Joly, C.A.: 2002, 'Os avanços, sutis mas significativos, da Rio + 10', *Biologia, Jornal do Conselho Federal de Biologia* **95**, 9.

Laurance, W.F., Cochrane, M.A, Bergen, S., Fearnside, P.M., Delamônica, P., Barber, C., D'Angelo, S. and Fernandes, T.: 2001a, 'The future of the Brazilian Amazon', *Science* **291**, 438–439.

Laurance, W.F., Fearnside, P. and Cochrane, M.A.: 2001b, 'Response to development of the Brazilian Amazon', *Science* **292**, 1651–1654.

Laurance, W.F. and Fearnside, P.M.: 2002, 'Issues in Amazonian development', *Science* **295**, 1643–1644.

Leis, H.: 1996, 'Globalização e democracia após a Rio-92: a necessidade e oportunidade de um espaço público transnacional', in L.C. Ferreira and E. Viola (eds.), *Incertezas de sustentabilidade na Globalização,* chapter 3, Ed. da Unicamp, Campinas, pp. 67–91.

Lewis, N.: 2002, WSSD 'must embrace knowledge diversity'. Indigenous knowledge key documents, www.SciDev.Net.

Lima, D.M.: 1999, 'Equity, sustainable development, and biodiversity preservation: some questions about ecological partnerships in the Brazilian Amazon', in C. Padoch, J.M. Ayres, M. Pinedo-Vasquez and A. Henderson (eds.), *Várzea – Diversity, Development and Conservation of Amazonia's Whitewater Floodplains,* New York, New York Botanical Garden, pp. 247–263.

Manson, P.: 1898, *Tropical Diseases: a Manual of the Diseases of Warm Climates,* London, Cassel.

Meadows, D.H., Meadows, D.L., Randers, J. and Behrens, W.W.: 1972, *The Limits of Growth,* New York, Universe Books.

Moutinho, P., Nepstad, D., Santilli, M., Carvalho, G. and Batista, Y.: 2002, As oportunidades para a Amazônia com a redução das emissões de gases do efeito estufa, www.ipam.org.br.

Nath, B., Hens, L. and Devuyst, D. (eds.): 1996, *Sustainable Development.* Brussels, VUB University Press.

Nepstad, D., McGrath, D., Alencar, A., Barros, A.C., Carvalho, G., Santilli, M. and Diaz, V.M. Del C.: 2002, 'Frontier Governance in Amazônia', *Science* **295**, 629–631.

Nuffield Council on Bioethics: 2002, *The Ethics of Research related to Healthcare in Developing Countries,* London, NCB.

Piso, G.: 1658, *De Indiae utriusque re naturali et medica,* Amstelaed, Elsevirii.

Sawyer, D.: 2002, 'Population and sustainable consumption in Brazil', in D.J. Hogan, E. Berquó and H.S.M. Costa (eds.), *Population and Environment in Brazil: Rio +10,* CNPD (National Commission on Populations and Development), ABEP (Brazilian Association of Population Studies), NEPO (Population Studies Center, Unicamp), chapter 10, Campinas, pp. 224–254.

Schiermeier, Q.: 2002, Traditional owners 'should be paid'. Indigenous knowledge Key documents, www.SciDev.Net.

Sener, B.: 2002, *Biomolecular Aspects of Biodiversity and Innovative Utilization,* Dordrecht, Kluwer.

Simpson, G.G.: 1964, *This view of life,* New York, Harcourt, Brace & World, pp. 108–120.

Sydenstrycker-Neto, J.: 2002, 'Population and environment in Amazônia: from just the numbers to what really counts', in D.J. Hogan, E. Berquó and H.S.M. Costa (eds.), *Population and Environment in Brazil: Rio +10*, CNPD (National Commission on Populations and Development), ABEP (Brazilian Association of Population Studies), NEPO (Population Studies Center, Unicamp), chapter 3, Campinas, pp. 55–75.

Togeiro de Almeida, L.: 2002, Comércio e Meio Ambiente nas negociações multilaterais, in *Comércio e Meio Ambiente, uma agenda para a América Latina e o Caribe*, chapters 4 and 5, Ministério do Meio Ambiente, Brasília, pp. 97–133; 135–224.

Viana, J.P., Damasceno, J.M.B. and Castello, L.: *Manejo comunitário de recursos pesqueiros na Reserva de Desenvolvimento Sustentável Mamirauá (RDSM)- Amazonas, Brasil. 1as. Jornadas Amazônicas*, CDS-UnB, June 2002, Brasília, (in press).

Viola, E.: 1996, 'A multidimensionalidade da globalização, as novas forças sociais transnacionais e seu impacto na política ambiental no Brasil, 1989–1995', in L.C. Ferreira and E. Viola (eds.), *Incertezas de sustentabilidade na Globalização*, chapter 1, Ed. da Unicamp, Campinas, pp. 15–63.

WHO: 1997, *Health and Environment in Sustainable Development. Five years after the Earth Summit*, Geneva, World Health Organization.

World Bank: 1998, Indigenous Knowledge for development, a framework for action. Indigenous knowledge Key documents, www.SciDev.Net.

CHAPTER 11

SUSTAINABLE DEVELOPMENT AND THE ROLE OF THE FINANCIAL WORLD

HERWIG PEETERS

Ethibel asbl – Stock at Stake sa, Strategic Advice and Project Development,
Rue du Progrès 333/7, B-1030 Brussels, Belgium,
(e-mail: herwigpeeters@pandora.be; fax: C32(0)2 206 11 10; tel.: C32(0)2 206 11 14)

Abstract. The incapacity to finance sustainable development through philanthropic official assistance turned the Johannesburg Summit to business world and the financial industry.

Pioneering financial institutions – including development banks and private banks – have developed a wide range of innovations that can support sustainable development. This article highlights a few innovative products and markets and focuses on the progress made by financial players on the level of standards, metrics and guidelines to improve sustainability management systems, reporting and accounting practices and the multi-stakeholder dynamic.

The role of the socially responsible investing (SRI) community has been underexposed by the Summit. Through its voice and market success, SRI has moved from a green market niche to the mainstream, however not becoming mainstream. The invaluable levering effect of SRI has just been discovered by authorities and market regulators and is becoming instrumental.

In order to show the business case of Corporate Social Responsibility and to prove the financial viability of the People, Planet, Prosperity investing approach, the SRI community should critically reflect on its own quality assurance systems, sound disclosure and verification practices.

Key words: Codes of conduct for financial institutions, Corporate social responsibility in the financial sector, Financial instruments for sustainable development, Sustainable and Responsible Investing, Sustainable banking.

Abbreviations: 3P (Triple P) – People, Planet, Profit; or: People, Planet, Prosperity; AA1000 – AccountAbility 1000; ABI – Association of British Insurers; AsrIA – SRI in Asia; BASD – Business Action for Sustainable Development; BS7750 – British Standard 7750; CDP – Carbon Disclosure Project; CERCLA – Comprehensive Environmental Response, Compensation and Liability Act; CSR – Corporate Social Responsibility; CSRR – Corporate Sustainability and Responsibility Research; EMAS – The EU Eco-Management and Audit Scheme; EPI – Environmental Performance Indicator; ESI – Ethibel Sustainability Index; FCCC – (United Nations) Framework Convention on Climate Change; FDI – Foreign Direct Investment; GFT250 – Top 250 companies of the Fortune 500; GNP – Gross National Product; GRI – Global Reporting Initiative; GWI – Global Warming Indicator; ICC – International Chamber of Commerce; ISEA – Institute for Social and Ethical Accountability; ISO 14000 – An environmental management standard specification by ISO; ISO – International Organisation for Standardisation; NGO – Non-governmental organisation; ODA – Official development assistance; Rio + 10 – 10 years after Rio = Johannesburg 2002; SEE – Social, Environmental, Ethical; SIF – Social Investment Forum; SiRi – Sustainable Investment Research International Group; SRI – Socially Responsible Investing; Sustainable and Responsible Investing; UN – United Nations; UNCED – United Nations Conference on Environment and Development; UNCSD – United Nations Commission for Sustainable Development; UNEP FI – UNEP Finance Initiatives; UNEP FII – UNEP Financial Institutions Initiative; UNEP III – UNEP Insurance

---

Readers should send their comments on this paper to: BhaskarNath@aol.com within 3 months of publication of this issue.

L. Hens and B. Nath (eds.), The World Summit on Sustainable Development, 241–274.
© 2005 *Springer. Printed in the Netherlands.*

Industry Inititative; UNEP – United Nations Environment Programme; VfU – Verein für Umweltmanagement in Banken; VQS – The Voluntary Quality Standard; WBCSD – The World Business Council for Sustainable Development

1. Johannesburg and its means of implementation

The Plan of Implementation of the Johannesburg Summit 2002 is very much a pamphlet with a sustained call for action upon governments, international organisations and other relevant stakeholders (UN, 2002c). It is very much a repetitive appeal for support and efforts, including the provision of financial and technical assistance and capacity building for developing countries.

One of the key outcomes of the 2002 Summit was reported as "the broadening and strengthening of the understanding of sustainable development, particularly the important linkages between poverty, the environment and the use of natural resources" (UN, 2002a). The Summit wants "to take sustainable development to the next level, where it will benefit more people and protect more of our environment" (UN, 2002c).

'The gap between developed and developing countries points to the continued need for a dynamic and enabling international economic environment, supportive of international cooperation, particularly in the areas of finance, technology transfer, debt and trade, and full and effective participation of developing countries in global decision-making' (UN, 2002b).

Agenda 21 (UN CSD, 1992) suggested financial strategies, to fund socially and environmentally responsible development, a key concern of governments, corporations and multilateral development banks. Agenda 21 is a comprehensive plan of action to be taken globally, nationally and locally by organisations of the United Nations (UN) System, Governments, and Major Groups in every area in which human impacts on the environment. Agenda 21, was adopted by more than 178 Governments at the UN Conference on Environment and Development (UNCED) held in Rio de Janeiro, Brazil, 3–14 June 1992. The full implementation of Agenda 21, the Programme for Further Implementation of Agenda 21 and the Commitments to the Rio principles, were strongly reaffirmed at the World Summit on Sustainable Development (WSSD) held in Johannesburg, South Africa from 26 August to 4 September. To realise the internationally agreed development goals, included in Agenda 21, in the Millennium Declaration (UN, 2000), and in the Johannesburg plan of action, significant increases in the flow of financial resources from developed to developing countries are required. Support was given to establish a world solidarity fund for the eradication of poverty and to promote social and human development in the developing countries.

The original cost of implementing Agenda 21 has been estimated at USD600bn. To reach the Millennium Development Goals estimates indicate that it will take on the order of an additional USD40 to USD60 billion a year.

Thus considerable new funds and the creative utilisation of existing resources are needed. At the Summit, the role of the private sector and the individual citizens relative to governments in funding the endeavours, in order to achieve the implementation targets, a has particularly been stressed.

This paper looks at the material needs and the financial means for implementing Agenda 21 and the Johannesburg Earth Summit objectives. It observes the ever growing impact and the potential role of financial institutions and corporate world regarding social responsibility and sustainable development. It presents a short and incomplete picture of some existing key financial instruments, explaining their successes and shortages and points towards the imbalances in the multilateral trading systems and the insecurities in the global financial systems. It entails a personal commentary on the Johannesburg Summit regarding the role of finance for sustainable development.

The paper compares the initiatives and investments of governments *versus* corporations, Non-governmental organisations (NGOs) and other private bodies. It critically evaluates the role of "sustainable bankers", both in understanding their powerful intermediary role regarding globalization and sustainable development and their own "corporate sustainability". It finally highlights the 'forgotten' role of the sustainable and responsible investment community and encourages the sector of SRI research and rating agencies to critically strive to more convergence on the level of research processes, quality assurance systems.

2. Official development assistance and the development banks

Official development assistance (ODA) and Foreign Direct Investment (FDI) are two important tools for funding international sustainable development.

The amount of ODA has fallen steadily from 1990 to less than 0.3% of the Gross national product (GNP) of OECD member countries, under half of the UN global target of 0.7% of the GNP (Annan, 2002b). Even when this target would be met, this assistance level will never be sufficient to deliver the level of development that is required. The ambitious goal of bringing the levels of ODA from developed countries to USD125bn was never materialised.

Because of this deficiency, the role of development banks as financiers of sustainable development became more manifest. The World Bank has become a main financier and started looking beyond the economic considerations and into broader social issues and impacts of its projects. Vice president of environmentally and socially sustainable development Ian Johnson said: "We have shifted the way we work in support of Agenda 21, we have sharpened the poverty focus of our work, expanded support for social services, equitable broad-based growth, good governance and social inclusion and are integrating gender and environmental considerations into development efforts" (Hahn, 2002). The main themes within the World Bank's sustainable development projects – energy use, population growth and environmental impact – are affected by the politics of globalisation. Moreover, critics argue that the World Bank is politically captured by the G8 governments and therefore is a big part of the problem, which in turn stimulated the Bank to adapt its modus operandi.

In 1999 the "Comprehensive Development Framework" was launched, a "holistic approach to development that balances macroeconomic with structural, human and physical development needs" (World Bank, 2002a).

Another example is the introduction of a new measure for countries' sustainability in 1998, the "adjusted net savings" (previously "genuine savings") (Bolt et al., 2002). This indicator builds on the concepts of green national accounts. Adjusted net savings measure the true rate of savings in an economy after taking into account investments in human capital, depletion of natural resources and damage caused by pollution. The "adjusted net savings", together with the "green net national product" and "the wealth accounting" are prominent indicators to link the macroeconomy and the environment (World Bank, 2002b). By developing 'greener' national accounts, the World Bank intends to place environmental problems into a framework that key economic ministries in any government will understand.

During the Summit the world's four regional development banks (African Development Bank, Asian Development Bank, European Bank for Reconstruction and Development Bank, Inter-American Development Bank) stressed their continuing commitment to promoting sustainable development and their belief that the principles of sustainability will be implemented best through a partnership of governments, international institutions, sub-regional organisations, private enterprises, local populations and other stakeholders (Kabbaj et al., 2002). The banks intend to leverage their expertise towards sustainable prosperity either by helping countries to build and undertake sustainable development agendas that will reduce poverty and preserve the environment, or by enhancing transparency, or by financing the private sector to invest responsibly towards development. They also pointed out the need to refine systems to monitor effectiveness of their programmes.

3. Open and equitable multilateral trading and financial systems

A number of problems interfere however with the multilateral projects and efforts of the development banks and are faced in the Plan of Implementation.

A first problem is the radical divergence in the North-South view on main themes as industrial development, population growth and the urgency of the environmental agenda. Next to this the trade barriers block many developing countries from access to lucrative markets. Agricultural subsidies in developed countries are still standing at USD350bn a year (seven times what these countries spend in ODA). Furthermore, there are the problems of corruption, market instability, political insecurity and governance risks.

"Globalisation offers opportunities and challenges for sustainable development. Globalisation and interdependence are offering new opportunities to trade, investment and capital flows and advances technology, including information technology, for the growth of the world economy, development and the improvement of living standards around the world. At the same time, there remain challenges, including serious financial crises, insecurity, poverty, exclusion and inequality within and among societies. Globalisation should be fully inclusive and equitable . . . " (UN, 2002c).

Debt relief or debt cancellation, duty-free and quota-free market access for the least developed countries trade facilitation, technology transfer, education, . . . are just a few other "sustainability investing tools" which the Johannesburg Plan of Implementation mentions.

3.1. PROBLEMS REGARDING SHORT TERM FINANCIAL SECURITY

In preparation to the Earth Summit 2002, investigations have started up toward enhancing global financial security (Gardiner, 2001). "The current financial system has not attained the objectives of poverty eradication, social equality, environmental sustainability and economic growth. Although financial markets profit from a certain degree of volatility, the markets do not take account of the consequential impacts of crises and financial risk on poor communities and the environment, especially in developing countries".

"Financial security is considered as a global public good: no one is required to maintain it but everybody can benefit from it. A reform of the financial system is needed to prevent further recurrence of financial crises and providing better mechanisms for financial crisis management . . . to support and reinforce sustainable development" (Gardiner, 2001).

Some problems associated with short-term financial volatility include:

(i) Escalating financial uncertainty: highly dangerous speculation; leveraged derivatives or hedge funds; dotcom bubble; . . .

(ii) National economic exposure: huge escalation of global capital flows exposing domestic economies to greater financial volatility; sudden outflow of foreign investment; . . .

(iii) Moral hazard: imprudent financial risk-taking by corporations, incited by government support in financial crises;

(iv) Market saturation: competitiveness is increasingly tough, causing cross-border consolidations of financial institutions to boost profits and dividends, with an increased rate of innovation of new financial products; derivatives markets deal now with 47 000 different kinds of options.

3.2. PROBLEMS REGARDING LONG TERM FINANCIAL SECURITY

Other factors impact the long-term financial security:

(i) Misgovernance: both corruption and internal conflicts not only undermine financial institutions but also the international credibility of a national financial system, deterring both domestic and foreign investment;

(ii) Debt: the on-going debt burden continues to undermine the long-term development of many developing and transitional countries;

(iii) Omission of social and environmental factors: financial insecurity studies focus on the financial and macroeconomic conditions related to crashes. There are few integrated assessments of the social and environmental factors underlying and affected by financial insecurity. The environment is a key component of financial security. Environmental degradation and pollution have implications for social welfare and the economic conditions underpinning domestic financial stability. Educated, skilled, healthy and happy people are key drivers of economic growth and financial stability. This again requires broad and long-term investments in social needs.

Table I lists some of the key financial instruments that are currently used to maintain short and long term financial security (Gardiner, 2001). Some of these instruments are commented in this text.

4. Foreign direct investments

Kofi Annan, UN Secretary-General, vented his frustration with the slow government decision-making at the Johannesburg Summit by urging business to press ahead with development initiatives. He told corporate delegates not to wait for governments to make decisions and laws to promote development in the world's poorest countries and environmental protection. "We realise that it is only by mobilising the corporate sector that we can make significant progress" (Annan, 2002a).

4.1. THE ROLE OF CORPORATIONS

The corporate community was most visible at the Johannesburg Summit. The presence of corporations was large compared to the Rio Summit. Corporations, showed active participation in partnerships. Business Action for Sustainable Development (BASD) refers to the promotion of corporate responsibility and accountability through "development and implementation" of intergovernmental agreements (BASD, 2002; WBCSD, 2001). BASD is an initiative of the International Chamber of Commerce (ICC) and the World Business Council on Sustainable Development (WBCSD). The concept of partnerships between governments, business and civil society was given a large boost by the Summit and the Plan of Implementation.

Business representatives did advocate a free trade approach to alleviate environmental problems (such as the Kyoto protocol) and did favour voluntary measures. NGOs did favour a regulation-based framework and binding rules, not impressed by voluntary and non-committal initiatives that do not have any monitoring or enforcement mechanisms.

A number of instruments facilitate the corporate financial and non-financial world to take up their role.

TABLE I. Existing financial instruments for maintaining financial security in the short and the long term (Gardiner 2001).

Mechanism	Target		Scope	
	Stakeholder	Aim	Short term	Long term
Supplement reserve facility (IMF)	Government	Capital liquidity	+	
Special drawing rights (IMF)	Government	Capital liquidity	+	
Exchange rate mechanisms	Government	Efficient currency transaction		+
Official Development Assistance	Government	Social, economic, environmental development		+
Contingent credit lines	Government Corporate	Debt relief, capital liquidity, economic growth	+	+
Debt workouts (rescheduling, cancellation, standstills)	Government Corporate	Debt relief	+	
Debt for nature swaps	Government Corporate	Debt relief, environmental protection	+	+
Tradable permits (e.g. carbon credits)	Government Corporate	Environmental efficiency		+
Financial screening	Corporate	Social responsibility, environmental management, risk management	+	+
Credit guarantees (government & international sources)	Corporate	Reduced financial risk, economic growth	+	+
Micro-credit (government, NGO, international sources)	NGO SME	Economic growth, poverty eradication		+
Alternative currencies (e.g. LETS, Time banks)	NGO SME	Domestic economic growth, poverty eradication	+	+
Global Environment Facility (WB, UNEP, UNDP)	Government NGO	Environmental enhancement		+
Foreign private investment (direct; portfolio investments, bonds)	Government Corp., NGO	Economic growth	+	+
Domestic private investment	Government Corp., NGO	Economic growth	+	+

4.2. FDI VERSUS ODA

In the decade since the Rio Summit, commercial credits and FDI indeveloping economies have grown substantially, to the point where the influence of the private sector eclipses that of public institutions such as the development banks and the International Monetary Fund, worrying many that they focus on large-scale industrial projects that do not meet the principles of sustainable development.

In 1990, for every dollar of long-term ODA, less than one dollar in long-term capital did flow from the private sector (Gardiner, 2000). Today, for every dollar ODA there are four dollars FDI.

This is the working field of the International Finance Corporation, the private-sector lending arm of the World Bank. The IFC did examine its own experiences on the level of sustainability sensitisation and produced a "practical guide for change" in order to prove the business case for its corporate clients in developing countries and to help them to couple profits with sustainability (Anon., 2002b; Armstrong, 2002).

The vast majority of the FDI have gone to just ten middle-income countries and is heavily concentrated in a limited number of industries. The African countries received only about 1% of the global flows. The capital flow was highly volatile, especially during the financial crash in the Asia-Pacific region. Political risksremain as the primary risk for private sector investment. It has been suggested that combining ODA and FDI approaches could lower political risks, which could in turn vitalise FDI (Vitalis, 2002). Generally spoken projects aim to improve institutional infrastructure and governance (Ribbans, 2002) e.g., by capacity building programs to improve the skills of civil servants to attract sustainable investment, by promoting tax incentives for developing countries' investment funds, by providing technical assistance and seed funding for project development, by strengthening managerial skills of micro-credit institutions, . . .

The challenge anyway is to attract more private sector resources to developing countries and to channel it to activities that support sustainable development efforts.

5. Micro-finance

Poor people have a variety of financial needs and the informal sector has been quicker to respond to such needs than traditional banks by providing tailor made remittances, leasing, and insurance services as well. The experience of many countries shows that micro-financing can empower individuals and the informal business sector, thought to be uncreditworthy (Anon., 2002a). Micro-finance therefore can well foster sustainable livelihoods and generate substantial non-financial benefits. This means that micro-finance not only proved that borrowers are able to pay their credit, but also that a dynamic cycle of sustainable livelihoods is created within a community, facilitating the collective ability to pay for clean water, to secure electricity and to significantly save time, benefit education and study during evenings, creating health and enabling better economic opportunities and productivity. Once marginalised communities achieved independence through economic empowerment, resulting in significant socio-economic and environmental benefits. Today micro-credit loans reach nearly 30 million borrowers and are growing rapidly, benefiting of its high repayment rates.

A variety of financing approaches could assist micro-finance institutions: the introduction of public or private equity capital (e.g. Deutsche Bank Microcredit Development Fund) or cost recovery systems compensating the extra costs of making small loans, are just two examples.

6. Generating resources in the public sector

It is often more effective to improve the efficiency of existing resources than to look for additional revenues coming from the private sector In some cases removing subsidies that are expensive and often environmentally harmful is one way to do this. Energy subsidies e.g. do encourage wasteful consumption and have a negative impact on the local environment and the global climate. By 2001, the combination of subsided tariffs and subsequent wasteful use of energy, together with high technical losses and widespread non-payment resulted in total losses of more than USD5bn. every year. It was calculated that if losses were reduced by only one-third, the subsequent savings would be sufficient to fill every teacher vacancy in India and to provide every school with running water and toilets (Anon., 2002a). Similar calculations and conclusions apply to reducing the water subsidies, both for irrigation and domestic use, although much more sensitive and complex to handle (Anon., 2002a).

Generating additional resources has a huge potential, depending significantly on the local context. A few mentioned mechanisms are the so-called "capturing of natural resource rents", especially timber rents in a limited number of countries; the "charging for services", e.g. based on tourism.

The potential of environmental taxes and charges to generate revenue for the public sector while simultaneously discouraging environmentally harmful behaviour has been a topic of growing interest. Earmarking the revenues, by using the green levies to provide basic environmental services, is also gaining increasing acceptance. Barriers to introduce environmental taxes, such as the lack of capacity for designing and administering these taxes in developing countries, should not be neglected however.

7. Sustainable bankers and insurers pushing the codes

Although the financial sector responded more slowly than other sectors to the Rio call for action – perhaps considering itself an environmentally friendly industry – quite some achievements exist in the field of environmental legislation, liability, accountancy and reporting, and in the field of socially and environmentally responsible investing.

It seems however that the banking sector, the insurance industry, the capital market, and the stock market have just begun to fully understand their powerful intermediary role regarding globalisation and sustainable development.

7.1. UNEP STATEMENT BY FINANCIAL INSTITUTIONS

Back in 1992, the UN Environment Programme (UNEP) Financial Institutions Initiative on the Environment was founded (originally known as the 'Banks Initiatives') to engage a broad range of financial institutions – from commercial banks to investment banks to venture capitalists to multilateral development banks and agencies – in a constructive dialogue about the nexus between economic development, environmental protection, and sustainable development (UNEPFI, 1997b) (see Box 1). The initiative promotes the integration of environmental considerations into all aspects of the financial sector's operations and services, starting

Box 1. UNEP Statement by Financial Institutions on Environment and Sustainable Development (Revision 1997, abbreviated) (UNEPFI, 1997b).

Commitment to sustainable development

1. We regard sustainable development as a fundamental aspect of sound business management.
2. We believe that sustainable development can best be achieved by allowing markets to work within an appropriate framework of cost-efficient regulations and economic instruments.
3. We regard the financial services sector as an important contributor towards sustainable development, in association with other economic sectors.
4. We recognise that sustainable development is a corporate commitment and an integral part of our pursuit of good corporate citizenship.

Environmental management and financial institutions

1. We support the precautionary approach to environmental management.
2. We are committed to complying with local, national, and international environmental regulations applicable to our operations and business services. We will work towards integrating environmental considerations into our operations, asset management, and other business decisions, in all markets.
3. We recognise that identifying and quantifying environmental risks should be part of the normal process of risk assessment and management, both in domestic and international operations.
4. We will endeavour to pursue the best practice in environmental management, including energy efficiency, recycling and waste reduction. We will seek to form business relations with partners, suppliers, and subcontractors who follow similarly high environmental standards.
5. We intend to update our practices periodically to incorporate relevant developments in environmental management.
6. We recognise the need to conduct internal environmental reviews on a periodic basis, and to measure our activities against our environmental goals.
7. We encourage the financial services sector to develop products and services which will promote environmental protection.
   (...)

from producing a formal environmental policy statement and making this publicly available, to building environmental risk assessments into theircredit decisions and paying close attention to their own corporate ecology. The original six signatories have grown to almost two hundred financial institutions. It is striking however, that some banks are signatories to the UNEP declaration, but do not even report any environmental policy objectives.

7.2. UNEP STATEMENT BY THE INSURANCE INDUSTRY

In collaboration with UNEP FII a group of leading insurance, reinsurance and pension fund companies has developed a Statement ofEnvironmental Commitment for the Insurance Industry (UNEPFI, 1997a) (see Box 2). Ninety-two signatories representing 27 countries did commit to "achieve a balance between economic development, human welfare and the environment".

Box 2. UNEP Statement by the Insurance Industry on Environment and Sustainable Development (abbreviated) (UNEPFI, 1997a).

General principles of sustainable development

(…)

3. We regard a strong, proactive insurance industry as an important contributor to sustainable development, through its interaction with other economic sectors and consumers.
4. We believe that the existing skills and techniques of our industry in understanding uncertainty, identifying and quantifying risk, and responding to risk, are core strengths in managing environmental problems.
5. We recognise the precautionary principle, in that it is not possible to quantify some concerns sufficiently, nor indeed to reconcile all impacts in purely financial terms.

Environmental management

1. We will reinforce the attention given to environmental risks in our core activities. These activities include risk management, loss prevention, product design, claims handling and asset management.
2. We are committed to manage internal operations and physical assets under our control in a manner that reflects environmental considerations.
3. We will periodically review our management practices, to integrate relevant developments of environmental management in our planning, marketing, employee communications and training as well as our other core activities.
4. We encourage research in these and related issues. Responses to environmental issues can vary in effectiveness and cost. We encourage research that identifies creative and effective solutions.
5. We support insurance products and services that promote sound environmental practice through measures such as loss prevention and contract terms and conditions. While satisfying requirements for security and profitability, we will seek to include environmental considerations in our asset management.
6. We will conduct regular internal environmental reviews, and will seek to create measurable environmental goals and standards.

(…)

It is known that the climate change threatens the financial viability of this industry. Especially since 1998 there was an enormous increase in the insured damage. Because of the relationship between these losses and climate change, the insurance industry became involved in the Kyoto debate and is in constant contact with a broad range of stakeholders who share similar concerns, namely environmental groups and climate scientists. The Kyoto Protocol is the global climate treaty that was signed and approved by 159 countries in December 1977 at the UN Framework Convention on Climate Change (FCCC) in Kyoto (UN FCCC, 1997). Today, the total number of ratifications, accessions or acceptances rose to only 97.

The non-binding, voluntary policy of the UNEP Insurance Industry Initiative attracts criticism that it is merely a public relations tool by which signatories can be seen to be doing something without having to commit to change. UNEP III however has been promoting a corporate carbon dioxide indicator for insurers and has initiated the concept of standardised corporate greenhouse gas inventories, which has lead to the development of a Global Warming Indicator (GWI), replacing the wide array of environmental reporting standards across the industry (Dunstan, 2000). A number of initiatives have been taken, including the development of the Greenhouse Gas Indicator (Thomas et al., 2000) and the Carbon Disclosure Project (CDP) (see further).

8. A survey on the state-of-the-art sustainability and banking

Through their lending, investment and insurance practices, and through their intermediary position in the economy, the impact of banks and insurance companies is potentially very high in promoting sustainable economic growth. One of the observers of the changing attitudes of financial world in challenging sustainability is Marcel Jeucken, a senior economist at Rabobank (Jeucken, 2001; Bouma, 2001; Jeucken, 2002; UNEPFI, 2002).

Jeucken explored the current state-of-the-art on the sustainability activities of thirty-four leading banks around the world. He scored them on criteria in five broad categories: communication, information, environmental finance, special products and social issues. The survey did highlight important differences between regions, countries and banks. The majority of the banks adopts a defensive position towards the environment (the prevailing indicator). He observed ten pro-active "front-runners", six 'followers' and eighteen 'stragglers'. In terms of sustainability strategy Jeucken distinguishes between "heavily defensive" (ignorant), "slightly defensive" (starting to act), 'preventive' (reducing costs by environmental care and lowering risks by environmental risk assessment), 'offensive' (preventive banking, designing requested products), 'sustainable' (trans-commercial, holistic objectives). With an environmental report observation period from 1998 to 2000, the overview might be outdated and the geographical spread might be limited, but Jeuckens conclusions seem not to lack validity. "Though the banking sector has been slow to pick up the challenge if sustainability, change is underway. However, listening to and focussing on the pro-active banks – such as Crédit Suisse, UBS, Rabobank and Deutsche Bank – will gain the wrong impression that the banking sector is well underway. This is a faulty observation: a large group of banks still do not see the role they can play and maybe should play towards a sustainable development. The main herd of banks is largely inactive. The business case still needs to be proven to these banks" (Jeucken, 2002). Jeucken's listing of "proactive banks" remains arguable however, due to a limited scope of indicators on the sustainability responsibilities of financial institutions and is not in line with the analysis of CSR research groups.

8.1. THE LONDON PRINCIPLES

The London Principles project was presented at the Summit as one of the Type 2 partnerships (Pearce, 2002) (see Box 3). Eleven European companies and investment institutions (have signed a set of seven principles intended to clarify how financial markets can encourage sustainable development. The London Principles on Sustainable Finance, launched by the Corporation of London, commit

---

Box 3. The London Principles (Pearce, 2002).

The London Principles . . .

. . . for Economic Prosperity:

Principle 1. Provide access to finance and risk management products for investment, innovation and the most efficient use of existing assets.
Principle 2. Promote transparency and high standards of corporate governance in themselves and in the activities being financed.

. . . for Environmental Protection:

Principle 3. Reflect the cost of environmental and social risks in the pricing of financial and risk management products.
Principle 4. Exercise equity ownership to promote efficient and sustainable asset use.
Principle 5. Provide access to finance for the development of environmentally beneficial technologies.

. . . for Social Development:

Principle 6. Exercise equity ownership to promote high standards of corporate social responsibility (CSR) by the activities being financed.
Principle 7. Provide access to market finance and risk management products to businesses in disadvantaged communities and developing economies.

signatories to provide greater access to financial products for the socially excluded and to "use equity ownership to promote CSR".

Aside from the principles themselves, the Corporation of London has also published case studies of best practice in sustainable development by financial firms. Also, a mechanism to ensure signatories demonstrate 'continual progress' against the principles is in development.

Criticisms on the London Principles point to the fact that the seven principles are vague, that, a lot of best practice and another set of voluntary principles are not needed, and that they coincide with a duplication of the UN Environment Programme's Financial Initiative. The UNEP initiative however focuses more on environmental matters, where The London Principles intend to look specifically at the role of the financial services sector in terms of sustainable development.

The London Principles apply to all aspects of finance and not just value-based investments and banking niches. They are aspirational and seek to encourage continuous improvement. Signatories will report annually on progress towards their implementation.

8.2. WBSCD JOINT STATEMENT AT JOHANNESBURG SUMMIT

At a side event organised by the WBCSD and UNEP, another eleven chairmen of financial companies issued a joint statement urging the sector to become more involved with sustainable development "By taking the environmental and social aspects into account when conducting our business, we can reduce risks, further improve our bottom line and create long-term value" (Smith, 2002). The coherence and relevance of the firm statement broke apart, where the declaration was questioning itself to which extent the financial sector should involve itself in sustainable development. "Is the financial sector merely the intermediary in creating finance and development or is it an agent of change?"

8.3. RIOC10 FINANCE COMMITMENTS

A set of ambitious proposals by Tessa Tennant, the "Rio+10 Finance Commitments", was prepared for the Summit (Baue, 2002) (see Box 4). These guidelines would open SRI – in principle, to every citizen of the planet through responsible lending policies, high impact community investments, micro-finance schemes and a fine blend of government and financial sector commitments regarding sustainability statements, publicly availability of social and environmental data, harmonised disclosure and accounting standards.

8.4. THE SiRi GROUP SUGGESTED ACTIONS FOR DISSEMINATING SRI-PRACTICES

Offstage another action agenda has been proposed by the Sustainable Investment Research International Group (SiRi) and Fundación Ecología y Desarrollo, calling

---

Box 4. The Rio+10 Finance Commitments (Baue, 2002).

1. A social investment option available to every saver on the planet.
2. Not less than 3% of assets to community investment by every SRI fund.
3. Micro-finance schemes available to every citizen on the planet.
4. The adoption of borrowing principles by lending institutions and their clients.
5. Ambitious investment programmes for renewable energy, mass transit infrastructure and sustainable housing and agriculture.
6. Pensions law in every nation requiring a sustainability impact statement of fund investments.
7. Environmental and social indicators to be disclosed as a legal requirement in company accounts.
8. Universal methodologies for calculating these indicators adopted as international accounting standards.
9. Country ratings for sustainability.

---

governments, civil society organisations, trade unions and financial institutions to adopt a wide range of policies to encourage SRI and CSR and to effectively decide on the sustainable use of their own financial resources (Peyo, 2002) (see Box 5). The SiRi Group was formed in 2000 and comprises eleven specialised SRI research organisations based in Europe, North America and Australia, a research base of over 100 specialist SRI analysts worldwide. SiRi Group members provide SRI research on corporations based in their own home markets, but with consistent content, in a standardised format and with harmonised quality standards, giving clients the benefits of global coverage based on local knowledge.

9. Environmental care, CSR and accountability

There is a growing awareness in the financial sector that the environment brings risks and opportunities. In the United States, under the Comprehensive Environmental Response, Compensation and Liability Act (CERCLA), banks can be held directly responsible for environmental pollution of clients, which made them alert for the environmental factor within the credit risks. European banks focused on the issue in terms of designing and marketing of environmental friendly financial products. The UNEP statements reflect these concerns.

9.1. SUSTAINABILITY REPORTING, GRI AND AA1000

About 45% of the top 250 companies of the Fortune 500 (GFT250) publish environmental, social or sustainability reports in addition to their financial reports (Molenkamp, 2002). Moreover, the number of reporting companies within the financial sector increased dramatically by around 60% in the period 1999–2002.

Box 5. The SiRi Group suggested actions for disseminating SRI-practices (abbreviated) (Peyo, 2002).

Civil society organisations are encouraged to:

- implement formal SRI policies to their own investments and savings, according to their own values;
- select their financial services suppliers according to the same SRI criteria;
- promote among their sympathisers the socially responsible savings and investments products;
- engage with financial institutions to broaden their supply of sustainable and responsible financial products.

Trade unions are encouraged to:

- implement formal SRI policies to their own investments and savings, according to their own values;
- select their financial services suppliers according to the same SRI criteria;
- promote among their associates the orientation of their savings and investments towards socially responsible business practices;
- engage with financial institutions to broaden their supply of sustainable and responsible financial products;
- play an active role in promoting a SRI approach among pension funds managers.

Financial institutions are encouraged to:

- commit to the integration of environmental considerations into all aspects of their operations, as stated by the UNEP FII commitments;
- include, driven by self interest, social responsibility criteria in all their financial products;
- pay attention to client motivations and concerns about sustainability, in designing and offering saving and investment products;
- complement their financial analysis with social and environmental assessments;
- use their power as institutional investors to engage with companies in order to promote socially responsible business practices.

Regulators are encouraged to:

- condition public subsidies to companies to the adoption of formal CSR policies;
- stimulate companies' adoption of transparency criteria regarding environmental and social practices, so that investors and consumers can take investing or purchasing decisions according to CSR standards;
- include disclosure regulation in the existing legislation, requiring that financial products disclose to which extent social, environmental or ethical considerations are taken into account;
- encourage and support, through multi-stakeholder partnerships, research and education with regard to SRI and CSR.

"The finance sector is increasingly assessing companies on their sustainability risks as part of their decisions", says Molenkamp. "Many companies are eager to be listed on new indices such as the Dow Jones Sustainability Index or the FTSE4Good Index, which in turn raises their profile on the financial markets. The SRI sector has experienced dramatic growth during the past few years. If this trend continues, sustainability will become a major deciding factor for access to equity capital and investments. Corporate sustainability reporting will therefore from a business point of view become increasingly important. Initiatives like the Global Reporting Initiative (GRI) to harmonise corporate sustainability reporting and make performance of companies better comparable will enhance this".

The GRI was established in late 1997 with the mission of developing globally applicable guidelines for reporting on the economic, environmental, and social performance, initially for corporations and eventually for any business, governmental, or NGO (GRI, 2002). The aim of the guidelines is to enable them to prepare comparable "triple bottom line reports". The third version, the new 2002 Sustainability Reporting Guidelines, was formally released on the Johannesburg Summit.

Founded by the Coalition for Environmentally Responsible Economies (CERES), the GRI is an official collaborating centre of the UNEP and works in cooperation with the UN Secretary-General Kofi Annan's Global Compact (UN, 2002b).

The GRI incorporates the active participation of corporations, NGOs, accountancy organisations, business associations, and other stakeholders from around the world. GRI is now established as a permanent, independent, international body with a multi-stakeholder governance structure and located in Amsterdam. Its core mission is the maintenance, enhancement, and dissemination of the GRI Guidelines through a process of ongoing consultation and stakeholder engagement.

Specific GRI-guidelines for the financial sector are being produced.

The Global Compact is a UN-sponsored platform for encouraging and promoting good corporate practices and learning experiences in the areas of human rights, labour and the environment. It is an entry point for the business community to work in partnership with UN organisations in support of the principles and broader goals of the UN, and provides a basis for structured dialogue between the UN, business, labour and civil society on improving corporate practices in the social arena.

AccountAbility 1000 (AA1000) is an accountability standard focused on securing the quality of social and ethical accounting, auditing and reporting. At its launch in Denmark in 1999, AA1000 provided the first systematic stakeholder based approach to accountability. Since then it has been used by businesses, nonprofit organisations and public bodies in framing corporate responsibility policies, stakeholder dialogue, auditing and verification of public reports and professional training and research.

In June 2002 AA1000 published a consultation document of the first of the five modules, known as the AA1000S, which form an updating of the entire AA1000 process (ISEA, 2002). This first module, the AA1000 Assurance Standard, Guiding principles, is under discussion, with the final document scheduled for publication in early 2003.

9.2. The frameworks of EMAS, ISO, VFU, EPI, and ABI

Formal environmental management systems, that systematically reduce the environmental impact of the internal processes are generally lacking. No bank has achieved The EU Eco-Management and Audit Scheme (EMAS) or British Standard 7750 (BS7750), a few banks have an International Organisation for Standardisation (ISO) 14000 certification on a local, national or global level, for environmental management systems. The first insights in the new ISO standards on CSR are expected for the year 2003.

VfU (Verein für Umweltmanagement in Banken), initiated by a few German and Swiss banks, developed a standard for internal environmental care (Schmid-Schönbein et al., 2002), later complemented by a standard for external environmental care, the Environmental Performance Indicator (EPI) Finance 2000 (Schmid-Schönbein and Braunschweig, 2000). Additionally, social performance indicators have recently been launched (SPI-Finance, 2002).

Within about the same period of time two British banks were involved in the development of a series of guidelines on environmental management and reporting, together with the FORGE Group (FORGE, 2000). Recently, this FORGE Group has published detailed guidelines on CSR management and reporting for the financial services sector in response to growing demands for financial services companies to demonstrate CSR (FORGE, 2002). The guidance centres provide a practical toolkit to address CSR by identifying priority CSR issues for the sector, providing best practice guidance on developing and implementing a CSR management and a reporting framework. They equally present action plans for incorporating CSR consideration into the design, management and delivery of financial products. The Guidance is often called the Association of British Insurers "(ABI)-Guidelines", referring to the project partner, the ABI.

10. Sustainable and responsible investments

Since Rio, the global asset management industry has been booming due to the IT expansion, has been collapsing due to the Asia crisis and has been imploding due to the dotcom bubble, before loosing a lot of its credibility due to the corporate governance crisis. One of the ever growing market niches however has been SRI. In fact, SRI has been the biggest success story related to sustainability in financial world since the Rio Summit.

Despite its success, SRI has not even been mentioned in the Plan of Implementation of Johannesburg. The potential of SRI had well been emphasised in Rio. Back in 1992, the "Rio Resolution on Social Investment" (Miller et al., 1992) (see Box 6) communicated by the international investment community, described their role in achieving sustainable development.

With a Plan of Implementation, calling for "mobilizing international and domestic financial resources", and "looking for mechanisms that do not distort international trade and investment", and "stressing the role of international financial

---

Box 6. The 1992 Rio Resolution on Social Investment (Miller et al., 1992).

We call on individuals to question whether the investment policies of their mortgage, loans, insurance, savings and/or pension plans support the objectives of sustainable development.

We call on companies and NGO to ensure that their policies enable investments to be managed for environmental and social benefit.

We call on financial institutions to begin the process of integrating environmental and social considerations into the investment analysis process and to have particular regard for investments in emerging capital markets.

We call on governments to introduce incentives for private capital to invest in community enterprises that support low-cost housing, small business start-ups, education, sustainable agriculture and other projects, which enhance the common good.

---

institutions", while "employing market-based incentives", this negligence is hard to believe. With its plea for "Changing unsustainable patterns of consumption and production", for "Enhancing corporate environmental and social responsibility and accountability", for "Encouraging dialogue between enterprises and the communities in which they operate and other stakeholders" and for "Encouraging financial institutions to incorporate sustainable development considerations into their decision-making processes", one by one key issues in SRI world, it was a surprise.

What exactly are the achievements of SRI?

10.1. SUCCESS IN THE MARKET

In the United States nearly one out of every eight dollars under professional management is reported to be involved in "social and environmentally responsible investing" (SIF, 2001). This means that nearly 12% of all assets under management reside in a professionally managed portfolio utilizing one or more of the three strategies that define SRI in the United States: screening, shareholder advocacy and community investing. The screening strategy is most often based on negative criteria. It is avoidance investing. According to the biannual Social Investement Forum (SIF) Industry Research Program there were 230 mutual funds incorporating social screening (SIF, 2001). The growth rate for socially screened portfolio assets was more than 1.5 times that of all professionally managed assets.

On December 31st 2001 there were 280 green, social and ethical funds operating in Europe, with a particular high growth rate since two years. SRI in Europe is one of the most dynamic and rapidly growing activities in the funds industry (Avanzi/SiRi Group, 2002). In terms of number of funds four leading countries – the UK, Sweden, France and Belgium – account for 68% of the SRI market. In terms of asset under management five countries – the UK, Sweden, the Netherlands, Italy and Belgium – reach 80% of the total SRI market (Figure 1).

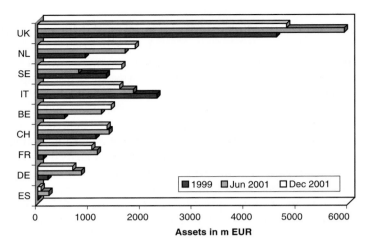

Figure 1. Green, social and ethical funds in Europe: assets under management (m EUR) per country (Avanzi/Siri Group, 2002. In cooperation with CSR Europe).

The relative weight of SRI funds in relation to the total European market of UCITS funds (collective investment funds complying to the EU UCITS Directive) is only 0.40% and this relative weight has decreased sharply during 2001. In Belgium, the leader in this ranking, the relative weight grew from 1.47% to 1.70%, followed by the Netherlands where the weight grew from 1.45% to 1.61% (Avanzi/SiRi Group, 2002).

SRI has a limited impact on FDI. SRI however is rapidly expanding in Latin America and especially through the Asia Pacific, with the launch of SRI in Asia (ASrIA), the Association for Sustainable & Responsible Investment in Asia (ASrIA, 2002). By this growth some influence and the introduction of social criteria in emerging markets may be expected. The growth of the SRI market in itself is not the most important part of the success. The SRI community did have a much larger indirect impact through the fertilisation of the CSR debate by providing terminologies, methodologies and metrics in order to study, analyse, evaluate and rate CSR behaviour, by providing full risks and opportunities assessments to financial analysts, by enriching the discussion concerning reporting guidelines, and the development of self assessment tools, management audit systems . . .

SRI does contribute to the promotion of CSR, through the development by rating agencies of criteria and indicators that identify the factors of competitive advantage and business success of socially responsible practices.

10.2. PERFORMANCE

For most investors a financial underperformance of sustainability investment products is unacceptable. A key issue within the SRI community, for a number of years, is the question if screening companies against social, environmental and ethical criteria ("SEE screening") helps asset managers to outperform the broader market.

Recent evidence (Pearce et al., 2002) provides a detailed and balanced picture. In theory SEE screened funds are subject to higher risks, because they limit the number of stocks in which they invest. On the other hand they use important information not well understood by the broader market, which may result in out-performance. Evidence shows that the potential "SEE effect" appears to balance for the cost of lower diversification in an SRI portfolio. SRI funds do not show lower returns, as shown by dozens of research reports produced during the last few years. Quite some studies poorly discriminated between different 'generations' and qualities of SRI screening methodologies and most of the research was carried out on passive screened indexes and funds.

The more forward-looking, "new generation" approaches, show that a shareholder-value focus makes out-performance well possible. The question is not does being green and socially pay, but when does it pay? This would be the advantage to investors of 'engaging' with companies to encourage them to behave as responsible corporate citizens. A number of arguments are seen by analysts, which explain the advantages of spending on CSR.

(i) There might be "first-mover" advantage for companies that are proactive in CSR issues. (ii) CSR is considered as an important indicator for good overall management. (iii) CSR can create competitive advantage through reputation and stakeholder management. (iv) CSR can create value by stimulating process and product innovation and better market knowledge.

According to Pearce (Pearce et al., 2002) the majority of studies carried out in the 1970s, 1980s and 1990s found a correlation between CSR performance and financial performance, however, without proving a causal relationship. Recent studies would suggest that the financial out-performance has been caused by non-CSR factors after all. Past research rejects the claim that being green and socially responsible always pays. Recent evidence shows that CSR can create shareholder value for some issues, in some industries, with some firms and for some management strategies. This would suggest that a discretionary approach in engaging with companies to improve their CSR performance could improve financial performance (Pearce et al., 2002).

10.3. MAINSTREAMING AND CONVERGENCE

"SRI research puts pressure on companies. Constant requests of information by rating agencies leads to improved data collection and data management within the companies. This in turn strengthens the companies' abilities to identify and tackle social and environmental problems" (Kahlenborn, 2002). The recent examples of abuses of power show the need to expand this role into mandatory driven systems to control corporate governance issues.

Next to this role as corporate watchdog, the SRI community has fertilised the CSR debate by providing terminologies, methodologies and metrics in order to study, analyse, evaluate and rate CSR behaviour. SRI did contribute to the promotion of CSR, through the development by rating agencies of criteria and indicators that identify the factors of competitive advantage and business success of socially responsible practices.

The coin has two sides. Where investors and SRI rating agencies call companies to improve disclosure and transparency practices . . . companies call for more transparent research and evaluation methodologies.

There is a huge divergence in SRI research standards. Standardisation of SRI rating methodologies however is considered to be desirable to allow users to compare different companies' ratings. CSR practices and instruments will be more effective if they are part of a concerted effort and based on clear and verifiable standards. A few studies have been screening companies and have expressed concern, while defining best practices characteristics (Mistra, 2001). "It is recommended that SRI research should achieve greater independence, standardisation and transparency. SRI research should be independent because SRI analysts should represent all stakeholders. Only truly independent, objective information about corporate social and environmental performance will allow all stakeholders to ascertain the true value of companies" (van den Brink, 2002).

Quality assurance is one of the steps to be taken towards the mainstream. The launch of The Domini Index in 1990, the first social index raised the visibility of SRI considerably and enforced greater confidence by showing good performance. In 1999, Dow Jones and Sustainable Asset Management launched the first series of sustainability indices, the Dow Jones Sustainability Indices. In 2000, the FTSE4Good Series indices, taking CSR as a core investment value was launched. The mentioned indices have often been criticised for their non-transparent and very lightish green approach. Another series of more fundamental sustainability benchmarks, the Ethibel Sustainability Index (ESI), was launched in 2002 and groups the ESI Global, ESI Americas, ESI Europe and ESI Asia Pacific (Box 7). The ESI is building on more than ten years research and experience with the Ethibel label system, a European collective quality label system for sustainable investing (Ethibel, 2002) (Box 7).

"Although SRI has reached the mainstream, it is still not mainstream" (UNEPFI, 2002).

10.4. ENGAGEMENT

A survey on the way in which investors in the UK and US markets engage with companies about their SEE performance showed that this is considered as associated with financial performance and good corporate governance. This is why SEE performance is integrated in the engagement process, both by large investors and their advisors. Protection of the shareholder value is the principal objective. Soft, behind-the-scenes dialogue and collaboration is perceived as the most effective option; public campaigns are avoided. Improving information disclosure is seen as an important aim. The relational investing model (long-term shareholdership to

Box 7. The ESIs, the corresponding label and the financial constituents (Ethibel, 2002).

The Ethibel Sustainability Index

- The ESI provides a comprehensive perspective on the financial performance of the world's leading companies in terms of sustainability for institutional investors, asset managers, banks and retail investors. This set of indexes was first published on June 27, 2002.
- Standard & Poor's is maintaining and calculating the ESIs. The Indexes are designed to approximate the sector weights on the S&P Global 1200, but the selection remains the exclusive responsibility of ETHIBEL.
- They are free-float weighted indexes containing the pioneer and best-in-class companies with respect to sustainability and CSR across sectors and regions.
- The ESI groups four regional indexes: ESI Global, ESI Americas, ESI Europe and ESI Asia Pacific. Each of the ESI indexes is calculated as price and total return indexes in both USD and EUR giving a total of 16 indexes.
- All the constituents in the Index are included in the Ethibel Investment Register, which is a broader list of sustainability leaders around the world that have passed Ethibel's proprietary screening methodology and criteria. As of December 2002 the ESI Global included 162 components.

The Ethibel Label for Sustainable and SRI funds

- The Ethibel Label is only attributed to 4t generation investments funds using investment values out of the Ethibel Investment Register.

(Continued)

Box 7. *Continued*

- Stock selections are conducted following a harmoni-ed but dynamic methodology with 64 fully detailed criteria, 97 indicators and 320 precise benchmarks and ratings (see e.g. Charlier et al., 2001).
- The research processes are under strict quality control. Independent expert advice and evaluations are externally organised.
- All the aspects of the sustainability and the social responsibility of a company, including its social, environmental and economic-ethical policy, are taken into account.
- A permanent dialogue with all the stakeholders is maintained, through the data collecting process, the evaluation process and the periodical reviews of the research methodology.

The ESI 'Financial' Constituents (20/12/02)

- Best of class "Financials – Banks" include: Abbey National (UK); Bank of America (US); Bank of Montreal (CA); Bank of Nova Scotia (CA); Barclays (UK); Commerzbank (DE), Dexia (BE); FleetBoston Financial Corporation (US); HBOS (UK); Lloyds TSB Group (UK); National Australia Bank (AU); Sumitomo Mitsui Bank (JA); Swedbank (FöreningsSparbanken) (UK); Unicredito Italiano Group (IT); Westpac Banking Corporation (AU).
- Best of class "Financials – Diversified Financials" include: Federal National Mortgage Association (US); ING Group (NL).
- Best of class "Financials – Insurance" include: Münchner Rück (Munich Re) (DE); Réassurances (Swiss Re) (De); Rentenanstalt/Swiss Life (CH); Skandia (Forsakrings) (SE); Unumprovident Corp. (US).
- Additional "Financials" in the Ethibel Investment Register include: Hachijuni Bank (JP); Millea Holdings (JP).
- No constituent "Financial" has been awarded a "first class" or 'pioneer' nomination based on CSR performance up till now.

---

increase influence) is not seen as an important tool for engaging on SEE performance (Pearce and Ganzi, 2002).

10.5. THE CARBON DISCLOSURE PROJECT

A potentially powerful engagement initiative has been taken by thirty institutional investors, that drove up the pressure on companies to improve disclosure on their social and environmental impacts. The "Carbon Disclosure Project" (CDP) wrote to 500 largest corporations asking them to quantify their greenhouse gas emissions, to identify the business implications of their exposure to climate-related risks and to communicate their plans for reducing them (CDP, 2002).

This, the investors claim, is needed to allow them to make long-term investment decisions, not only for their SRI portfolios, but also for other investments they manage. According to CDP climate change presents shareholders with a prospective material risk to the value of their investments. Due to fiduciary duty, investment institutions are legally obliged to ensure that everything they can do to maximise shareholder value is being done. Therefore, the institutions must engage on the issue. In this perspective, CDP sees four key climate change related risks: extreme weather events (severe droughts or floods); political and regulatory momentum to constrain emissions; shifts in consumer sentiments; climate change sensitive technologies, products and services superseding those existing today.

Swiss Re reviews the companies. They document what they are doing to manage climate change risk and consider the exclusion of companies that are not addressing the issue (Cortese, 2002).

11. SRI: global objectives, local divergences

Investors can choose among a number of SRI products that include negatively screened, positively screened, best in class, stakeholder activist and engagement based portfolios.

The gap between the various approaches and backgrounds can be documented by examining two (amongst many more!) definitions given to SRI:

"SRI is an investment process that considers the social and environmental consequences of investments, both positive and negative, within the context of rigorous financial analysis" (SIF, 2001).

"The integration of personal values with investments decisions is called SRI" (SIF, 2002).

Where a majority of Anglo-Saxon fund managers take an avoidance approach and a moralistic stance, the European SRI movement is referring to SRI, thus clearly incorporating the broadly discussed, science based and operationally detailed concepts "Sustainable Development" and 'CSR'.

To indicate the ethical depth and the quality of SRI criteria, Ethibel offers a classification in four generations of SRI funds and CSR research methodologies (Peeters, 2001) (Box 8).

(1) Socially responsible funds of "the first generation" are only based on negative criteria. This means that the fund manager when drawing up the portfolio will exclude companies that are involved in specific activities, and/or products or services. The investor gets a guarantee that his/her money is not, for instance, being spent in the arms trade or nuclear energy production but that's as far as it goes. These types of funds offer the investor a chance to protest but this formula is less suited to providing a positive stimulus to the corporate world.

(2) The second generation of SRI funds applies positive criteria and focus on a specific sector or theme. Researchers for this type of funds actively look for

Box 8. Some schematic elements and features of the Ethibel Fund Typology and Research Morphology classification.

Fund typology	1st generation	2nd generation	3rd generation	4th generation
Research morphology				
Research framework				
Criteria system	Negative criteria	Some positive criteria	Full range of positive criteria; possibly some exclusions	Full range of positive criteria; possibly some exclusions
Sustainability scope	Avoiding non-sustainable products or activities	Focus on thematic aspects of sustainability and CSR	Full sustainability and CSR scope, some stakeholder dialogue	Full sustainability and CSR scope, full stakeholder dialogue
Scope of analysis	Issue analysis	Thematic analysis	Corporate Sustainability & Responsibility Research (CSRR)	CSRR + risks & opportunities assessment
Moral dimension	Rather moralistic approach	<	>	Science based and risk rating driven
Framework orientation	Rather client driven	<	>	Referring to broadly discussed concepts of SD and CSR
Cultural context taken into account	No	No	Some	Ideally yes
Data collection process				
Desktop research	Yes	Yes	Yes	Yes
Company consultation	?	?	Yes	Yes
Review right by company	?	?	?	Yes
Monitoring/review	Yes	Yes	Yes	Yes
Stakeholder consultation				
On basic framework	No	No	?	Yes
On sector specific issues	No	No	?	Yes
For data collection	?	?	?	Full range
For conclusions/ ratings	No	No	?	Yes
For quality control	No	No	No	Yes
Rating/selection				
Benchmarks or other normative system	Full exclusion based on qualitative or quantitative norms	Some process of positive selection	Miscellaneous systems of best-in-class selection	Misc. systems of positive rating + expert advice
Quality management				
Internal QMS	Yes	Yes	Yes	Yes
External verification				
Output control	?	?	?	?
Process control	?	?	?	?

companies performing well in a specific field, for instance, by implementing a comprehensive social policy or by making considerable efforts to produce ecologically responsible products. For these funds, companies are screened for only one or some aspects of sustainable entrepreneurship.

(3) Third generation investment funds can be called 'sustainable' in the sense that investigations into these funds comprise all areas of sustainable entrepreneurship. Based on this comprehensive approach, companies that are suited to sustainability are selected. Investigations focus on internal staff policy and the relationship with the social environment as well as efforts made in the environmental domain and the ethical aspects of the company's economic policy.

(4) Fourth generation sustainable and responsible investment funds invest in sustainable enterprises in the widest sense of the word. The added value, in this case, is in the quality and the method of evaluation. Vital to fourth generation evaluation is a consistent inclusion of stakeholders' views on the company. Stakeholder communication takes place at three (separate) levels: (a) input into the research methodology, (b) the data collecting process and within the evaluation procedure. Finally, a transparent screening approach must be a guarantee for quality. This means that strong internal procedures must assure the accuracy, completeness and verifiability of sources and data. The various steps of the investigation must be verifiable by external auditors. This approach is considered to be most needed in order to produce full risks and opportunities assessments.

12. The impetus of the European definition on CSR

Two months before Johannesburg, the European Commission (EC) proposed a multi-stakeholder Forum as a means to advance CSR (EC, 2002). The launch of the Forum is the result of a comprehensive consultation process stemming from the Green Paper (EC, 2001). The Forum aims at promoting transparency and convergence of practices and instruments on CSR. The EC will not regulate, noting that by definition CSR is voluntary. A voluntary approach is favoured, as the main proposal is to set up a multi-stakeholder platform to develop commonly agreed guidelines on issues such as reporting, assurance and codes of conduct and to put forward the "business case for CSR" to large and small companies, amongst other targets. The EC will review the Forum's progress in 2004.

Critics consider the proposals in the EC communication not strong enough to tackle the issues of e.g. corporate governance and claim that the EC definition of CSR as a voluntary activity based on a strong business case is not a convincing foundation for policy development.

CSR is considered by the EC as a business contribution to Sustainable Development (EC, 2002) and the Green Paper defines CSR as: "Taking up the triple bottom line approach ... by going voluntarily beyond legal requirements ... in offering fair deals to stakeholders ... while having a dialogue with them" (EC, 2001). Considering this definition as the emanation of the European views on CSR, some indications can be given of the possible implications on CSR research methodologies. What might be the operational consequences on the future quality standards?

12.1. TRIPLE BOTTOM LINE APPROACH...

The triple bottom line refers to the idea that the overall performance of a company should be measured based on its combined contribution to economic prosperity, environmental quality and social capital.

Research processes within the field of CSR and sustainability should address four main areas of analysis: the internal social policy, the external social policy, the environmental policy and the ethical-economic policy. Within these domains a number of themes are 'generally' accepted as being key issues to be addressed. Within the area of internal social policy, themes as strategy, employment, job content, terms of employment, working conditions and industrial relations seem to be recognised by academics, NGOs and other stakeholders as being essential for any social audit of a company. Within the area of environmental policy, issues related to strategy, management, production and products are essential elements for any CSR analysis. Similarly, a minimal scope on the level of external social policy, including labour rights and human rights and the ethical and economic policy can be defined. The detailing into topics, indicators and ratings can be left to the professionalism and skills of the rating agencies.

12.2. ... VOLUNTARILY BEYOND LEGAL REQUIREMENTS...

Any CSR assessment should at a minimum level reflect on the attitude towards legislation and identify the seriousness and absence of infringements. It should present data and interpretations that facilitate the comparison or benchmarking of companies and disclose a best of class principle (whether best in class, best in sector or any other relevant reference) and document it in the assessments. It should be transparent in its output level and on all procedures and methodologies regarding rating, benchmarking or any other comparative analysis.

12.3. ... OFFERING FAIR DEALS TO STAKEHOLDERS...

In describing the external dimension of CSR the Green Paper states: 'CSR extends beyond the doors of the company into local community and involves a wide range of stakeholders in addition to employees and shareholders: business partners and suppliers, customers, public authorities and NGOs representing local communities, as well as the environment. In a world of multinational investment and global supply chains, CSR must also extend beyond the borders of Europe' (EC, 2001). Minimal areas of research should be defined and minimal requirements on stakeholders issues be addressed. Any CSR assessment should include information on the degree to which the company is transparent for its stakeholders about its societal impacts and is engaged in stakeholder dialogue.

12.4. ... HAVING A DIALOGUE WITH STAKEHOLDERS

The "in depth quality" of CSR analysis may differ dramatically, often provoked by different engagements in active and dialogue based screening processes. There is a huge difference in doing "desktop-research" based on questionnaires and electronic information on the one hand and audit-like visits and conversations with management on a variety of levels and responsibilities on the other hand.

Furthermore, the "stakeholder involvement" practices differ dramatically. Relatively few groups do integrate stakeholders opinions in the data collecting process, in refining the methodological framework or in facilitating and improving the assessment process.

Four levels of stakeholder engagement can be seen: (i) passive incorporation of stakeholders view, by desktop research; (ii) active dialogue with stakeholder groups representatives in order to prepare/refine sector or company specific methodologies or audits; (iii) active dialogue with a company's stakeholders or stakeholder groups representatives in order to incorporate – in a balanced and interpreted way – their views into the assessments; (iv) active participation of stakeholder representatives in evaluation/rating processes.

In order to conduct valuable, coherent and honest risks and opportunities assessments, any research quality standard should reflect on minimal requirements concerning the engagement of stakeholders in the data collecting process.

13. A quality standard for CSR/SRI research and rating ... processes

The European Commission calls for "convergence and transparency of SRI rating methodologies" and for "transparency on the level of investment management of SRI funds and pension funds" (EC, 2002). "CSR practices and instruments will be more effective if they are part of a concerted effort and based on clear and verifiable standards".

The CSR and sustainability concepts are not uniform and sometimes limited in scope. It does seem that every bank has a different concept and every investment fund has another view and even every compartment of those funds. Nobody can claim the final concept of CSR and discussion about Sustainable Development should in no way be stopped. However, the great variety of concepts is confusing for consumers and investors.

13.1. HARMONISATION OF THE DATA COLLECTING PROCESS?

There have been quite some efforts to harmonise the data collecting process.

Harmonisation based on the same framework of questions and indicators is very valuable, depending only on the intrinsic quality of the indicators. The comparability, exchangeability and validity of the data will be very much improved by standardisation. CSR/SRI research however can never be fully harmonised due to the different cultural and socio-political structures and backgrounds. The only solution here is the establishment of networks that are global and local and multidisciplinary and multicultural.

13.2. No harmonisation of the evaluation and rating process

These "collective data collecting processes" however do not lead to uniform evaluations or ratings. They do not necessarily lead to the same conclusions. This has often led to frustration by corporations and analysts. This is because processing data and comparing them with the evaluation criteria are part of a different process, than the data collection process itself.

No standardisation of the criteria seems desirable. No discussions should be killed that way. No evaluation procedures should be harmonised on a European level either. However, in order to be transparent, all methodologies, criteria and evaluation procedures should be disclosed.

13.3. Maximum transparency

Very strict internal as well as external quality control systems are needed, not only as a matter of service guarantee to fund managers, but also as a matter of credibility towards investors and integrity towards the screened companies.

CSR/SRI research and evaluations processes should be clearly defined and fully detailed, including the sustainability standards and research methodology, the data collecting activities, the way these analyses are reported, the evaluation and rating, the disclosure and communication of the output, the integrity assurance and the internal and/or external verification.

13.4. The European VQS

A number of independent CSR research organisations have started to work out a Voluntary Quality Standard for CSRR (VQS, 2002). The purpose of setting this up is to provide a transparent framework for professional quality research. This framework will only focus on undertaking SRI research and will not attempt to define or harmonise criteria.

14. Conclusions

During the decade following the Rio Summit, a substantial gap in funding for sustainable development issues, a lack of implementation, governance and political will has been observed, although the Rio approach was generally accepted as 'correct' and has been approved in Johannesburg.

The financing of sustainable development is clearly deficient when it would rely on official philanthropic development assistance only. The Johannesburg Summit's Plan of Implementation is highlighting business – including not at the least the financial industry – as a key partner in many aspects to forward sustainable development. Financial institutions are encouraged to further take up their social responsibilities, through their intermediary role in the economic system and through partnerships and product innovation. There is a huge interest in facilitated access to the public and private market finance.

There is a critical need for a more integrated and sustainable financial system, with sound new mechanisms toward enhancing global financial security.

Within the financial sector a limited number of progressive players are raising pressure to move things forward.

Significant progress has been made on standards, metrics, and guidelines to improve sustainability management systems, reporting, accounting and multi-stakeholder dynamic. Numerous voluntary initiatives, effectively monitored codes of conduct and partnerships with governments and civil society show the tendency to grow out of the infancy stage.

The role of the sustainable and responsible investment community has been invaluable and will continue to be crucial by critically rewarding sustainability leaders, assessing corporate sustainability risks and opportunities and by continuously pleading for transparency and good governance practices.

In order to prove the business case of CSR and of the triple bottom line to the mainstream, the sector of SRI research and rating agencies should critically investigate itself and strive to more convergence on the level of research processes, quality assurance systems and sound disclosure and verification practices.

References

Annan, K.: 2002a, *Both Business and Society Benefit from Working Together*, speech at 'Business Day', Business Action for Sustainable Development, Johannesburg, 1 September 2002.

Annan, K.: 2002b, *Implementing Agenda 21, Report of the Secretary-General*, New York, United Nations Economic and Social Council.

Anon.: 2002a, *Creative Financing for Sustainable Development*. An input to the World Summit on Sustainable Development (non-official consultation paper), World Bank Environment Department, consultation draft (http://www.worldbank.org/devforum/forum_financing.html).

Anon.: 2002b, *Developing Value. The Business Case for Sustainability in Emerging Markets*, SustainAbility/IFC/Ethos, Herndon, London.

Armstrong, G., 2002, *The Contribution of Private Investment to Sustainable Development: A Framework*, Herndon, London, International Finance Corporation.

Asria: 2002, www.asria.org.

Avanzi/SiRi Group: 2002, *Green, Social and Ethical Funds in Europe 2002*, Sustainable Investment Research International Group (with support of CSR Europe), Milano.

BASD: 2002, *Comments and Key Business Messages*, Business Action for Sustainable Development (http://www.basd-action.net/docs/releases/20020904_keybus.shtml).

Baue, W.: 2002, *SRI Contributions to the Earth Summit Blocked from Within and Without*, SocialFunds.com Rio + 10 Series, August 16, 2002, SRI World Group (http://www.socialfunds.com/news/article.cgi/article909.html).

Bolt, K., Matete, M. and Clemens M.: 2002, *Manual for Calculating Adjusted Net Savings*, Washington DC, World Bank, Environment Department.

Bouma, J.J., Jeucken, M. and Klinkers, L. (eds.): 2001, *Sustainable Banking, The Greening of Finance*, Sheffield, Greenleaf Publishing.

Charlier, A., De Smedt, E., Peeters H., Van Braeckel, D., Vandenhove, J. and Zeilinger, I: 2001, *Benchmarking Sustainability in the European Car Industry 2001–2002*, Stock at Stake & Ethibel, Brussels.

CDP: 2002, *Institutional Investors Collaborate on Greenhouse Gas Emissions Questionnaire*, The Carbon Disclosure Project May 31, 2002, (http://www.cdproject.net).

Cortese, A.: 2002, *As the Earth Warms, Will Companies Pay?*, New York, *The New York Times*, August 18, 2002.

Dunstan, S.: 2000, *Insurers can kick-start Kyoto*, *Environmental Finance*, May 2000, London (http://www.environmental-finance.com/2001/0105may/featmay2.htm).

EC: 2001, *Promoting a European Framework for Corporate Social Responsibility – Green Paper*, Luxembourg, The European Commission, Office for Official Publications of the European Communities.

EC: 2002, *Corporate Social Responsibility. A Business Contribution to Sustainable Development*, Luxembourg, The European Commission, Office for Official Publications of the European Communities.

Ethibel: 2002, *The Ethibel Sustainability Index, The ESI Rulebook, The Ethibel Research Methodology, Brussels, The Ethibel Label* (www.ethibel.org).

Forge: 2000, *Guidelines on Environmental Management and Reporting*, London (www.abi.org.uk).

Forge: 2002, *Guidance on Corporate Social Responsibility Management and Reporting for the Financial Services Sector. A Practical Toolkit*, London (www.abi.org.uk).

Gardiner, R.: 2000, *Briefing Paper on Foreign Direct Investment: A Lead Driver for Sustainable Development?*, Towards Earth Summit 2002, Economic Briefing Series No. 1, London, UNED International Team. (www.earthsummit2002.org).

Gardiner, R.: 2001, *Briefing Paper on Sustainable Finance: Seeking Global Financial Security*, Towards Earth Summit 2002, Economic Briefing Series No. 2, London, UNED International Team (www.earthsummit2002.org).

GRI: 2002, *2002 Sustainability Reporting Guidelines*, Global Reporting Initiative, Amsterdam (http://www.globalreporting.org/).

Hahn, T.: 2002, Who will finance Agenda 21?, *Tomorrow*, August 2002, pp. 40–42.

Inter-departmental Group on CSR: 2002, *Business and Society. Corporate Social Responsibility Report 2002*, London, Department of Trade and Industry (DTI).

ISEA: 2002, *The AA1000 Assurance Standard*, Institute of Social and Ethical Accountability (http://www.accountability.org.uk/default.asp).

Jeucken, M.: 2001, *Sustainable Finance & Banking. The Financial Sector and the Future of the Planet*, Earthscan, London.

Jeucken, M.: 2002, Banking and sustainability – slow starters are gaining pace, *Ethical Corporation Magazine* **11** (November 2002), pp. 44–48 (www.ethicalcorp.com).

Kabbaj, O., Chino, T., Lemierre, J. and Igledias, E.: 2002, *Statement of Heads of Regional Development Banks at the World Summit for Sustainable Development*, Johannesburg, 2 September 2002.

Kahlenborn, W.: 2002, *The Social and Environmental Impact of SRI. Results of Past Research*, Summary report of the Eurosif Conference 'SRI: Task or Tool for Public Policy', Frankfurt.

Miller, A., Schultz, J., de Sousa-Shields, M., Blom, P. and Rosen, R. (eds.): 1992, *The 1992 Rio Resolution on Social Investment* (http://www.uksif.org/publications/archive-rio-1992-05/content.shtml#press) (posted on the web in March 2000), UK Social Investment Forum, US Social Investment Forum, The Canadian Social Investment Organisation, The International Association for Investors in the Social Economy, The Australian Social Investment Forum.

Mistra.: 2001, *Screening of Screening Companies. An assessment of the quality of existing products/services*, Mistra, Stockholm and London, The Foundation for Strategic Environmental Research (www.mistra.org/eng).

Molenkamp, G.: 2002, *Corporate Social Responsibility. Sustainability Reporting becoming mainstream*, KPMG Global Sustainability Services, De Meern, The Netherlands.

Pearce, B.: 2002, *A Principled Approach*, Environmental Finance, July–August 2002, 22–23.

Pearce, B. and Ganzi, J.: 2002, *Engaging the Mainstream with Sustainability. A Survey of Investor Engagement on Corporate Social, Environmental and Ethical Performance*, London, Forum for the Future's Centre for Sustainable Investment and Finance Institute for Global Sustainability.

Pearce, B., Roche, P., Chater, N. and PIRC: 2002, *Sustainability Pays*, Co-operative Insurance Society (produced by the Centre for Sustainable Investment at Forum for the Future), London.

Peeters, H.: 2001, *Assessing Corporate Social and Environmental Performances*, Corporate Social Responsibility on the European social policy agenda, Conference of the Belgian Presidency of the European Union, 27-28/11/2001, Workshop CSR and the investor: promoting socially responsible investing, Brussels, (http://www.socialresponsibility.be/pdf/speakers/B4-H.Peeters.pdf). See also (http://www.ethibel.org/subs_e/1_info/sub1_2.html).

Peyo, R.: 2002, *The SiRi Group Suggested Actions for Disseminating SRI-practices*, Fundación Ecología y Desarrollo/SiRi Group, Zaragoza (www.sirigroup.org).

Ribbans, E. (ed): 2002, *Investing for Sustainable Development. Getting the Conditions Right*. WBCSD, IUCN, The World Conservation Union, World Bank Institute, LEAD International, Deutsche Bank, Conches-Geneva.

Schmid-Schönbein, O. and Braunschweig, A.: 2000, *EPI-Finance 2000. Environmental Performance Indicators for the Financial Industry*, November 2000 (www.epifinance.com).

Schmid-Schönbein, O., Furter, S. and Oetterli, G.: 2002, *VfU Indicators 2002: Revision and Further Development of the VfU Indicators on In-house Environmental Performance for Financial Institutions*, VfU Verein für Umweltmanagement in Banken, Sparkassen und Versicherungen e.V., Bonn (www.vfu.de).

SIF: 2001, *2001 Report on Socially Responsible Investing Trends in the United States*, Washington, The Social Investment Forum.

SIF: 2002, *2002 Directory of Socially Responsible Investment Services*, Washington, The Social Investment Forum.

Smith, B.: 2002, A different drummer. Sustainable development in tough times. *Sustain* **20** p. 4 (November 2002), World Business Council for Sustainable Development.

SPI-Finance: 2002, SPI-Finance 2002, Social Performance Indicators for the Financial Industry, E2 Management Consulting, Zurich (www.spifinance.com).

Tennant, T.: 2002, *RioC10 Finance Commitments (incl. a set of social and eco- performance indicators)* (http://www.asria.org/pro/news&events/rioC10FinanceCommitments.htm).

Thomas, C., Tennant, T. and Rolls, J.: 2000, *The GHG Indicator: UNEP Guidelines for Calculating greenhouse Gas Emissions for Business and Non-Commercial Organisations*, United Nations Environment Programme, NPI Global Care Investments, Genève/London, Centre for Environmental Technology.

UN: 2000, *United Nations Millennium Declaration*, document (A/55/L.2), The United Nations (http://www.un.org/millennium/declaration/ares552e.pdf).

UN: 2000a, *Key Outcomes of the Summit*, The United Nations Johannesburg Summit, Division for Sustainable Development, United Nations Department of Economic and Social Affairs (http://www.johannesburgsummit.org/html/documents/summit_docs/2009_keyoutcomes_commitments.doc).

UN: 2002b, *The Global Compact*, New York (http://www.unglobalcompact.org/Portal/).

UN: 2002c, *UN Johannesburg Summit (2002) Plan of Implementation of the World Summit on Sustainable Development*, Division for Sustainable Development, United Nations Department of Economic and Social Affairs (http://www.johannesburgsummit.org/html/documents/summit_docs/2309_planfinal.htm).

UN CSD: 1992, *Agenda 21, Environment and Development Agenda*, Rio de Janeiro, United Nations Conference on Environment and Development (UNCED), 3-14/06/92; United Nations Division for Sustainable Development 23/10/2002 (http://www.un.org/esa/sustdev/agenda21.htm).

UNEPFI: 1997a, *Statement of Environmental Commitment by the Insurance Industry*, UNEP Insurance Industry Initiative steering committee, Geneva (http://unepfi.net/iii/statemen.htm).

UNEPFI: 1997b, *UNEP Statement by Financial Institutions on the Environment & Sustainable Development*, UNEP's Finance Industry Initiatives, Geneva (http://unepfi.net/fii/english.htm).

UNEPFI: 2002, Sector report: *Industry as a Partner for Sustainable Development. Finance and Insurance*, UNEP's Finance Industry Initiatives, Geneva.

UN FCCC: 1997, *Kyoto Protocol to the United Nations Framework Convention on Climate Change*, United Nations Framework Convention on Climate Change (http://unfccc.int).

van den Brink, T.: 2002, *Guide Screening and Rating Sustainability, How companies are Screened and Rated on Sustainability*, Amsterdam, Triple P Performance Centre.

Vitalis, V.: 2002, *Improving the Synergies between ODA and FDI. OECD Roundtable on ODA and FDI*, Paris, OECD.

WBCSD: 2001, *The Business Case for Sustainable Development, Making a Difference towards the Johannesburg Summit and beyond*, Conches-Geneva, The World Business Council for Sustainable Development.

World Bank: 2002a, *Comprehensive Development Framework, What is CD?*, Washington, The World Bank Group (www.worldbank.org/cdf).

World Bank: 2002b, *Environmental Economics and Indicators: Green Accounting*, The World Bank Group. Washington   (http://lnweb18.worldbank.org/ESSD/essdext.nsf/44DocByUnid/E700AB1A6A90B2BD 85256B500063D42A).

CHAPTER 12

EDUCATION FOR SUSTAINABLE DEVELOPMENT: THE JOHANNESBURG SUMMIT AND BEYOND

BHASKAR NATH

European Centre for Pollution Research, London
(e-mail: BhaskarNath@aol.com)

Abstract. The World Summit on Sustainable Development (WSSD), held in Johannesburg during 26 August and 4 September 2002, was a truly remarkable event, not least because it identified and committed the world community to what has to be done to realise Agenda 21 objectives.

Discussion begins with the "means of implementation" of the Johannesburg Plan of Implementation (JPI). Education for, and raising awareness of, sustainable development are the key commitments in the "means of implementation". The issues central to these commitments are discussed.

The crucial role of moral philosophy in education for sustainable development is then discussed. Defining the "problem" as lack of progress (in fact negative progress between Rio and Johannesburg) towards global sustainable development, a cause–effect relationship of the "problem" is developed based on a systematic and logical analysis. It shows that the "cause" is West's profoundly materialistic, environment-degrading and exploitative attitude and activities to satisfy grossly unsustainable, hedonistic and insatiably avaricious Western life-styles – life-styles that are held up by the West as "ideal" fruits of economic "development" to be aspired by all. The "effects" are pollution of air, water and soil; mounting loss of biodiversity, ecosystems and species; relentlessly widening north–south divide, etc. It is argued that while science and technology can address some of the "effects", they cannot address the "cause". Only moral philosophy can by fundamentally re-orienting moral values *genuinely* to respect nature and the environment.

Based on sound and tested principles of Educational Psychology, a proposal is then made for including moral philosophy in the formal *curricula* (content and pedagogy) of primary, secondary and higher education for instilling in children and young people genuinely environment-respecting moral values. To this end a generic syllabus for the secondary level is proposed.

Finally, it is argued that if the scientific community really believes that science or technology *alone* can radically change the pervasive environment-degrading moral values to those that *genuinely* respect the environment, thus paving the way to *real* global sustainability, then it must demonstrate how this could be done and explain why, despite their abundant science and technology, the developed nations are the biggest polluters and consumers with grossly unsustainable life-styles. Certainly, examples would be much more convincing than rhetoric or tired old clichés about how science and technology alone could deliver global sustainable development.

Key words: curricula, development, education, environment, moral, philosophy, psychology, science, sustainable, technology.

1. Introduction

The World Summit on Sustainable Development (WSSD), held in Johannesburg during 26 August and 4 September 2002, reaffirmed sustainable development as

---

Readers should send their comments on this paper to: BhaskarNath@aol.com within 3 months of publication of this issue.

L. Hens and B. Nath (eds.), The World Summit on Sustainable Development, 275–298.
© 2005 *Springer. Printed in the Netherlands.*

a central element of the international agenda, giving renewed impetus to global action to eradicate poverty and protect the environment. The Summit broadened the understanding of sustainable development and strengthened it by focusing on the important linkages that exist between poverty, the environment and the use of natural resources. An important achievement of the Summit was that governments agreed to and reaffirmed a wide range of concrete commitments and targets for action to achieve more effective implementation of sustainable development objectives than hitherto.

The *Johannesburg Declaration on Sustainable Development*, which is a political declaration, and the *Johannesburg Plan of Implementation* (JPI), are the two most important documents to emerge from the WSSD. The former sets out a vision of the future by tracing the evolution of sustainable development from Stockholm to Rio de Janeiro to Johannesburg, and focuses on the challenges faced by the international community in implementing it. It makes political commitments on ways in which to implement sustainable development more effectively than hitherto. The JPI, on the other hand, reaffirms the international community's commitment to the Rio principles, the full implementation of Agenda 21, and the Programme for the Further Implementation of Agenda 21. To these ends the JPI contains a total of 170 Paragraphs committing the international community to a wide range of issues that directly or indirectly impinge on the implementation of Agenda 21.

The focus of this paper is on education for sustainable development. More precisely, it is on the kind of education needed to make meaningful progress towards global sustainable development (we note that negative progress has actually been made between Rio and Johannesburg). For this an assessment is made of the "means of implementation" in the JPI and the *Ubuntu Declaration*, both of which emerged from Johannesburg. Assessment is also made of some parallel initiatives on education for sustainable development. It is demonstrated that although science and technology can help the process of sustainable development, they cannot be exclusively relied upon to deliver it. It is then argued that moral philosophy must be included as an important element of formal educational curricula at all levels, if the world community is at all serious about achieving global sustainable development. Such curricula (content and pedagogy), based on sound principles of Educational Psychology, are briefly discussed along with some key issues of teaching and learning.

2. "Means of implementation" of the JPI and some parallel initiatives

2.1. "MEANS OF IMPLEMENTATION" OF THE JPI

Paragraphs 81–136 incl. of the JPI elaborate on the "means of implementation" for realising the Agenda 21 objectives. A distillation of these Paragraphs produces 3 key commitments as means of implementing the JPI, and, not surprisingly, they are concerned with education for, and awareness of, sustainable development.

They are to

(a) Ensure that, by 2015, all children will be able to complete a full course of primary schooling and that girls and boys will have equal access to all levels of education relevant to national needs.
(b) Eliminate gender disparity in primary and secondary education by 2005.
(c) Recommend to the UN General Assembly that it consider adopting a decade of education for sustainable development, starting in 2005.

Item (a) above is reaffirmation of a *Millennium Development Goal*, while (b) is reaffirmation of the *Dakar Framework for Action on Education for All.*

Taken together, the presumption in commitments (a) and (b) would appear to be that meaningful progress towards global sustainable development is contingent upon raising universal literacy rate and level of education without gender discrimination. Certainly, a society's disposable income and purchasing power increases as it becomes more literate, and to this extent access to all levels of education without gender discrimination is a very desirable objective and a basic human right too. However, there is ample evidence worldwide to show that as a society becomes more affluent, it consumes more and pollutes more. China provides a good and contemporary example of this, while the highly developed countries illustrate this phenomenon very well. These countries are characterised by high or very high literacy rates and highest educational achievements on a wide front. And yet, they are the biggest polluters and consumers with matching life-styles that are grossly unsustainable, and demonstrably so. And so it is not clear how commitments (a) and (b), if fulfilled, could or would help the process of global sustainable development. In fact, the contrary would appear to be true, although we are not advocating that people anywhere should be denied their basic right to education.

The proposal in item (c) above is much to be welcomed, for at the very least it would raise public awareness of the need to achieve sustainable development and inform people about what they ought to do individually, collectively and institutionally for the practical realisation of Agenda 21 objectives.

2.2. THE *Ubuntu* DECLARATION

In the *Ubuntu Village*, which is about thirty minutes' drive from the main WSSD conference venue at *Sandton* in Johannesburg, a series of parallel events (workshops, special meetings, conferences, etc.) were held during the Summit. An important outcome of one of these events was the *Ubuntu Declaration on Education and Science and Technology for Sustainable Development* (www.unesco.org/iau/tfsd_unbutu.html).

In this Declaration, signed by eleven of the World's foremost learning and scientific organisations, a call was made for an initiative to strengthen science and technology education for sustainable development. To that end it called on

Governments of the WSSD and Post-Summit Agenda to (among other things):

- Designate educators as the tenth stakeholder group in the WSSD process, and
- Review the programmes and curricula of schools and universities, in order to better address the challenges and opportunities of sustainable development, with focus on:

    (a) Plans at the local, regional and national levels.
    (b) Creating learning modules which bring skills, knowledge and reflections, ethics and values together in a balanced way.
    (c) Problem-based education at primary and secondary levels in order to develop integrated and non-instrumental approaches to problem-solving at an early stage in the education cycle.
    (d) Problem-based scientific research in tertiary education, both as a pedagogical approach and as a research function.

As will be seen later, reference to "ethics and values" in (b) above is of particular significance in the context of sustainable development and its realisation.

2.3. SOME OTHER INITIATIVES ON EDUCATION FOR SUSTAINABLE DEVELOPMENT

2.3.1. The ICSU initiative

Aware of the need to generate the capacity to apply science and technology to meet the challenge of sustainable development, the International Council of Science (ICSU) convened a small group to draw up a document outlining some of the steps that are necessary for building up such capacity. Recommendations of the group for further action are given in its report (ICSU, 2002). Some of the key recommendations for implementation during the follow-up to the WSSD in Johannesburg are listed below (from ICSU, 2002):

Primary and secondary education

- The teaching of science through inquiry-based, hands-on approaches at primary and secondary levels needs to be incorporated as a fundamental component of basic education in both developing and developed countries. Its essential role must be recognised, as it prepares children to live and work in a world increasingly defined by science and technology, equipping them with personal decision-making and for their roles as citizens. Policies that affect finances, curricula, teacher preparation, materials development and assessment to support this critical goal must be established.
- Topics important for sustainable development, such as the relationship between science and technology to health, energy, food production and the environment, should be included while providing basic conceptual frameworks for lifelong learning. Scientists should assist curriculum developers in identifying relevant topics and creating appropriate materials for teaching and learning.

- It is especially important to ensure that girls and young women (as well as boys and young men) receive high quality education, given their current under-participation in basic education and in scientific and technical courses at all levels, in addition to their roles within the family, community, society and economy.
- The scientific community must build meaningful partnerships with governments and schools to support quality science and technology education in primary and secondary education. The roles that scientists and engineers can play include: serving as advocates to governments and to donor agencies to support quality approaches to learning and teaching science, mathematics and technology; and working with teachers and educational administrators to support the development, implementation, scaling-up and sustaining of quality, hands-on science instruction in schools.

Tertiary education and research

- For universities to take the lead in the changes required for science to respond to the challenges for sustainable development, they must revise their curricula, the organisation and assessment of research, and their working links with different sectors of society.
- This is especially true in those areas where sectors such as non-governmental organisations, local communities, small enterprises, and so on, are to play a role as partners in sustainable development efforts.
- Universities and higher education institutions need adequate and stable funding to maintain their capacity to engender innovation and provide quality education. Cuts in the budgets of public universities are therefore a threat to capacity building for sustainable development, and should be a matter of serious consideration by national governments.
- Group work among students and multidisciplinary training should be promoted. Students pursuing studies in any single discipline should be required to take at least a course in another discipline or a multidisciplinary subject of relevance to sustainable development.
- Research of high quality and relevance related to sustainable development, especially that carried out in developing countries, should be recognised in specific ways, for instance through international awards. The nomination of developing country scientists to international committees should be encouraged.

2.3.2. Global Higher Education for Sustainability Partnership (GHESP)

Formed in 2000, the GHESP came about as a result of the work programme of the Commission on Sustainable Development and in anticipation of the Johannesburg Summit. Its partners are convinced that the leaders of higher education institutions and their academic colleagues in all disciplines must make sustainable development a central and academic focus in order to create a just, equitable and ecologically sound future. This requires the generation and dissemination of knowledge through interdisciplinary research and teaching, policy-making, capacity building and technology transfer. To the partners of the GHESP it is critical that higher education institutions understand and accept their responsibility within the broader context of social and economic development, and the building of democratic, equitable and ecologically-minded societies (GHESP, 2002; www.unesco.org/iau).

2.3.3. Comments on the ICSU and GHESP initiatives

Successful pursuit of the ICSU recommendations, it is claimed, would enlarge and enhance both intellectual resources and capacity of educational institutions, especially in developing countries, for the application of science and technology to achieve sustainable development. The emphasis seems to be almost exclusively on scientific and technological education and research. This is a pity. Because, as will be gathered from Sections 3.1 to 3.3, science and technology cannot deliver sustainable development, and so it is not clear why such exclusive emphasis is placed on science and technology.

The scope of "education" in the GHESP initiative, though not explicit, would appear to be wider. In particular, ". . . all disciplines must make sustainable development a central and academic focus . . ." and ". . . it is critical that higher education institutions understand and accept their responsibility within the broader context of social and economic development . . ." are much to be welcomed. However, in the light of what we have said in this paper, we very much hope that moral philosophy is given the importance it deserves in the GHESP paradigm of research and education for sustainable development.

3. The problem and an analysis of its cause–effect relationship: can science and technology deliver sustainable development?

3.1. THE PROBLEM AND SOME OF THE ISSUES CENTRAL TO IT

Serious lack of progress towards global sustainable development is the "problem" to be addressed. In the true scientific tradition we will analyse the problem systematically in what follows with reference to some of the issues central to it. Our purpose is to establish the cause–effect relationship of the problem in order to develop (in Section 5) a heuristic for addressing it.

The earth is a complex geo-biochemical entity whose precise functioning, as well as the complex interactions that occur among its myriad elements, we are yet to understand fully. Its fragile self-regenerative systems (e.g. the carbon cycle) have limited capacity for processing anthropogenic environmental contamination. When, as now, the quantity, chemical complexity or toxicity of contaminants discharged to the environment exceed the limits of self-regeneration, the excess accumulates to disrupt or disable the systems themselves, or to cause adverse environmental impacts with serious implications for health and/or nature's environmental integrity (Nath, 2002). And this is the main reason for the mounting environmental predicament confronting us today.

An objective analysis of the "problem" will show that it is caused and exacerbated by relentlessly rising production and consumption of goods and services, and that it is the main obstacle to the realisation of even a modest degree of global sustainable development. According to the Brundtland Commission Report, *Our Common Future*, during 1950 and 1985 world production of consumer goods rose by a factor of seven (WCED, 1987), and it is safe to assume that the factor today (2003) is significantly greater. Greater production means greater consumption of energy and natural resources notwithstanding recycling and reuse efforts, matching amounts of both production and post-consumption wastes to be disposed of, and the environmental consequences of all these. Indeed, this open-ended and mainly avarice-driven consumption to satisfy the "wants" of materialistic lifestyles, increasingly characterised by vanity and hedonism, is proving to be the nemesis of global sustainable development.

Two points are relevant in this context. First, while the authoritative definition of sustainable development, given in the Brundtland Commission Report (WCED, 1987), is in terms of "needs" and not "wants", the prevailing economic system is increasingly preoccupied with supplying the "wants" of avarice and hedonism. And second, the Report also states that adoption of less-consumptive and less-polluting life-styles, especially by the rich, is a necessary pre-condition for progressing towards global sustainable development. However, this is unlikely to happen, not least because the prevailing economic system, which works only when there is uninterrupted growth in production and consumption, would cease to function if it did. Moreover, the rich (and therefore powerful) are as unlikely to willingly renounce their hegemony of wealth and power as are the poor to curb their developmental ambitions for a better life. And so the prospects of global sustainable development are caught ever more firmly between a rock and a hard place.

With regard to the above, the following excerpt from the *Living Planet Report 2000* (WWF, 2000) does much to concentrate the mind:

> Man has wiped out a third of the natural world in the last thirty years and soon will have to start looking for a new planet to live on . . . . The scale of devastation is so great that man will have used up all the Earth's natural resources by 2075 . . . If every human alive today continues to consume resources and produce carbon dioxide at the same rate as the average Briton, we will need to colonize at least two Earths to survive . . . . Our current rate of consumption is eroding the very fabric of our planet and will ultimately threaten our long-term survival.

It is sobering to consider the logical implications of the above. If we fail to colonise at least two earths by around 2075 – and we have yet to find even one in the unimaginable vastness of the Cosmos let alone colonise it – the already huge income disparity between rich and poor nations is likely to persist and widen. A logical corollary to this is that if we fail to colonise at least two earths by around 2075, and a "cap" has to be put on global anthropogenic pollution in order to maintain some degree of homeostasis, then the poor must become poorer in order for the rich to become richer, or *vice versa*. However, the *vice versa* is unlikely to happen, because the rich and powerful will not willingly renounce their hegemony of wealth or power. And so, either way, it is a "no-win" scenario for the World's poor.

3.2. CAUSE–EFFECT RELATIONSHIP

Consumption, and the desire for it, is open-ended – there is no end to it. Modern consumerism (marketing in particular) is based on the manipulation of the masses using Freudian psychoanalytic techniques. Pioneered in the USA in the 1950s (BBC, 2002) and global in its scope today, consumerism thrives on people's insatiable greed for material things and lust for hedonism. It has brought about the pervasive "throw-away" culture and the absurd concepts of "Retail therapy" and "Conspicuous consumption".

The pervasive Western culture of unfettered production and consumption to satisfy the "wants" of unsustainable life-styles increasingly characterised by avarice and hedonism is the "cause" of the "problem".

This open-ended consumer culture, which we may characterise as a "deep malaise", is born of the profoundly anthropocentric Western world-view, whose origins go back to Aristotle (see Section 4.2) and the environment-degrading moral values that stem from that view. Consequently, Western attitude to the environment is grossly exploitative – an attitude that characterises and underpins West's highly materialistic, highly consumptive and grossly environment-degrading and unsustainable life-styles. Amazingly, such life-styles and those environmentally dysfunctional values have been, and are being promoted by the West as "ideals" to be emulated by developing nations as coveted "fruits" of their economic development (to be realised, of course, with expensive Western technical assistance they can ill afford, together with loans from Western donors that have an unfortunate habit of turning into crippling debt-burden on many). Yet, more amazingly, many of the developing nations have been adopting those values and abandoning their own much older values that taught them how to live contended lives in harmony with nature.

The "effects" ("symptoms") of this "deep malaise" are all too obvious to see. They have been manifesting as relentlessly increasing contamination of air, water and soil; mounting loss of species, ecosystems and biodiversity; ozone layer depletion; global warming; etc. The cause–effect relationship is shown schematically in Figure 1.

According to Behavioural Psychologists (Gross, 2001; Eisenberg, 1982) – and it follows from common sense too – how we treat the environment (or anything else for that matter) is fundamentally determined by our attitude to it, and, our attitude, in turn, is shaped by the moral values we hold. In other words, the "cause" is not extraneous. It is within us, in our psyche, and so too is the remedy to it. Clearly, education in moral philosophy is needed to radically change our attitude to nature and the environment, from one of gross exploitation as at present to that of genuine respect and prudent husbandry. It is hard to see how science or technology, however clever, could be helpful in this matter.

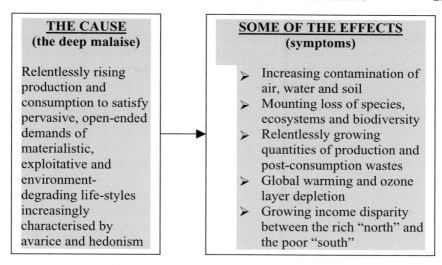

THE CAUSE **(the deep malaise)**	**SOME OF THE EFFECTS** **(symptoms)**
Relentlessly rising production and consumption to satisfy pervasive, open-ended demands of materialistic, exploitative and environment-degrading life-styles increasingly characterised by avarice and hedonism	➢ Increasing contamination of air, water and soil ➢ Mounting loss of species, ecosystems and biodiversity ➢ Relentlessly growing quantities of production and post-consumption wastes ➢ Global warming and ozone layer depletion ➢ Growing income disparity between the rich "north" and the poor "south"

Figure 1. Schematic of the cause–effect relationship of the problem of little or no progress towards global sustainable development.

3.3. CAN SCIENCE AND TECHNOLOGY DELIVER SUSTAINABLE DEVELOPMENT?

Science[1] and technology are so pervasive today that it is hard to find any aspect of modern life untouched by them. Directly or indirectly they have brought immense benefits to human societies, enriched our lives materially, brought us creature comforts and longevity, and given us the means to understand how the physical world around us works. Indeed, achievements in all the branches of science and technology bear ample testimony to humankind's genius and incessant quest for knowledge, and how to use that knowledge for the benefit of human societies.

It is to be noted, however, that science and technology are by themselves neutral in the sense that their impacts are determined by how they are applied, why they are applied, and whether or not we choose to apply them in the first place (Nath and Talay, 1996). That is, by themselves they are neither good nor bad. However, in so far as the natural environment is concerned, whether they turn out to be good or bad is determined by how they impact on the environment.

Following the industrial revolution, economic development through industrialisation based on science and technology became the preferred, if not the only approach to wider socio-economic development. Thanks to the efforts of international structures such as the World Bank and the International Monetary Fund, this particular model of development, which happens to be environment-degrading, is now universal (George and Sabelli, 1994).

---

[1]Comprising the exact sciences (e.g. Biology, Chemistry, Physics and Mathematics) and the social sciences (e.g. Economics, Geography and Sociology). Technology is defined as practical application of the sciences, or the study or use of mechanical arts.

Driven largely by industrial and commercial interests that are concerned mainly, and sometimes exclusively, with making profit, application of science and technology for economic development has brought human societies to the cross-roads of history where our very survival in the long-term is put at serious risk. How could we then trust science and technology to solve the environmental problems they have created in the first place?

No doubt science and technology have brought immense benefits. However, we are paying a high environmental "price" for it in terms of the "effects" of Figure 1, and the "price" is escalating to thwart the achievement of even a modest degree of global sustainable development. And this has serious implications for future generations.

An analysis would show that while science and technology can offer economically viable solutions to small-scale environmental problems, such as those for treating municipal wastewater or restoring relatively small areas of contaminated land, they cannot be applied to solve large-scale or global problems, or even to alleviate their impacts. Consider the following problems among many others that could be cited:

- As a result of relentless industrial development in Europe, the Baltic, the Mediterranean and the North Seas have been contaminated by all kinds of chemicals discharged into them. These contaminants, and compounds deriving from their mutual chemical reactions, are causing increasingly serious and adverse impacts on the marine life of those Seas.
- Global warming is no longer a myth. It is already here, caused mainly by relentlessly increasing carbon dioxide emissions to the atmosphere.

Obviously, application of science and technology has created these problems. Can science and/or technology offer politically acceptable and economically viable solutions to them? It is hard to see how. Even if they could, we will still be treating, like an incompetent physician, the "symptoms" of Figure 1 and not the "deep malaise" causing them in the first place. Indeed, an examination will show that science and technology are almost exclusively concerned with treating the "symptoms" and not the "cause". We argue that this conventional approach, which focuses almost exclusively on the symptoms, cannot and will not bring meaningful progress towards global sustainable development.

And so we are compelled to conclude that although science and technology can help the process of global sustainable development in a limited way, they cannot deliver it.

The situation in the rich, developed countries strongly supports this conclusion. These countries are abundantly endowed with latest science and technology, as well as ample financial and skilled manpower resources. So, if science and technology alone could deliver sustainable development, they should be the most sustainable. But they are not. On the contrary, they are the biggest consumers and polluters. The USA illustrates this well. With only about 4% of the World's population, that nation consumes an estimated 25% of the World's resources and produces an estimated 26% of the global pollution (Pilger, 2002). Such a nation, or the life-style it maintains, cannot be said to be sustainable by any stretch of the imagination.

4. Evolution of human attitude to the environment

In order to develop a heuristic for addressing the "problem" defined in Section 3.1, it would now be both instructive and helpful to examine how human attitude to nature and the environment has evolved through the ages in both Eastern and Western philosophical traditions.

4.1. IN THE ANCIENT CIVILISATIONS

Whether out of awe, love or respect, many of the older societies had successfully established and maintained a harmonious relationship with nature and their environments – notably those founded on *Buddhism*, *Daoism*, and the *Vedic* philosophy. The last, which flourished in ancient India, endures even today as the foundation of that country's culture and way of life. In that culture divine status is afforded to many of the natural and cosmic entities (e.g. the Sun, wind, seas, the universe, etc.) in human or terrestrial forms to which it is easy for humans to relate. The rationale for this is obvious to see. For example, the Sun God was worshipped (and still is) because without Him the earth would be an icy, sterile wasteland. Even today planet earth is *always* referred to as *Dharitri Mata*, which in Sanskrit means Mother Earth, and venerated with deep respect for her abundant benediction without which life on earth cannot exist.

Such moral values, which engendered *genuine* love and respect for nature, encouraged one to take from nature only that which one needed to live, and no more. To do otherwise, and especially to exploit Mother Earth in any way, was considered a cardinal sin, just as it is for one to exploit or abuse his or her biological mother. It was recognised, of course, that humans must kill in order to live. Even when we breathe, we kill countless micro-organisms through inhalation. It was understood, however, that in order to qualify as a kind, compassionate and genuinely environment-respecting human being, one must minimise one's killing activities.

> Everything animate or inanimate that is within the universe is controlled and owned by the Lord. One should therefore accept only those things necessary for himself, which are set aside as his quota, and one should not accept other things, knowing well to whom they belong.
>
> Mantra One, *Sri Isopanisad*

A common thread running through all the classical philosophical texts of ancient India is that all things in creation, animate or inanimate, are parts of the *same* continuum of existence, differentiated only by the level of consciousness of each as compared to the others; and that the consciousness of each, animate or inanimate, is inextricably linked to the *universal consciousness*, meaning The Divine. Interestingly, the concept of the *universal consciousness* is also to be found in the theory of *many worlds* in Quantum Physics (Nath and Talay, 1996; Rae, 1993). This concept of oneness of all things engenders respect for all things animate and inanimate, and especially for the planet earth, *always* referred to and venerated as Mother.

Islam, which is a much younger religion by comparison, also teaches its followers to respect nature and the natural environment. In The Holy Koran there are many passages to this end, of which the following is typical:

> Uncorrupted water is the sign of paradise. If one wants to improve his way of life to match the high quality of the Most Perfect, one should stop polluting water.
>
> Verse 47, Chapter 15 of The Holy Koran

4.2. IN WESTERN CIVILISATIONS

In Western (Occidental) civilisation human attitude to nature and the natural environment evolved in a very different way. The origins of this evolution can be traced back to Aristotle whose philosophical world-view (Allan, 1970) eventually shaped the foundation of modern science, technology and economics. According to Aristotle, nature has no intrinsic value. It is of value *only* if it benefits humans. Thus, for example, a rare plant in the tropical rain forest is valuable and worth preserving *only* if some useful drug could be made from it, or if it serves a useful purpose to benefit the humankind. Clearly, it is a highly utilitarian and exploitative attitude to nature and all non-human things within it. It is also a profoundly anthropocentric view which does not acknowledge the right of nature, or of anything non-human within it, to exist for its own sake. Historically this exploitative attitude, which is all too common in pervasive Western cultures, has driven the evolution of both science and technology and still continues to do so.

The Platonic world-view, on the other hand, acknowledges the intrinsic value of nature, and of all things within it, for its own sake (Lesser et al., 1997). That is, nature and all things within it have their own intrinsic values independently of humans and regardless of what humans thought those values might be. We humans may not know about or understand those values or their intrinsic qualities because of our own limitations, ignorance or selfishness. Clearly, it is an eco-centric world-view which is benign to nature at the very least. It is tempting to speculate on how human societies would have evolved with the Platonic world-view as the foundation of economics, science and technology rather than the Aristotelian world-view which prevailed.

The exploitative and profoundly anthropocentric world-view of Aristotle pervades the Judaeo-Christian tradition too, as will be gathered from the following: "And God blessed them, and God said unto them, Be fruitful and multiply, and replenish the earth, and subdue it; and have dominion over the fish of the sea, and

over the foul of the air, and over every living thing that moveth upon the earth"
(Genesis 2.28, The Holy Bible). The words "subdue" and "dominion" had been
interpreted to mean taking licence to exploit nature and all things within it for the
benefit and pleasure of man.

> In Western terms, one of the underlying factors which may have contributed (by being taken lit-
> erally) to the desire to dominate nature, rather than live in harmony with it on a sustainable basis,
> is to be found in the Book of Genesis where it records that "God said unto man, be fruitful and
> multiply, and replenish the Earth and *subdue it: and have dominion over the fish of the sea and
> over the fowl of the air and over every living thing that moveth upon the Earth.*" To me, that Old
> Testament story has provided Western man, accompanied by his Judaeo-Christian heritage, with
> an overbearing and domineering attitude to God's creation.
>
> HRH The Prince of Wales (Porritt, 1991)

In the Seventeenth Century this utilitarian and highly exploitative attitude
was reinforced by Francis Bacon (1561–1626) and René Descartes (1596–1650),
among others, in the secular context (Anderson, 1948; Clarke, 1982). Their thesis
was that nature and everything within it was for the sole benefit, well-being and
pleasure of man. In other words, man had *carte blanche* to exploit nature as he
pleased for his own benefit and pleasure. However, as it has now become clear,
this attitude more than any other factor, has been responsible for the continuing
degradation of earth's natural environmental capital, thus bringing us to the cross-
roads of history where our long-term survival as a species is put at risk *vis-à-vis*
the environment and nature's life-support systems.

Then the Western scientist and technologist arrived on the scene with their characteristic arrogance and superiority complex to proclaim that respecting the earth
as "mother" was sentimental nonsense. The earth and all its resources, they proclaimed, was there to be exploited for the sole benefit and pleasure of man. If ever
there were to be an environmental catastrophe of apocalyptic proportions – and
heavens forbid it – this will surely be its most fitting epitaph.

Soon after the end of World War II, exploitation of nature and her resources
took off as never before. In the 1950s a number of large, state-sponsored projects
were undertaken in the USA to explore if, or how, Sigmund Freud's psychoanalytic techniques could be employed to promote and achieve the twin objectives
of capitalism – uninterrupted growth of consumption, and effective social control (BBC, 2002). Led by Anna Freud and Edward Bernays (daughter and nephew
respectively of Sigmund Freud), the central thesis of these projects was that individuals harbour dark and powerful forces repressed in their unconscious minds
which, if not kept in check, could rise to the conscious mind to destabilise society
itself. Therefore, could the hedonistic pleasures of open-ended consumption (ultimately leading to the absurd manifestations of "retail therapy" and "conspicuous
consumption") be relied upon to keep those forces in check? If so, it would serve
the aforementioned twin objectives. We note in passing that open-ended consumption, and its uninterrupted growth, is the cornerstone of the capitalist (*laissez-faire*)
economic system; and that it is diametrically opposed to sustainable development
whose achievement is contingent upon people adopting less consumptive and less
polluting life-styles (WCED, 1987).

Researched response to the above question was a resounding "yes", and so began a veritable orgy of consumption leading to the pervasive "throw-away" culture of today which is proving to be the nemesis of both environmental protection and sustainable development. The high priests of this orgy are the captains of Western multinationals who worship at the altar of Mammon. They make uplifting pronouncements on the need to protect the environment, as and when necessary, largely to enhance their corporate image and environmental credentials, and, most importantly, to improve their "bottom lines". Assisted by legions of public relations consultants, psychoanalysts and others they spend vast sums of money to produce clever advertising that bombards (brainwash?) people relentlessly to fuel their greed for the hedonistic pleasures of open-ended consumption. It is indeed depressing to note that while all major religious and philosophical traditions are disdainful of avarice, greed and gluttony, it is precisely these that are now the mainstays of capitalism.

> The world manufactures seven times more goods today than it did as recently as 1950. Given population growth rates, a five- to ten-fold increase in current manufacturing output will be needed just to raise developing-world consumption of manufactured goods to industrialised world levels by the time population growth rates level-off next century.
>
> (WCED, 1987, page 15)

For many, the latest strategy for further exploitation of the world's poor being pursued by the powerful Western business interests, morally if not actively supported by their respective governments and international structures such as the WTO, World Bank and the IMF, goes by the name of "globalisation". To a large and growing body of people, called "globaphobes", it is little more than an updated old ploy with which to further impoverish the poor nations, claim their resources, and ensnare them even more tightly in a culture of debt and dependency. And in the process to degrade their quality of life, their environment, and eventually the global environment even further.

There is a very good reason for this pessimistic view of globalisation. It is this. Ever since it became unfashionable for the rich and powerful nations to exploit the poor and weak nations through colonisation, the former has been adopting trade and aid as preferred, modern and politically-correct instruments with which to exploit the latter and to control their resources.

> In their rhetoric, governments of rich countries constantly stress their commitment to poverty reduction. Yet the same governments use their trade policy to conduct what amounts to robbery against the world's poor. When developing countries export to rich country markets, they face stiff tariff barriers that are four times higher than those encountered by rich countries. Those barriers cost them $ 100 billion a year – twice as much as they receive in aid.
>
> (Oxfam, 2002; page 5)

There is no reason to believe that the innate exploitative Western mindset, born of a profoundly exploitative world-view that characterised the colonial era, has now changed for the better. On the contrary, consumed by a pervasive culture of greed and driven by apparently insatiable lust for hedonism and grossly unsustainable materialistic life-styles, their exploitative mindset is as strong today as it has ever been, probably stronger. Evidence for this, if any was needed, is typically and amply provided by recent spectacular financial scandals at giant US corporations such as ENRON, TYCO and WorldCom, among many others, that have recently been unravelling to expose their breathtakingly corrupt behaviour. These scandals typically show that Western multinationals do not have a moral compass, and neither do the political establishments or the international structures that support them.

One may reasonably ask: if giant US multinationals can deceive their own shareholders on such a grand scale despite a plethora of laws, regulations and agencies (e.g. the powerful Securities and Exchange Commission) to monitor and punish such behaviour, what can they not get up to for maximising profit in poor countries that have little or no bargaining power, and where laws and regulations to control such behaviour are lax or non-existent and the enforcement regime pliable? And, can their profit-oriented activities be trusted to respect and protect the environment? One may also be forgiven for thinking that the list of corrupt multinationals exposed to date merely represents the tip of the proverbial "iceberg", and that such avaricious behaviour is, has been, and will continue to be endemic notwithstanding efforts to control it (unless convincing proof to the contrary could be provided).

A factual account of how globalisation is further impoverishing the world's poor and degrading their environment, and yet how it is being vigorously promoted as being "good" for them by duplicitous, greedy and morally bankrupt Western governments and business interests, is given in Pilger (2002).

Amid this encircling gloom and pessimism there is something encouraging to report, however. It is that the need to respect the environment is now beginning to be acknowledged officially even in important political documents. Article 2 of the Treaty of the European Union (1992) states that EC's environmental policy objectives should include the goals of "sustainable and non-inflationary growth respecting the environment" (Lee, 1995). Although it is reticent about what "respecting the environment" is supposed to mean in practice, it is a good start in the right direction nonetheless.

5. Heuristic for a solution

5.1. NEED FOR A MORAL RENAISSANCE

As observed in Section 3.3, science and technology cannot deliver sustainable development, and it is an illusion to think they could. We argue, therefore, that in order to address the "problem" of Section 3.1 effectively, the pervasive Western anthropocentric and grossly exploitative attitude to nature and the environment must give way to an eco-centric attitude fostering *genuine* respect and care for nature and her abundant benediction that makes life on earth possible. For it is only then that meaningful progress towards restoring global environmental integrity and sustainable development can be made. However, in order for this to happen, and considering how difficult it would be to turn the tide of rampant consumerism to satisfy avarice-driven and open-ended "wants", nothing short of a moral renaissance is needed to re-orient our moral values genuinely in ways that would substantially restore the homeostasis that must exist among nature, all non-human things within her, and the humankind.

The main objective of the proposed renaissance, to be pursued mainly through education, is two-fold:

- To instil in children a set of moral values deeply and genuinely respectful of nature and the environment in the classic mother–child relationship model – morals that teach them to respect all animate and inanimate things in nature, and fellow human beings, for their intrinsic values, not what they could be profitably exploited for.
- To integrate the best that science and technology can offer for human welfare with the moral values mentioned above, without inflicting excessive damage on the environment.

The focus of moral education is deliberately on children, from preschool through to secondary and higher education. This is because, from the standpoint of educational psychology, whether we take the *empiricist* view of the child as an "empty vessel to be filled with knowledge and information", or the favoured *interactionist* view as "partly empty vessel and partly biologically pre-programmed to behave in certain ways", early childhood (1–8 years) is the ideal time to instil moral values and basic notions of good and bad in order for such values and notions to mature in adulthood and endure throughout their lives (Kohlberg, 1981; Fontana, 1995; Nath and Talay, 2003). As we grow old, on the other hand, we become set in our ways and comfortable with what is familiar to us, and so it becomes more and more difficult and often impossible for us to change our values or attitudes fundamentally.

It would be less than satisfactory for this renaissance to be spearheaded by the West for two reasons:

- Western philosophical thought has evolved from an anthropocentric, materialistic and profoundly exploitative world-view that does not acknowledge the intrinsic value of nature or of anything non-human within it (also see Section 4.2). It is shallow, mechanical in its approach, utilitarian in essence, and incapable of providing the spiritual bond crucial to an enduring mother (earth)–child (people) relationship.

> [Western] philosophers reduced the scope of their inquiries so much that Wittgenstein, the most famous philosopher of this century, said, "The sole remaining task for philosophy is the analysis of language". What a comedown from the great tradition of philosophy from Aristotle to Kant!
> Stephen Hawking (1988), page 191

- Western approach is essentially utilitarian, characterised by immense propensity for making profit. So, if this sacred mission is spearheaded by the West, it is likely to be turned into a profane "business opportunity" for profit, as has happened in the case of sustainable development for example.

Considering the geo-political landscape of today and the quality of philosophical input needed, the ancient *Vedic* philosophy of India is to be highly recommended to spearhead the moral renaissance for the following reasons:

- The ancient *Vedic* philosophy of India (enshrined in numerous texts such as the *Vedas, The Bhagavad-Gita, The Upanishads*), which endures today as strongly as it ever did, is eco-centric and deeply respectful of the earth, always referred to as "mother earth", as well as of all animate and inanimate creations (also see Section 4.1). Even today the moral values that engender and sustain such respectful behaviour are deeply embedded in Indian psyche and culture.
- In Sanskrit the generic name of India's ancient philosophy is *Sanatana Dharma*, which in English means *Eternal Religion* of man ("religion" is the nearest translation of *Dharma*). *It is a philosophy and not a religion, organised or otherwise. Sanatana Dharma* is secular in two different senses. First, it is *Sanatana* (meaning eternal, without beginning or end) and all-inclusive – meant for all of mankind. And second, the meaning of *Dharma* is very different from the meaning of "religion" in English. *Dharma* means intrinsic value or property of man, or righteous moral duty of man, in both cosmic and terrestrial senses, and it does not refer to a set of top-down commandments or codes of social behaviour to be obeyed.

Popular perception of the philosophy can be different, however. This is because, as the philosophy is so profound, it can only be understood with a fine intellect which is very often lacking in those responsible for disseminating it to the masses. As a result, the ritualistic aspects are emphasised at the popular level, as are the manufactured dogmas.

> Every morning I bathe my intellect in the stupendous and cosmogonal philosophy of the *Bhagavad-Gita*, in comparison with which our modern world and its literature seem puny and trivial.
>
> Henry David Thoreau (1817–1862)

> In the whole world there is no study so beneficial and so elevating as that of the *Upanishads*. They are destined sooner or later to become the faith of the people . . . It has been the joy of my life – it will be the solace of my death.
>
> Arthur Schopenhauer (1788–1860)

5.2. METHODOLOGY

Shortage of space does not allow us to describe in detail how curricula (content and pedagogy) for genuinely environment-respecting education are to be developed and implemented. We will therefore give below only an outline for preschool to secondary school levels. A comprehensive treatment of the subject, up to and including undergraduate university level together with important issues of educational psychology, will be found in a collection of contributions by the author and his colleagues, soon to be published in UNESCO's *Encyclopaedia of Life Support Systems* (EOLSS).

5.2.1. Basic criteria for curriculum development for school children

Recalling our own school days, most of us have memories of how an incompetent or dull teacher made an interesting subject boring, and, conversely, how an enthusiastic and competent teacher made an otherwise boring subject interesting and worthwhile. Like the social sciences themselves, the "art" of teaching is imprecise and made even more so by the complex and varied personalities of both teachers and pupils. And so there is no such thing as a perfect teaching-learning method or process. The effectiveness of the method or process is fundamentally determined by the ability and motivation of pupils to learn on one hand, and by the competence and motivation of the teacher(s) to teach on the other.

Following the basic principles of educational psychology (e.g. Fontana, 1995; Gross, 2001), the development of curricula (content and pedagogy) must be based on certain criteria for maximising effectiveness. With reference to the stated objectives (Section 5.1), some of the basic criteria common to preschool, primary and secondary levels are these:

(a) Environmental education must be a mandatory and an integral part of curricula, and it should be adequate in terms of both depth and scope.

(b) The teaching objective must be to instil in children a caring and deeply respectful attitude to nature and all her creation that would endure throughout their lives.

(c) The method of teaching must be one that makes the subject interesting and relevant to children. In order for this to happen, the teacher must be enthusiastic, knowledgeable and able to present the subject-matter in a way that accords with the realities of the child's world as he or she perceives them. This is especially important for young children.

(d) If deemed necessary or desirable, a reward-oriented scheme of constructive competition should be introduced to stimulate or enhance children's extrinsic motivation to learn. In the case of young children reward may be in the form of gold or silver stars (made of paper) prominently displayed in the classroom, while for older children a reward is usually in the form of prizes awarded at an important school ceremony.

(e) Today there is a wealth of accessible material on nature and the environment on video, television, CD-Rom and the internet. There are also beautifully illustrated large-print children's books that children of all age groups are likely to find interesting. These materials should be used to educate children through entertainment.

(f) Children should be given the opportunity to participate in interesting activities, both in and out of school, that excite their imagination and enhance their knowledge of nature and the environment. For preschool and primary school children such activities should be closely supervised and include painting of pictures of trees, animals, etc.; and use of post-consumption waste such as paper and cardboard, cardboard boxes etc. to make toys and models (this activity will instil the rudimentary concepts of recycling and reuse in young minds). At the secondary level such activities should translate to locally-relevant joint environmental projects – involving laboratory and field work, discussion, and report writing – each to be done jointly by a team of 2–4 children as appropriate. Experience shows that such projects generate motivation and give children a sense of purpose, responsibility and satisfaction of doing something of value.

(g) Visits to ecologically interesting sites should be organised for the older children under the supervision of adults knowledgeable about the site and able to respond enthusiastically and competently to the questions children invariably ask during such visits. In the case of secondary school children such visits could be gainfully linked to their project work. Experience in Costa Rica and elsewhere shows that such visits bring enormous benefits even to primary school children.

(h) As children are a part of the culture in which they grow up, importance of the socio-cultural context to education generally cannot be over-stated. In many cultures there are deeply embedded religious and/or philosophical traditions and associated values that are highly conducive to engendering respect for nature and the environment. Over the ages these traditions and values have generated a wealth of myths, legends, rituals, folklore and even lullabies encapsulating those values. They should be invoked to instil or reinforce nature- and environment-respecting moral values in children directly, indirectly or even subliminally.

> Every ancient society seems to have been governed and influenced by mythology. While to the modern way of thinking the myths of the past may seem primitive and irrelevant to a technological society, I would contend that it is precisely because we have lost sight of those myths, and failed to see their true significance in unconscious terms, that our whole approach to life and to our natural surroundings has become so unbalanced.
>
> HRH The Prince of Wales (Porritt, 1991)

(i) Before its implementation, a newly developed curriculum should be scrutinised by independent or government-appointed experts to ensure that it is of the required quality standard and meets the teaching objectives of Section 5.1. Subsequently it should be reviewed periodically for upgrading and/or updating if necessary.

The teacher(s) must never lose sight of the teaching objectives (Section 5.1) and be constantly aware that psychologically the teacher–class relationship is greatly conditioned by children's understanding (and expectation) of what the school is able to offer them.

5.2.2. Curriculum development for preschool and primary levels

An effective and realistic approach to curriculum development for preschool and primary school children is proposed by Bartholomew and Bruce (1993). This approach, valid for secondary schools too, has three distinct but interacting aspects:

- The *child*, referring mainly to the *affective factors* of personality.
- The *content*, which refers to what the child already knows, what he/she needs to know, and what he/she wants to know.

- The *context*, defined by people, culture, race, gender, special educational needs, access, materials and physical environment, indoors, outdoors, places and events.

In order for the teaching-learning process to be effective, for all age groups – preschool, primary and secondary – the development of curricula and the process itself must address what are called the *affective factors* of personality (Fontana, 1995). As a part of the human condition itself, these factors (e.g. interest and relevance, self-esteem, attitude and behaviour) can and do influence, in varying degrees depending on the individual's psychological make-up, both the ability of pupils to learn and the ability of teachers to teach.

With regard to content, "what the child already knows" constitutes his or her initial knowledge-base on which to build further knowledge. "What he/she needs to know" is a judgement made by the teacher, or curricula developer(s), on what is considered to be of value for the child to know. "What he/she wants to know" is determined by the child's intrinsic motivation to explore his/her sur-roundings through the senses for deriving hedonistic pleasure characteristic of children (Eisenberg, 1982). However, with the development of children's extrinsic motivation, they become increasingly interested in what is being taught.

At both preschool and primary levels the "what he/she needs to know" element should cover the following with an environmental theme: supervised painting; making toys and models; games and music; read-ing recommended children's books, and watching recommended televi-sion programmes. Clearly, at each level the design of curricular activi-ties must be commensurate with the cognitive ability of the child at that level. Teaching must be through entertainment and pleasure-giving activi-ties, because both preschool children and younger primary school children tend to have predominantly hedonistic and self-focused orientation (Eisen-berg, 1982).

The *context* constitutes the personal *space* or *world* of the child, within which he or she perceives his or her own reality.

With regard to the above, curricula should be developed based on the criteria listed under Section 5.2.1. Experience shows that best results are achieved when curricula are balanced. That is, when all three aspects – *child, content* and *context* – are given equal emphasis. If one of the aspects is over-emphasised, curricula can become unbalanced resulting in loss of quality (Bartholomew and Bruce, 1993).

5.2.3. Curriculum development for the secondary level
In many societies there is, and has been, a tendency for secondary school pupils to take those subjects that are advisedly more likely than others to lead to a university degree, diploma or vocational qualification (in accountancy, medicine, engineer-ing, information technology, etc. depending on market demand) likely to secure a comfortable "meal ticket" on completion of formal education. Although this behaviour is understandable, it often relegates environmental and moral studies (if included in the curriculum at all) to the "optional" status with the result that only a small minority of pupils opt for them.

However, if we are at all serious about environmental protection and sustainable development, education engendering genuine respect for nature and the environment must be a mandatory part of formal curricula, not optional. To that end the outline of a generic syllabus is proposed in Box 1. It is to be taught as five or six different subjects throughout secondary education, each commensurate with pupils' growing cognitive skills.

Emphasis on non-technical topics in the syllabus of Box 1 is deliberate. It is because environmental subjects based on the "hard" sciences are almost exclusively concerned with "end-of-the-pipe" strategies (for example, chemistry in the case of waste water treatment). That is, dealing with the consequences of pollution *after* it has been created. However, what is really needed for meaningful environmental protection, and especially sustainable development, are "before-the-pipe" strategies *to prevent or eliminate, if possible, the creation of pollution in the first place*. Indeed, prevention is at the top of the European Union's hierarchy of waste management options (Powrie and Robinson, 2000). Although technology can play a significant role in pollution reduction (e.g. with clean or cleaner technology, which only the rich countries can afford as it happens), prevention and elimination strategies, based on the social sciences and moral philosophy, are arguably more important.

6. Conclusion

The "means of implementation", agreed at the WSSD in Johannesburg to reinvigorate the Agenda 21 process aiming at global sustainable development, is the starting point of discussion in this paper. Focusing on raising both universal literacy rate without gender discrimination and awareness of sustainable development, these "means" are expected to push forward the Agenda 21 process on a wider front than hitherto.

Given that access to primary education is a basic human right, and that widest public awareness of sustainable development is a priority, these "means" are much to be applauded. However, as pointed out in Sections 2.1 and 2.3.3, it is not clear how raising literacy rates would or could advance the cause of sustainable development. In fact, the opposite would appear to be true; compelling evidence for this is provided by the rich, developed nations themselves. With high or very high literacy rates and high educational achievement, they are the biggest consumers and polluters with life-styles that are grossly unsustainable and increasingly so.

In Section 3.3 it is demonstrated that although science and technology can help the process of global sustainable development, albeit in a limited way, they cannot deliver it. And in Section 5 that, if the international community is at all serious about achieving even a modest degree of sustainable development, then the moral values that support and promote the pervasive, exploitative and grossly environment-degrading Western life-styles must be radically changed to those that *genuinely* respect both nature and the environment. In order for this to happen,

---

Box 1. Outline of a proposed generic environmental syllabus for the secondary level.

Objective: To impart knowledge of earth's natural environment, how it works, how human activities have been relentlessly degrading its integrity, and how such adverse impacts could be minimized or eliminated. To emphasise the one-ness of all living and non-living entities inhabiting the same continuum of existence and consciousness and nourished by Mother Earth and her abundant life-sustaining benediction.

Environmental study: Ecology, nature and the environment. Earth's geo-biochemical processes and natural cycles (e.g. the carbon cycle). Environmental degradation caused by domestic, industrial and agricultural activities. Pollution of air and water. Contamination of land. Effects of growing consumerism on the environment. Global warming, ozone depletion and growing fresh water deficit, and their likely impacts on environment and quality of life.

Studies on sustainable development: The concept of sustainable development (SD), its importance for both present and future generations, and what needs to be done to achieve it. Political, economic, demographic and other obstacles to SD. What individuals can do to facilitate SD. Consequences of pervasive global-isation for nature and the environment. The widening "north–south divide" and its consequences for the global environment. Case studies.

Moral studies: Moral and ethical basis of determining what is good and what is bad. Morality, rights and obligations with regard to nature and the environment. Psychological basis of morality. Religious basis of morality. Western moral values *vis-à-vis* nature and the environment advocated by Albert Schweitzer, Alfred North Whitehead, Vladimir Vernadsky, Pierre Teilhard de Chardin, James Lovelock and Arne Naess among others. Eastern moral values advo-cated by Ming Dao Deng, *The Ramayana, The Mahabharata, Sri Isopanisad* and *The Bhagavad-Gita.*

Coursework: Emphasis should be given to coursework in the form of projects as mentioned in (f) and (g) in Section 5.2.1. In addition, group discussions should be organised under teacher supervision on topical local or national environmen-tal issues and problems with the objective of engaging children in articulating practical ways in which they and society at large could help protect the environ-ment and promote SD.

Invited lectures (extra curricular): From time to time experts from academia, business and industry, and policy-makers from local (municipal) governments should be invited to make presentations to children aged 14 and over on topical and locally-relevant environmental issues and problems. And children should be given ample opportunity to ask questions and express their views during such events. Experience shows that these events enhance children's motivation to learn and reinforce the relevance of the subject in their eyes.

the case for including moral philosophy as an important (mandatory) element of formal curricula at all levels of education – primary, secondary and higher – is strongly argued.

Curiously, the scientific community appears to have a deep but demonstrably misplaced conviction that science and technology *alone* can deliver global sustainable development. Those who are so convinced need to contemplate and reflect on the following points:

- Philosophy is unquestionably the foundation of all knowledge. As Albert Einstein famously observed, *"Knowledge without philosophy is just mechanics"*. Surely, the objective of scientific or technical education (or any education for that matter) cannot be, and must not be, to produce scientists or engineers who think or act mechanically. Yet, today philosophy does not seem to appear as an integral part of formal scientific or technical curricula.
- Unlike animals of lower species, humans have an innate need to rationalise all their actions and thoughts. And it is this need that sets us apart from other earthly creatures. Philosophy, moral philosophy in particular, provides this rationale, and by doing so it gives us our humanity.
- If science and technology alone could deliver global sustainable development, re-orient moral values or alter attitudes for achieving it, then how is it that the rich, developed nations, that are abundantly endowed with latest science, technology, and financial and skilled manpower resources, are the biggest polluters and consumers with grossly unsustainable life-styles?

Given the international community's exclusive but demonstrably misplaced reliance on science and technology to deliver global sustainable development, as well as anticipated resistance from various vested interest groups that stand to benefit from the *status quo*, we have no illusion about the difficulty of educating the young in moral philosophy as proposed herein. It has to be done, nonetheless, in the interests of both present and future generations. Otherwise the global environment will continue to deteriorate, and, as a result, future generations will remain at serious risk of inheriting a highly polluted world denuded of its resources, a world of science without civilisation, and insatiably avaricious societies bereft of humanity.

It is hoped that this paper, deliberately written in a slightly provocative style, will generate constructive discussion on this important issue.

References

Allan, D.J.: 1970, *The Philosophy of Aristotle*, Second Edition, Oxford, UK, Oxford University Press.

Anderson, F.H.: 1948, *The Philosophy of Francis Bacon*, USA, University of Chicago Press.

Bartholomew, L. and Bruce, T.: 1993, *Getting to Know You*, London, Hodder & Stoughton.

B.B.C.: 2002, 'The Century of the Self', a documentary broadcast by the British Broadcasting Corporation, London, during 29 April and 2 May, 2002 (www.bbc.co.uk/bbcfour/documentaries/features/century_of_the_self).

Clarke, D.: 1982, *Descartes Philosophy of Science*, UK, Manchester University Press.

Eisenberg, N.: 1982, 'The development of reason regarding prosocial behaviour', in N. Eisenberg (ed.), *The Development of Prosocial Behaviour*, New York, Academic Press.

Fontana, D.: 1995, *Psychology for Teachers*, Third edition, Basingstoke, United Kingdom, Palgrave.

GHESP: 2002, *Global Higher Education for Sustainable Partnerships*, Paris, International Association of Universities (IAU), Unesco House.

George, S. and Sabelli, F.: 1994, *Faith and Credit: The World Bank's Secular Empire*, London, Penguin Books.

Gross, R.: 2001, *Psychology: The Science of Mind and Behaviour*, Fourth edition, London, Hodder & Stoughton.

Hawking, S.: 1988, *A Brief History of Time*, New York, Bantam Books.

ICSU: 2002, ' Science education for capacity building for sustainable development', *Series in Science Education for Sustainable Development*, No. 5, Paris, International Council for Science.

Kohlberg, L.: 1981, *Essays on Moral Development*, New York, Harper & Row.

Lee, N.: 1995, 'Environmental Policy', in M.J. Artis and N. Lee (eds.), *The Economics of the European Union*, UK, Oxford University Press.

Lesser, J.A., Dodds, D.E. and Zerbe, R.O.: 1997, *Environmental Economics & Policy*, Addison-Wesley, New York.

Nath, B. and Talay, I.: 1996, 'Man, science, technology and sustainable development', in B. Nath, L. Hens and D. Devuyst (eds.), *Sustainable Development*, Brussels, VUB Press, pp. 17–56.

Nath, B.: 2002, 'Environmental regulations and standard setting', in *Knowledge for Sustainable Development – an Insight into the Encyclopaedia of Life Support Systems*, Paris, UNESCO.

Nath, B. and Talay, I.: 2003, 'The importance of teaching environmental education at an early age', in *Encyclopaedia of Life Support Systems (EOLSS)* Paris, UNESCO (to be published in 2004).

Oxfam: 2002, *Rigged Rules and Double Standards – Trade, Globalisation, and the Fight Against Poverty*, London, Oxfam International.

Powrie, W. and Robinson, J.P.: 2000, 'The sustainable landfill bioreactor – a flexible approach to solid waste management', in B. Nath, S.K. Stoyanov and Y. Pelovski (eds.) *Sustainable Solid Waste Management in the Southern Black Sea Region*, Dordrecht, the Netherlands, Kluwer Academic Publishers, pp. 113–140.

Pilger, J.: 2002, *The New Rulers of the World*, London, Verso.

Porritt, J.: 1991, *Save the Earth*, London, Dorling Kinderseley.

Rae, A.I.M.: 1993, *Quantum Mechanics*, Third Edition, UK, Institute of Physics Publishing, Bristol.

WCED (World Commission on Environment and Development): 1987, *Our Common Future*, Oxford, UK, Oxford University Press.

WWF: 2000, *The Living Planet Report 2000*, London, World Wildlife Fund.

CHAPTER 13

SCIENCE, RESEARCH, KNOWLEDGE AND CAPACITY BUILDING

ALFRED W. STRIGL

*Austrian Institute for Sustainable Development, c/o University of Natural Resources and
Applied Life Sciences, Vienna Lindengasse 2/12, A-1070 Vienna, USA
(e-mail: alfred.strigl@boku.ac.at, www.boku.ac.at/oin)*

Abstract. A small part of the scientific community is seeking hard to enhance the contribution of science, knowledge and capacity building to environmentally sustainable and socially fair human development around the world. Many researchers over the globe share the same commitment – anchored in concerns for the human condition. They believe that science and research can and have influenced sustainability. Therefore their main goals are to seek and build up knowledge, know-how and capacity that might help to feed, nurture, house, educate and employ the world's growing human population while conserving its basic life support systems and biodiversity. They undertake projects, that are essentially integrative, and they try to connect the natural, social and engineering sciences, environment and development of communities, multiple stakeholders, geographic and temporal scales. More generally, scientists engaged in sustainable development are bridging the worlds of knowledge and action. This pro-active, heavily ethics- and wisdom-based "science for sustainability" can be seen as the conclusion of all dialogues and discussions amongst scientists at the World Summit on Sustainable Development (WSSD) 2002 in Johannesburg. The "Plan of Implementation" after WSSD will be based on political will, practical steps and partnerships with time-bound actions. Several "means of implementation" are going to be proofed and initiated: finance, trade, transfer of environmentally sound technology, and, last but not least, science and capacity building.

Some characteristics of working scientific sustainability initiatives are that they are regional, place-based and solution-oriented. They are focusing at intermediate scales where multiple stresses intersect, where complexity is manageable, where integration is possible, where innovation happens, and where significant transitions toward sustainability can start bottom-up. And they have a fundamental character, addressing the unity of the nature – society system, asking how that interactive system is evolving and how it can be consciously, if imperfectly, steered through the reflective mobilization and application of appropriate knowledge and know-how. The aims of such sustainability-building initiatives conducted by researchers are: first to make significant progress toward expanding and deepening the research agenda of science and knowledge-building for sustainability; secondly to strengthen the infrastructure and capacity for conducting and applying science, research and technology for sustainability – everywhere in the world where it is needed; and thirdly, to connect science, policy and decision-making more effectively in pursuit of a faster transition towards real sustainable development. The overall characteristic is, that sustainability initiatives are mainly open-ended networks and dialogues for the better future. A world society that tries to turn towards sustainable development has to work hard to refine their clumsy technologies, in "earthing" their responsibility to all creatures and resources, in establishing democratic systems in peace and by heeding human rights, in building up global solidarity through all mankind and in commit themselves to a better life for the next generations.

Key words: capacity building, implementation, knowledge, science, sustainable development.

Abbreviations: CGIAR – Consultative Group on International Agricultural Research; COMSATS – Commission on Science and Technology for Sustainable Development in the South; ICSU – International

Readers should send their comments on this paper to: BhaskarNath@aol.com within 3 months of publication of this issue.

L. Hens and B. Nath (eds.), The World Summit on Sustainable Development, 299–317.
© 2005 *Springer. Printed in the Netherlands.*

Council for Science; IFS – International Foundation for Science; IMF – International Monetary Fund; IPCC – Intergovernmental Panel on Climate Change; ISP – International Science Programme; MEA – Millennium Ecosystem Assessment; NGO – Non-governmental Organization; ODA – Official Development Aid; PrepCom – Preparatory Commission for the World Summit on Sustainable Development; SCRES – The Standing Committee for Responsibility and Ethics in Science; SRISTI – Society for Research and Initiatives for Sustainable Technologies and Institutions; TWAS – Third World Academy of Sciences; TWNSO – Third World Network of Scientific Organizations; UN – United Nations; UNDP – United Nations Development Programme; UNEP – United Nations Environment Programme; UNESCO – United Nations Educational, Scientific and Cultural Organization; UNFPA – United Nations Populations Fund; WB – World Bank; WCED – World Commission on Environment and Development; WGBU – German Advisory Council on Global Change; WSSD – World Summit on Sustainable Development; WTO – World Trade Organization.

> *To what avail are all our pieces of knowledge,*
> *if we do not care about what holds them together.*
>
> Dalai Lama

1. Introduction: what does a fair world mean in respect to finite environmental resources?

For several years the concept of "Sustainable Development" has occupied a major role on the global agenda. But although the idea spread over the world, concrete results so far are poor. The World Summit on Sustainable Development (WSSD) in Johannesburg 2002 reflected and reassessed this ambitious goal again. And at this occasion, the international community once addressed the challenges posed by chronic poverty and resource-hungry affluence. Socially just and ecologically embedded development is high on the agenda for the coming decades. This can be fully understood, given the systematic neglect of justice, equity and fairness in world politics. But in contrast to that, the scientists in Johannesburg argued that it was about time that the South along with economies in transition embraced the environmental challenge. They claimed that responsible care for the environment is one of the keys for ensuring livelihood and health for the marginalized sections of the world's citizenry. In fact, many studies and observations show that there can be no poverty eradication without saving the ecological function of nature. In this respect a prominent overview is given in the UNFPA-Report *"The State of the World Population 2001: population and Environmental Change"*. Moreover, an environmental strategy is indispensable for moving beyond the hegemonic shadow of the North. And both North and South have to leapfrog beyond fossil-based development patterns.

Sustainability at all scales is now historically the greatest challenge. In particular, economic globalization has largely washed away gains made on the micro level, spreading an exploitative economy across the globe and exposing natural resources in the South and in Russia to the pull of the world market. For the Johannesburg Agenda several background themes have been identified which ought to run through all the debates: water, energy, health, agriculture and biodiversity. Above all, the following question was seen as crucial: what does a fair world mean in respect to finite environmental resources? The answer sounds easy and wise: real fairness, on the one hand, that entails enlarging the rights of the poor to their own habitats. And on the other hand, it calls for a fair economic and financial system to cut back the claims of the rich to the resources exploited in the South.

The question of global fairness has much to do with life-patterns of the rich population in the North which are spreading out globally. Interests of local communities in maintaining their livelihoods often collide with the interests of urban classes and corporations in expanding consumption and profits. These resource conflicts will not be eased unless the economically well-off around the globe move towards resource-productive patterns of production and consumption. Sustainable production and consumption therefore inherently means fair and many times more resource-efficient technologies and life-styles. The greatest necessity is to achieve sustainable societies through decoupling economic growth and environmental impact. Addressing issues related to clean technology, zero emission or eco-efficiency it is possible, as a point of departure, to generate sustainable energy and material use.

Secondly sustainability needs fair prices and a fair financial system. Due to the burden of debt, the South has to pay interest several times higher than what it gets from Official Development Aid (ODA). The winners of globalization are mainly found in the transnational corporations and financial business sector located in the North. The institutional protagonists for this kind of economic and financial globalization, the International Monetary Fund (IMF), the World Bank (WB) and the World Trade Organization (WHO) are heavily criticized for their liberalization policies and blamed for widening the gap between world's poor and rich citizens. We all should be aware that these organizations are ruled and controlled by us – the rich people's nations like the EU-states and the USA. Therefore the key for sustaining world's development is in our politician's hands (Strigl, 2001).

2. Which grand challenges do earth's societies face?

The United Nation (UN) Secretary General Kofi Annan reflected a growing consensus towards worldwide sustainable development when he wrote in his Millennium Report "*We, the Peoples: The Role of the United Nations in the 21st Century*" to the General Assembly that freedom from want, freedom from fear and the freedom of future generations to sustain their lives on this planet are the three grand challenges facing the international community at the dawn of the 21st century (Annan, 2000). But there is great asymmetry in the resources and attention devoted to harnessing science and capacity building in the service of these three goals.

Efforts to achieve "freedom from want" have created and been supported by several effective research and development systems, for example those engaged in international agricultural research and in certain global disease campaigns. Efforts to achieve "freedom from fear" are supported by a mature, well-funded and problem-driven research and development system for instance in the world's military establishments. In contrast, efforts to achieve sustainability are relatively new because, in the words of the Secretary General, the "founders of the UN could not imagine that we would be capable of threatening the very foundations for our existence." (Annan, 2000) Science, research, knowledge and capacity building are increasingly recognized to be central to the Secretary General Annan's three challenges (United Nations Development Program (UNDP), 2001; World Bank, 1998). The WSSD represented the best opportunity in a decade to construct a global research and development system tailored to the particular needs and magnitude of the sustainability challenge (Sachs, 2000).

The WSSD conference in South Africa was conceived as a follow-up to the Earth Summit held in Rio de Janeiro 1992. But this time, delegates were encouraged to move beyond the environmental focus of Rio, and address the three pillars of sustainable development. The six billion people currently living on Earth are thought to consume 40% of the terrestrial biomass, between a quarter and a third of marine resources and about 50% of the planet's accessible fresh water (UNFPA, 2001). The *"Global Environment Outlook 3"* report from the United Nations Environment Programme warned that half of the world's population is likely to suffer water shortages by 2032 (UNEP, 2002). Yet, the global population is projected to climb throughout this century, stabilizing at atleast 9 billion by 2100 (UNDP, 2001). Mankind's trajectory is far from reaching sustainability. From a scientific point of view we can say: we are most definitely destroying many important ecosystems of the planet.

Mankind still has a long way to go in tackling this challenge. Many times in history, it was a tiny but tenacious minority of committed people (amongst them often scientists – like in the 1970s the Club of Rome) who first defined the misery and have done much to address it (Clark, 1986). Researchers put climate change and the Earth's finite resources on the political agenda (at Rio), and their voices were some of the loudest in calling for policies to be changed in response to these threats. Since the Earth Summit in Rio 1992 these researchers have been more closely in tune with both the public and policy-makers and have begun to coordinate themselves in order to pursue sustainable development in many more arenas. Yet, there remains much more that individual researchers and their institutions can do. In this respect this article gives some suggestions.

Gripping the opportunity of the Johannesburg summit will require a strategic approach that has often exceeded the interests of individual nations, policy initiatives and research programmes. Fortunately, important elements of the foundation for such a strategy have been laid out over the last several years through a rapidly expanding discourse on the relationships among science, knowledge and sustainability. Many of the earliest and most thoughtful contributions to this discourse have come from the developing world through the work of individual scholars and institutions such as the International Council for Science (ICSU), the Third World Network of Scientific Organizations (TWNSO), the Third World Academy of Sciences (TWAS) the Commission on Science and Technology for Sustainable Development in the South (COMSATS), the Society for Research and Initiatives for Sustainable Technologies and Institutions (SRISTI), and the South Center (see listed homepages).

European ideas and strategies of the late 1990s are exemplified in Schellnhuber and Wenzel's *Earth systems analysis: integrating science for sustainability* (1998) and the European Union's Fifth Framework Programme (European Commission, 1998). A synthesis of US views from the same period is given in the National Research Council's *Our common journey: a transition toward sustainability* (1999). Initial efforts to capture an international cross-section of perspectives include the special issue on Sustainability Science published by the International Journal of Sustainable Development (1999), and the World Academies of Science report on a Transition to Sustainability in the 21st Century (2000).

In addition, international environmental assessments are increasingly reaching out to connect with sustainability issues, as are research planning efforts for global environmental change programmes at both national and international levels (Watson et al., 1998; IPCC, 2001; United Nations Environmental Programme, 2002). A number of academies of science have also recently addressed the links between sustainability and global change (German Advisory Council on Global Change (WGBU), 1997; African Academy of Sciences, 1999; Kates et al., 2000; Rocha-Miranda, 2000). A great deal of work in the area of science for sustainable development in a WSSD context has been done by the "ICSU" which launched a series of ten brochures for Johannesburg. In particular the report 4 (*Science, Traditional Knowledge and Sustainable Development*), report 5 (*Science, Education and Capacity Building*) and report 9 (*Science and Technology for Sustainable Development*) can be addressed here. ICSU is still very active in the post-WSSD process.

3. Was the WSSD in Johannesburg a science summit?

Why is this question raised, when the answer is obviously no? Because it could have been also a science summit – much as it was somehow an non-governmental organization (NGO)-summit for the NGOs and a business-summit for the business community. But the funny thing is that the results of the summit, the papers and documents, give the impression that the WSSD was an ordinary science symposium.

One of the major outcomes of the WSSD is the "Plan of Implementation". Analysing this document just in counting the frequency of selected words gives the following result:

Science (and scientific):	46
Research:	35
Knowledge (and know-how):	42
Capacity building:	41
Values:	9
Ethics:	4
Wisdom:	0

Due to the role of science and capacity building that is seen in the implementation document many scientists are cautiously optimistic – even after this summit. But some scientists – more than bureaucrats and politicians – have the conviction, that science and capacity building for sustainable development, especially, have to foster value shifts to sustainable and local wisdom, since ethics-based wisdom combines both knowledge and values. In that respect the "Plan of Implementation" heavily neglects the latter.

From scientific point of view it's obvious that the Johannesburg Plan of Implementation, the Political Declaration and the Type II Partnerships are not enough. Sustainable development is not high in the agenda of national parliaments and policy-makers, therefore it is not well placed in the agenda of high international politics. But pointing fingers at the politicians and putting the blame on them would be foolish. The way scientists can realize what (they think) has to be done either can be managed through influencing political decision directly or via "civil society". With the help of the public, there is a lot that science has achieved in the past, for example the Montreal Protocol, signed in 1987. The problem of ozone depletion as such was initially identified by the scientific and environmental community. High-profile meetings and actions by NGOs then convinced the relevant (political and industrial) actors, even before there was conclusive scientific evidence.

There are parallels with the today's Kyoto Protocol discussion. Companies that use large amounts of fossil fuel are clamouring for solid political commitments on carbon emissions. This industry is understandably reluctant to invest in infrastructure that may be outmoded in a few years' time. Actions that change behaviour without firm commitments from government are known as "type-2 – partnership initiatives" in the arcane parlance of the Johannesburg meeting, and have great potential to allow scientists and industry to address some of the critical issues of sustainable development. Two questions challenge the validity of type II – measures. First, are they going to become a practical and active reality in terms of capacity and institution building? Second, will these partnerships be enough to address the vast, major needs of sustainability. The scientific community could be relatively optimistic with the first question. Many important, promissory initiatives, in terms of design and funding, that can comply with those objectives were announced at the WSSD in Johannesburg. It only has to be ensured that they are really contradictory to existing unsustainable policies and of subsequent long term effectiveness. About the second question we should be more sceptical. The environmental and social problems are so large and so significant that "coalitions of the willing" and "voluntary partnerships" cannot replace the role of public policies, directives, laws, governance and institutional frameworks that need to transform and enforce the behaviour of societies and individuals, the ultimate factor of change to address international, national and local unsustainable *status-quo* situations.

As implied above, much of the knowledge needed for advancing sustainability goals involves making sense of how multiple environmental stresses, social institutions and ecological conditions interact in particular places. This means that science systems for sustainability will need to give special emphasis to integration at intermediate or regional scales (National Research Council, 2002). From this base, they will need to be structured to facilitate "vertical" connections between the best research anywhere in the world and practical experience in particular field situations. At the same time, they will need to foster "horizontal" connections among regional research and application centres that might learn from one another like the Knowledge Network of Grassroots Green Innovators demonstrates (see homepage list).

There is much that scientific expertize can achieve, especially in deploying existing know-how in the places where it is most needed. There is an abundance of high- and low-tech solutions to water management and energy generation in countries that lack advanced infrastructure. Yet, they need to be put in place in a rational way. But also the rich countries have much to gain from sustainable-development research. For example, the development of rural or coastal areas is regulated largely by local authorities, whereas the impacts of such development are often felt in the capitals great distances away. Fishermen and farmers can be introduced to scientifically informed approaches to fisheries management and agriculture where policies are lacking.

Finally, effective science and research systems for sustainability will need to bridge the artificial and pernicious divide between "basic" and "applied" research (Bronscomb et al., 2001). Progress on some of the most urgent problems of sustainability will almost certainly require fundamental improvements in our understanding of nature–society interactions. Thus, sustainability science needs fundamental research. On other issues the requirement is less for new knowledge than for learning how to apply what is already known in an experimental, problem-solving and solution-oriented mode. Sustainability science needs to be learning-by-doing. Sustainability science needs to be pro-active, pro-participatory and highly involved in the development process. More generally, promoting sustainability needs integrated *knowledge systems* that connect what have too often been the "island empires" of research, monitoring, assessment and operational decision support. And finally science and knowledge systems will be measured and appraised by real success in achieve higher sustainability.

4. What is on the science agenda – before and after Johannesburg?

High on the researchers' agenda is the problem of getting existing scientific expertize to where it is needed. It is arguable whether the techniques, tools and experience to do the work on the spot already exist. Geographical information systems, which combine spatial information on a particular geographical area with environmental, social and political data, have become more advanced and less expensive over the past decade. Action and participatory research, monitoring technologies of the environment and the understanding of eco- and social systems have also matured since the Rio summit. Also on the agenda is to ensure global access to scientific data and information. It is necessary to have adequate global scientific information (and monitoring) systems in place. Remote (environmental) sensing plays an important role, but it is also necessary to develop adequate monitoring of social and economic variables and to provide this data to the public.

The advances represent a unique opportunity to study for the first time resource use, climate change and the relationship between health, the environment, social and economic development. As technological advances dramatically increase the ability to obtain, store and analyse data, the digital divide between the North and the South is increasing rapidly (Kates et al., 2000). As a lead-up to the UN World Summit on the Information Society (December 2003), ICSU has announced that it will probe issues that are making scientific data and information more difficult to access – especially for developing nations. For example, research is increasingly being funded by the private sector, and organizations are intent on retaining ownership of their findings or generating revenue from intellectual property rights. In other cases, governments are looking for ways to commercialize data collected using public funds. At the same time, there is no consensus on how to pay for international and global monitoring systems.

All these tools are rarely focused on producing results that politicians and decision-makers can use. A lot of data produced by researchers never finds its way to policy, governments or other authorities. Also major international projects commonly have problems creating such 'policy-relevant' science. The Intergovernmental Panel on Climate Change (IPCC) for example, is widely admired for having communicated a strong scientific message amid the lobbying of industrial groups and environmental activists. But because the IPCC's reports are global in scope and contain huge amounts of data, it is difficult for policy-makers concerned with local issues to make use of them. As a conclusion: the crucial synthesis is linking local action to the global level.

Sustainability researchers face an exciting journey: linking global necessity with the local scale. Many local projects focus on single issues and forget to consider all dimensions of sustainability. What is needed, according to sustainability experts, is the expansion of local research projects so that they address the sustainability of all resources at a particular location at once. Such projects are beginning to emerge. Last year saw the launch of the Millennium Ecosystem Assessment (MEA). MEA is an ambitious endeavour to assess the impact of factors such as shifts in land use and loss of biodiversity on the Earth's ecosystems. Information from fish stocks to nitrogen cycles is going to be produced. But the data generated are not just for specialists. The MEA focuses on services of those ecosystems that people actually care about. MEA is of evident local interest.

For instance an MEA study of Norwegian ecosystems should help their government to decide how its fishing and oil-exploration industries can be expanded or transformed without further damaging marine ecosystems. Other MEA members are providing technical support to studies in western China or in Mexico's Yaqui Valley. There, the MEA project goes several steps further. By involving local researchers and politicians, research teams hope to produce data that can directly influence the development of the region. The effects of land and chemical use, irrigation schemes and crop type on the local ecosystem are all being studied, as well as the impacts of external factors such as agricultural policies, globalized markets and drought. By combining this knowledge, researchers hope to reveal how these factors affect the terrestrial and aquatic environment and the income of farmers and city dwellers. Decisions made in one part of the system, concerning one sector, cascade through the system, affecting many other sectors. As with many other sustainability projects the Yaqui Valley project is a "work in progress". The scientific community hopes, that with more information – particularly on the human societies at local level – a truly integrated approach will be possible.

5. How many dimensions does capacity building have?

Capacity building is a critical element for sustainable development and has been highlighted during the WSSD preparatory process. At the suggestion of ICSU, one of the two dialogue sessions during PrepCom IV focused on capacity building. It has also been widely discussed at the WSSD science forum. Discussions over how to improve scientific expertize and infrastructure for enhancing sustainable development in developed and developing countries – "capacity building" in the sustainability jargon will be limited by the lack of new money made available at the political summit. Projects that focus on local sustainable development could serve as a template for sustainability studies in other areas, but implementing similar studies or initiatives in other parts of the world could prove more difficult. The rich North has a good science base, but many countries, e.g. in Africa, would have difficulties in collaborating with OECD countries.

Nations around the world, both in the South and in the North, are experiencing a dramatic decline in interest for the natural sciences amongst young students. At the same time, a large proportion of the current generation of scientists is approaching retirement. To complicate matters further, scientists are being pulled in opposite directions: they need to be specializing to compete in cutting-edge disciplinary research, yet broader approaches are needed to deal with problems relevant for society. Some would argue that the scientific divide is even greater than the digital divide. Some scientific initiatives are exploring various initiatives to strengthen and coordinate efforts amongst research partners to address the issues around developing scientific capacity in all countries. Amongst them there is ICSU and the RING Alliance, the Regional and International Networking Group, a global alliance of research and policy organizations that seeks to enhance and promote sustainable development through a programme of collaborative research, dissemination and policy advocacy. Topics of discussion include science teaching at primary and secondary levels, specialization *versus* inter- and transdisciplinarity, capacity building, especially at the third world level, scientific networks for sustainability sciences and the challenges of institutional support. Networks of scientists are one the most important ways to tackle capacity building. They provide sharing of produced scientific knowledge, identification of common interests, understanding of impacts, dissemination and information gathering and support through sharing of facilities (Swiss Commission for Research Partnerships with Developing Countries, 2001).

Political and institutional decisions must be translated to fellowships for highly qualified education, programmes for young scientists, networks of institutions of excellence, sharing of innovative experiences, clusters of centres of excellence (like the US cooperation with India) solving practical development problems, developing international programmes, strengthening national academies and pushing political action in intergovernmental bodies. There is a need for a strategy on showing successful stories and cases to politicians, telling them what other countries are doing and how countries can learn from each other. In Africa, some countries are devoting important resources to science and technology and their example must be showed to others and followed.

Scientific networking in the 21st century has to provide an essential further step: "institutional networks". Institutionalization is more than just sharing. It strengthens the process towards a common research agenda, provides more human capacity and more resources and gives synergies for institutional growth. Centres of Excellence have a role but must be integrated in solid institutional networks. They are catalysts of research and provide capacity building opportunities and peer revision. The institutional networks must also be networks for sustainable development, with a culture of science, with people's participation, based on local realities and addressing the common good, embedded in the society and forming partnerships with governments, the productive sector and the civil society. The objectives should be to develop the social contract with science, supporting endogenous capacities and using diversity to sustain development.

6. What about "best practice examples" in capacity building?

"Leadership for Environment and Development", the full name of the LEAD-Project, is a prominent example of international capacity building, not only on science but on all sustainable development fields. The objective is finding mid career talented people, in the fields of academia, government, NGO, media and business. The LEAD network is composed by 1500 individuals, from which 30% come from the scientific community. Such networks should not only be established on global scale. Since all sustainability has to have local anchors, local, regional and national networks have to be initiated. Another example is the "Actors-Network Sustainable Austria" consisting of around 140 highly engaged individuals of decision-makers from politicians, business managers to civil servants and university professors. One aim of this network is to build up a "human infrastructure" and "social capital" to bring Austria on the track towards sustainability. The "Trieste model", a pure scientific project in international capacity building and supported by United Nations Educational, Scientific and Cultural Organization (UNESCO) and IAEA, is driven by the work of the International Centre for Theoretical Physics, the International Centre of Genetic Engineering and Biotechnology and the TWAS. They provide for capacity building in the biggest sense, contributing to the return of scientists to their countries, and transferring know-how and technologies. Trieste is a best practice example of North/South and South/South cooperation.

As a consequence of active capacity building, some official sustainable development organizations and governments are aiding individuals or are running small projects in developing countries. The Swedish government, for example, is attempting to spread its country's expertize in biomass fuels to the Baltic states. It hopes that use of the fuels, which are made from agricultural waste or specially planted crops, could reduce the Baltic region's reliance on coal and oil as energy sources. One example on larger scale is the European and Developing Countries Clinical Trials Programme, funded by the European Union. The 200 million Euro project, which was unveiled in Johannesburg, aims to establish centres for clinical trials at several locations in Africa by 2006. The scheme's backers hope that generating the right infrastructure and expertize will encourage pharmaceutical companies to run high-quality trials of treatments for diseases such as HIV, tuberculosis and malaria where they are most needed.

A new paradigm of education for sustainable development is set by the "Educational Model Network for a Global Seminar on the Environment". Cornell University as the centre organizes this global network of universities from the United States, Costa Rica, Sweden, Netherlands, India and Australia. It consists in videoconferencing, multiconferencing and satellite communication systems that focus on specific problems, with the objective of transforming institutions and empowering global citizens to cooperatively sustain human, environmental and food systems. Undergraduates and graduate students, working together with faculty form a global learning network based on concepts and theory, using literature from education and social sciences, with an holistic view. The Global learning concept and theory is constructivist, experiential learning, "learning to learn," and uses cognitive psychology. The subject matters are global warming, biodiversity, food security and supply, water and population.

Such projects are significant steps forward, but many researchers still have doubts about whether concepts such as policy-relevant science and capacity building can be successful on a large scale. Some major challenges faced are still how to decide how individuals and institutes can best collaborate to study sustainability and whether, perhaps by the United Nations, science for sustainable development would benefit from coordination at a global level. Any attempt to clarify these issues could be enormously helpful, especially to the numerous scientists who are interested in sustainability but are unsure how they can contribute (Funtowicz and O'Connor, 1999). Even if scientific organizations within the UN-system or the ICSU and its partners can provide a blueprint for organizing the field, big political stumbling blocks remain.

It is not impossible to design and implement effective research and development systems to mobilize science and technology for sustainable development. Some relatively successful international programmes exhibiting many of the characteristics outlined here have already been developed to address problems ranging from increasing agricultural productivity, combating human disease to protecting the earth's ozone layer. Likewise, there already exist efforts that have made a good beginning in implementing integrated, solution-driven, place-based research and applications programmes in support of sustainability as for example the programme of the Southeast Asia START Regional Centre which is part of the global START Network. START, the Global Change SysTem for Analysis, Research and Training, is another "best practice example" for global networking to encourage multidisciplinary research on the interactions of human and environment affecting and being affected by global changes (see homepage list).

7. How to communicate the "scientific value" of sustainability?

A second challenge for sustainability research is that both politicians and science-funding agencies want to see a short-term return from most of their investments, and preferably in their own country or research area. Money for long-term multi- and transdisciplinary projects is hard to secure and often dries up after an initial outlay. Funding for big UN-research centres (e.g. Consultative Group on International Agricultural Research, (CGIAR)) from developed nations has, for example, been falling over the past decade. Other areas that are critical to sustainable development are also suffering. Renewable energy sources, for instance, were set high on the political agenda in Johannesburg, but funding for the research in this field has fallen deep by more than half as many developed countries slashing their energy-research budgets.

There are exceptions. Funding agencies in some developed countries are beginning to make greater provision for research that focuses on sustainable development, but the overall prospects are bleak. One explanation is that a 4 or 5-year electoral cycle is not equipped to deal with initiatives that will take 20 years to begin to take effect. And politicians are (by their very nature) unlikely to opt for large, long-term investments. Although individual researchers cannot overhaul political systems, they can ensure that politicians are aware of how much science can contribute to sustainable development. This could in turn lead to changes in funding policy.

One should not forget that the limited actions towards sustainability we already have are predominantly driven by scientific data. Otherwise we would not think about such a topic at all. Often the academic world is struggling over definitions and the reliability of the data. Moreover, science is not particularly good at selling itself, if compared to politics or economy. Communication about sustainability issues is very poor today – even among scientists! Often many scientists cannot properly explain what they are doing to other scientists. So it is too easy to complain that the decision-makers do not understand the problem and the urgency. The key to success will be avoiding such communication problems. A solid description of exactly what researchers can do for sustainable development may provide a boost to everyone involved: from the scientists in developed countries to administrators who control the purse-strings of funding organizations. Only a pro-active involvement of scientists and the commitment of researchers in initiatives for sustaining our world's development can help to break out of the ivory tower.

Science and technology are global but the applications could be very local and relevant, like removing illiteracy and malaria, as targets that we want to achieve. It is not going to solve all the problems but it's just a tool that can help with job creation, access to goods and services reduce mortality rates, improve literacy, access to safe water as a result of improved management, money transactions and logistic support in disaster recoveries, among others. All modern technologies, especially information technologies, must be sustainable (Strigl, 2001). If people cannot afford these technologies, there are no benefits. They must be accessible and affordable; they cannot be a charity; the poor people must desire it. They should be easy to use and trustable. The scientific community should contribute by placing their knowledge into the network, and the institutions from developing countries, like TWAS for example, could introduce copyright schemes such as charging small royalties for providing information. Intellectual property rights could produce win–win situations and concrete solutions. Multilingual software, support of local entrepreneurs, power development without grids and provision of wide band are greatly needed. Such access to information and know-how is essential for reducing the gap between the North and the South and essential for social welfare in a rapidly changing world.

8. Wanted: scientists with hearts and new ideas – all over the world

Since Rio some progress has been made in the development of codes of practice and guidelines within the science community. Chapter 31 of Agenda 21 presents clear principles on the role of science and sets out the need to develop, improve and promote international acceptance of codes of practice and guidelines recognized by the society (United Nations, 1992). Engineers and medical doctors are bound by professional codes of ethics that state categorically that the public interest, life, safety and property, overrides private interest in the practice of their profession. The World Federation of Engineering Organizations (WFEO) incorporated a "code of environmental ethics" into its engineering code of ethics. The engineering community also endorsed the "Earth Charter" which calls upon member governments, professionals and civil society to accept a moral and ethical guide of conduct and to commit to sustainable development. ICSU's "Standing Committee for the Responsibility and Ethics in Science (SCRES)" has completed an analysis of 115 codes of practice and standards from the science and technology community (SCRES, 2002).

Society depends on scientists and engineers as responsible individuals, to guard against negligence and misconduct and to safeguard mankind. Ethical challenges include: conflict of interest, whistle blowing, human rights, free migration of professionals and research funding. In addition, scientists and engineers are increasingly being called upon to become more engaged with the public and policy-makers on highly emotive issues such as food safety, GMOs, gene technology, stem cells, cloning, use of animals in research and nuclear energy – to name a few. The view of scientists and engineers solely as "independent" knowledge generators has been irrevocably altered by changes in society. Scientists now acknowledge they must take responsibility for the implication of their results, potential uses and abuses and impacts on people and societies (SCRES, 2002).

Scientific knowledge and new technologies continuously challenge and sometimes change society's values radically. Scientists and engineers have an obligation to contribute to this discussion. No sector of society has more knowledge about issues that generate ethical dilemmas and who also have the capacity to help to resolve them. For that reason, it is important to promote ethical sensitivities beginning with individual scientists and engineers.

Closely related to these ethical concerns is increasing awareness that cultural diversity is a factor that must be effectively integrated within efforts to achieve sustainable development. Each country faces its own challenges and requirements guided by their own culture and values. Most of the scientists welcome the opportunity to engage in an open and constructive dialogue with policy and decision-makers and society that will enable us to better reflect the wide diversity of culture and values throughout the world (Gupta, 1999).

A new generation of scientists is needed, particularly for sustainability needs, with a more holistic approach, but not only in the rich developed states. How can we ensure that science is done everywhere? Input and work coming from "Diaspora", or repatriating scientists can be a solution, but nothing replaces the need of developing a home-based scientific capacity (Binder, 2002). New programmes for Ph.D. training are needed taking into account special needs of sustainable development, as well as competitive research grants. It is very important to not forget young scientists. On an individual or institutional level, researchers can begin to foster relationships with their colleagues in poor countries and to look for ways to apply their research to sustainable development.

Scientists, working in concert with others, are showing that they can help to steer the world towards a more sustainable future. Several multidisciplinary projects that are well suited to informing sustainable policy decisions have been created over the last years – for example the mentioned MEA. Plans are also afoot to reinforce much-needed scientific knowledge and research capacity in the developing world and there already exist some capacity (Swiss Commission for Research Partnerships with Developing Countries, 2001). Efforts funded by the WB to introduce more fuel-efficient cooking stoves have begun to pay off by reducing biomass burning and respiratory disease in places such as China and India. The partnership between the University of California, Berkeley, and Nairobi-based Energy Alternatives Africa to establish a photovoltaic electricity industry in Kenya is now spilling over into other African countries. The EU-funded European and Developing Countries Clinical Trials Programme, which was unveiled at Johannesburg, could attract developed countries to carry out the high-quality clinical trials for locally important drugs that are so desperately needed in Africa.

Medical science should be developed in order to meet the threat of newly emerging diseases like AIDS and the return of old diseases such as tuberculosis and malaria that are becoming resistant to present treatments. It is important to improve, promote and spread appropriate agricultural methods, in particular where introduction of industrialized methods of farming have lead to health disasters, destruction of natural biodiversity and traditional sustainable agricultural practices and the impoverishment of rural populations. The manipulation of human, animal and plant DNA must be treated with the greatest caution (precautionary principle). Such developments are irreversible, and scientific methods may fail to predict all the consequences and side effects. The patenting of life forms and the privatization of the knowledge of indigenous people must be prevented. Here the sustainability-group amongst the scientific community has the obligation to find ways and models how to deal with individual property rights, patents, trade-mark and copy-right protection whilst the poorest are suffering or being exploited from "protected" products and technologies and the rich still prosper from this asymmetry.

Last but not least the spectrum of economic schools of thought to be heard in decision-making processes and being taught in schools and universities must be widened. Neo-classical economics, for all its merits and harm, is unable to grasp important aspects of sustainable development. In many instances, it is more a part of the problem than of the solution. Other economic schools of thought like ecological economics provide insights that are essential for any policy towards sustainability and these should be properly valued. An important resource scarcity of the future could be the brainpower of heterodox economists.

9. Conclusion: what does the new contract between science and the public looks like?

The magnitude of human impacts on the ecological systems of the planet is apparent. There is also increased realization of the intimate connections between these systems and human health, the economy, social justice and national security. The urgent and unprecedented environmental and social changes challenge scientists to define a new social contract (Lubchenco, 1998). This contract represents a commitment on the part of scientists to devote their energies and talents to the most pressing problems of the day. Addressing social equity, poverty reduction and other societal needs must be integral to scientific, engineering and technological endeavours. The historically new and yet unmet needs of society include more comprehensive information, understanding and technologies for society to move toward a more sustainable biosphere, which is ecologically sound, economically feasible and socially just. New fundamentally deep and accurate science, pro-active and committed research that contributes not only to knowledge accumulations but also to a sustainable change, faster and more effective transmission of new and existing knowledge to policy- and decision-makers, and better communication of this knowledge to the public will all be required to meet this challenge. In turn, society has a responsibility to provide adequate funding, up-to-date research facilities, and appropriate career structures, as well as opportunities to inform and participate in the decision-making process. Such an effort requires a new contract between science and society in which ethical dimensions play a central and guiding role.

There is no doubt that (harnessing) science and knowledge-building have to become vital forces for sustainable development. Such a development depends on processes that will ensure the involvement of all appropriate (scientific) input and expertize in problem identification and response. Scientific excellence and integrity needs to be combined with a close dialogue and cooperation with policy-makers, stakeholders and implementers, Funtowicz and O'Connor, 1999). This includes full participation by experts with local and regional knowledge and wisdom in developed and in developing countries, since sustainability is a global challenge. Many scientific institutions asked an indispensable question: can real sustainability be reached without involving stakeholders? Their findings confirm that first and foremost, effective research and knowledge-building systems for promoting sustainability will need to be structured so that they are driven by the most pressing problems of sustainable development as defined by the people themselves (and not only by experts). In this respect "sustainable development" has to act as the solution to these problems and is therefore highly vision-oriented driven. This will almost certainly result in a much different agenda from one that would be obtained by continuing to allow priorities to reflect primarily the most acute problems (in science, knowledge and capacity building) as defined by stakeholders in research and innovation. As suggestions for some key elements in the development of the necessary new quality assurance, science communication and public policy processes there are:

- new institutions, networks and public procedures for the social evaluation of science advances,
- a shift in emphasis from one-way technology transfer to participatory learning and capacity building,
- a reassessment of the forms and locations of the "centres of excellence" capable of contributing knowledge and judgement needed for sustainability,
- and a reassessment of funding and financial support of research programmes and centres.

Many of the challenges for sustainable development involve issues and triggering mechanisms that are global, long-term and complex. Yet solutions need to be, for a large part, concrete, simple, short-term and local. To overcome this gap the purpose of different programmes and initiatives towards sustainable development e.g. Local Agenda 21 (United Nations, 1992) is to build local capacity and private–public–citizens partnerships for action. These processes are not only observed and monitored but often initiated and facilitated by pro-active scientists, researchers and others. Such processes can be seen, on the one hand, through the development of sustainability visions, targets, action plans and indicators appropriate for different scales. On the other hand, they bring stakeholders together at local levels and across levels, to define options for collective action. This develops new forms of governance and participation and can be seen as a strong tool to reinvent democracy. Multisector involvements through local stakeholders' networks are complementary to formal democratic institutions for implementing sustainable development policy. In that respect sustainable development actors have the task of "reinventing" democracy.

Generating, sharing and utilizing science to improve and integrate policy is a question of: scientific communication, international cooperation and capacity building for sustainable development. These are some of the new challenges that the quest for sustainable development poses to scientific research and the interface between science and policy. One of the issues fundamental to both science and policy is that of integration. Integration of scientific research requires a systemic approach, inter- and transdisciplinary research style, and the consideration not only of the relevant quantitative data but also the relevant qualitative information (Scholz, 2001). Appropriate mechanisms for making science available to policy-makers must include team-based social and regional approaches.

However, in Johannesburg it was recognized that to achieve real sustainable development, true inter-paradigmatic dialogues are necessary. Things to do or to take into account by groups dealing with science in support policy were also discussed. The fact that the high complexity of natural and societal systems implies a degree of irreducible uncertainty should not be interpreted as total ignorance and a licence for "anything goes" or "never touch the American life-style" in the policy realm. Adaptive and participatory pro-active approaches contrast with command-and-control approaches. In many cases, scientific research does not produce the kind of policy-makers in the scientific enterprise. This World Summit showed however, that innovative experiments on how to generate a dialogue between science and policy are needed. ICSU has demonstrated its commitment to the political process, and is now intent on further defining the focus of new initiatives in this area that are focusing the real threats of food security, global environmental change, loss of biodiversity and geohazards. Some weeks after WSSD delegates from the scientific community are eager to advance their role in organizing a global action plan for science and technology.

"Following what was, for many, the disappointing political outcome of the WSSD, it is very exciting for the ICSU to reach agreement on the need to roll up our sleeves and generate an action plan for science for sustainability," said Professor Jane Lubchenco, ICSU's new president at the General Assembly in October 2002. "Our top priority is to take an integrated approach to addressing the economic, environmental and social pillars of sustainable development." As a common conclusion and basic result of the Johannesburg summit all delegates shared the strong hope that a world society that tries to turn towards sustainable development has to work hard in refine their clumsy technologies, in "earthing" their responsibility to all creatures and resources, in building up global solidarity through all mankind and commit themselves to a better live for the next generations.

The most intimate formula for sustainable development scientific community found is earthing responsibility in respectfulness, tolerance and solidarity. This might be the only way to create global wisdom.

References

African Academy of Sciences: 1999, *Tunis Declaration: Millennial Perspective on Science, Technology and Development in Africa and its Possible Directions for the Twenty-first Century*, Fifth General Conference of the African Academy of Sciences, Hammamet, Tunisia, 23–27 April 1999.

Annan, K.: 2000, *We, the Peoples: The Role of the United Nations in the 21st Century*, New York, United Nations (http://www.un.org/millennium/sg/report/full.htm.).

Binder, C.: 2002, 'Research in partnership with developing countries: application of the method of material flux analysis in Tunja/Colombia', in F. Moavenzadeh, K. Hanaki and P. Baccini (eds.), *Future Cities: Dynamics and Sustainability,* Dordrecht, Kluwer Publishers.

Bronscomb, L., Holton, G. and Sonnert, G.: 2001, *Science for Society: Cutting-edge Basic Re-search in the Service of Public Objectives* (http://www.cspo.org/products/reports/scienceforsociety.pdf).

Clark, W.C.: 1986, 'Sustainable Development of the Biosphere: Themes for a Research Program', in W.C. Clark and R.E. Munn (eds.), *Sustainable Development of the Biosphere*, Cambridge, Cambridge University Press, pp. 5–48.

European Commission: 1998, 'Fifth framework programme: putting research at the service of the citizen', in S. Funtowicz, and M. O'Connor, (eds.), 1999, *Science for Sustainable Development*, Special Issue of *Int. J. Sustain. Dev.* **2**(3) (http://www.cordis.lu/fp5/src/over.htm).

Funtowicz, S., O'Connor, M., Ravetz, J.: 1999, 'Scientific Communication, International Cooperation and Capacity building for Sustainable Development', *Int. J. Sustain. Dev.* **2**(3), 363–367.

Gupta, A.: 1999, 'Science, sustainability and social purpose: barriers to effective articulation, dialogue and utilization of formal and informal science in public policy', *Int. J. Sustain. Dev.* **2**(3), 368–371.

International Council of Science (ICSU): 2001, *Global Change and the Earth system: A Planet Under Pressure*, Paris, IGBP Science Series No. 4.

IPCC – Intergovernmental Panel on Climate Change: 2001, *Special Report on Climate Change and Sustainable Development*, Plenary Seventeenth Session, Nairobi, April 2001. (http://www.ipcc.ch/meet/p17.pdf).

Kates, R., Clark, W.C., Corell, R., Hall, M., Jaeger, C.C., Lowe, I., McCarthy, J.J., Schellnhuber, H.J., Bolin, B., Dickson, N.M., Faucheux, S., Gallopin, G.C., Gruebleer, A., Huntley, B., Jäger, J., Jodha, N.S., Kasperson, H.E., Mabogunje, A., Matson, P., Mooney, H., Moore, III B., O'Riordan, T. and Svedin U.: 2000, *Sustainability Science. Research and Assessment Systems for Sustainability, Program Discussion Paper 2000–33*, Cambridge, MA: Environment and Natural Resources Program, Belfer Center for Science and International Affairs, Kennedy Scool of Government, Harvard University. (also published in Science, **292**, 641–642 (http://sustainabilityscience.org).

Lubchenco, J.: 1998, 'Entering the century of the environment: a new social contract for science', *Science* **279**, 491–497.

National Research Council, Committee on Global Change: 2002, *The Science of Regional and Global change: Putting Knowledge to Work*, Washington, National Academy Press.

Rocha-Miranda, C.E. (ed.): 2000, *Transition to Global Sustainability: The Contributions of Brazilian Science*, Rio de Janeiro, Academia Brasiliera de Ciências.

Sachs, J.D.: 2000, 'A new map of the world', *The Economist* **355**, 81–83.

Schellnhuber, H.J. and Wenzel, V. (eds.): 1998, *Earth System Analysis: Integrating Science for Sustainability*, Berlin, Springer Verlag.

Scholz, R.W. and Tietje, O.: 2001, *Embedded Case Study Methods, Integrating Quantitative and Qualitative Knowledge*, Thousand Oaks, Sage.

SCRES – The Standing Committee for Responsibility and Ethics in Science: 2002, *Standards for Ethics and Responsibility in Science – an Empirical Study* (http://www.icsu.org/Library/Reviews/SCRES/SCRES-Standards%20report.pdf).

Strigl, A.W.: 2001, 'Limits and options for sustaining technological development through systems renewal', In *Proceedings of the Environment Informatics Conference 2001: Sustainability in the Information Society*, ETH Zurich, Switzerland.

Swiss Commission for Research Partnerships with Developing Countries: 2001, *Enhancing Research Capacity in Developing and Transition Countries, Experiences, Discussions, Strategies and Tools for Building Research Capacity and Strengthening Institutions in View of Promoting Research for Sustainable Development*, Bern.

United Nations: 1992, Agenda 21: Programme of Action for Sustainable Development. Rio Declaration on Environment and Development. *Final Text of Agreements negotiated by Governments at the United Nations Conference on Environment and Development (UNCED)*, 3–14 June 1992, Rio de Janeiro, Brazil.

United States National Research Council: 1999, *Our Common Journey: A Transition Toward Sustainability*, Washington, DC, Board on Sustainable Development, National Academy Press.

UNDP – United Nations Development Program: 2001, *Making New Technologies Work for Human Development: The Human Development Report 2001*, Oxford, Oxford University Press.

UNEP – United Nations Environmental Program: 2002, *GEO: Global Environment Outlook*, London, Earthscan Publisher.

UN-FPA – United Nations Populations Fund: 2001, *The State of World Population 2001: Footprints and Milestones: Population and Environmental Change*, Denmark, Phoenix-Trykkeriet AS.

Watson, R., Dixon, J.A., Hamburg, S.P., Janetos, A.C. and Moss, R.H.: 1998, *Protecting Our Planet, Securing Our Future*, UN Environment Programme, Nairobi (http://www-esd.worldbank.org/planet).

WB – World Bank (1998) *Knowledge for development: The World Development Report for 1998/9*, Oxford University Press, Oxford.

WCED – World Commission on Environment and Development: 1987, *Our Common Future*, Oxford, Oxford University Press.

WGBU – German Advisory Council on Global Change: 1997, *World in Transition: The Research Challenge, Annual Report 1996*, Berlin, Springer-Verlag.

World's Scientific Academies: 2000, *Transition to Sustainability in the 21st Century*, Tokyo Summit, May 2000 (http://interacademies.net/intracad/tokyo2000.nsf).

Selected homepages

Commission on Science and Technology for Sustainable Development in the South (COMSATS): http://www.comsats.org.pk/index.html

Consultative Group on International Agricultural Research (CGIAR): http://www.cgiar.org/

Global Change System for Analysis, Research and Training (START): http://www.start.org

International Council for Science (ICSU): http://www.icsu.org

International Foundation for Science (IFS): http://www.ifs.se/index.htm

International Science Programme (ISP): http://www.isp.uu.se/Home.htm

Knowledge Network of Grassroots Green Innovators: http://www.sristi.org/Nissat.htm

RING Alliance: http://www.ring-alliance.org/index.html

Society for Research and Initiatives for Sustainable Technologies and Institutions (SRISTI): http://www.sristi.org

Southeast Asia START Regional Centre: http://www.start.or.th

Third World Academy of Sciences (TWAS): http://www.twas.org

Third World Network of Scientific Organizations (TWNSO): http://www.ictp.trieste.it/~twas/TWNSO.html

United Nations Millennium Ecosystem Assessment: http://www.millenniumassessment.org/en/index.htm

CHAPTER 14

GOVERNANCE FOR SUSTAINABLE DEVELOPMENT AND CIVIL SOCIETY PARTICIPATION

KRIS BACHUS

Higher Institute for Labour Studies, Catholic University of Leuven,
Kapucijnenvoer 33 Block H 4th floor, B-3000 Leuven, Belgium
(E-mail: kris.bachus@hiva.kuleuven.ac.be)

Abstract. Governance and participation were designated as important issues to be discussed at the World Summit on Sustainable Development in Johannesburg. In this paper, the concepts of governance, participation and civil society are defined and discussed. Special attention is given to the close link between these three concepts, and to how they interact with each other.

In the second part of the paper, the focus is on the Johannesburg conference and its outcomes with respect to governance and participation. The tenor of the argument is that the outcomes are disappointing due to the multitude of compromises, agreed upon during the negotiation process. The third and last part of the paper summarises the rather modest achievements of the process, and identifies the future challenges.

Key words: Governance, participation, civil society, good governance, World Summit on Sustainable Development, Plan of Implementation, governance for sustainable development, environmental governance, global governance.

Abbreviations: CBO – Community-Based Organisations; CEC – Commission of the European Communities; CGG – Commission on Global Governance; CSD – Commission on Sustainable Development; ELCI – Environment Liaison Centre International; GEF – Global Environmental Facility; GRI – Global Reporting Initiative; ICFTU – International Confederation of Free Trade Unions; ICLEI – International Council for Local Environmental Initiatives; IMF – International Monetary Fund; IUCN – International Union for Conservation of Nature and Natural Resources; MEA – Multilateral Environmental Agreements; MSP – Multi-Stakeholder Processes; NAI – New African Initiative; NCSD – National Council for Sustainable Development; NEPAD – New Partnership for Africa's Development; NGO – Non-Governmental Organisation; OECD – Organisation for Economic Co-operation and Development; UN – United Nations; UNCED – United Nations Conference on Environment and Development; UNCSD – United Nations Commission on Sustainable Development; UNDP – United Nations Development Programme; UNEP – United Nations Environmental Programme; WBCSD – World Business Council for Sustainable Development; WEO – World Environmental Organisation; WHAT – World Humanity Action Trust; WSSD – World Summit on Sustainable Development; WTO – World Trade Organisation

1. Introduction

Governance and participation both sprung up at the same time as the breakthrough of sustainable development itself. It is a difficult task – if not impossible – to address two related subjects like governance and participation separately. As shown in the conceptual parts of this paper, the two concepts are closely connected and can be regarded as prerequisites for each other. In fact, we

319

L. Hens and B. Nath (eds.), The World Summit on Sustainable Development, 319–345.
© 2005 *Springer. Printed in the Netherlands.*

propose a framework in which the two concepts are intertwined, both directly and indirectly through the concept of civil society. Civil society, which will also be addressed in this paper, links the two other issues.

In this model, we can propose a simple conceptual framework, as illustrated in Figure 1.

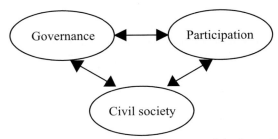

Figure 1. Conceptual framework linking governance, participation and civil society.

The paper is structured on the basis of this framework. First, the concept of governance is elaborated, followed by some thoughts on participation and civil society. Before we link the concept with sustainable development and the World Summit on Sustainable Development, the interconnections between the three concepts will be focused on.

2. The concept of governance

"Governance" and "good governance" are terms that have, in recent years, progressed from obscurity to widespread usage. Initially, "governance" and "government" were used interchangeably. Over the years, the need for a concept distinct from government appeared when people started to look upon government as an *organisation*, rather than a *process*. Today, government is seen as a set of institutions, designed by several actors.

For a number of years, governance has been the term used to describe the processes of how governments and other societal actors interact. Interest in public issues is not confined to government. Other actors can share an interest, like business organisations, trade unions, media, religious organisations and the military. These other actors are often referred to as "civil society", while the system involving many non-state actors in the policy process is sometimes referred to as "multi-actor governance".

There is no such thing as a generally accepted definition of governance. Many different angles on the concept exist; government, scientists and other actors mostly tend to define the issue starting from their own field of interest.

The lack of a generally accepted definition is not only a problem for the concept of governance in the general sense. It is even more problematic for the specific type of governance that is discussed in this paper: governance for sustainable development. Both governance and sustainable development are interpreted in numerous different ways, depending on the viewpoint of the protagonist. Environmentalists emphasise the importance of "environmental governance", while development workers would situate the concept in the sphere of development.

It is not the aim of this paper to give a comprehensive overview of the literature on the definition of governance; therefore, only some of the existing definitions will be dealt with.

UNDP (1997) defines governance as *"the exercise of political, economic and administrative authority to manage a society's affairs"*. It adds to this *"the term "governance" refers to the process by which society manages its economic, social and political resources and institutions – not only for development, but also for the cohesion, integration and well-being of its people"*.

The WHAT Governance Programme[1] uses the following definition: *"the framework of social and economic systems, legal and political structures within which humanity organises itself"*.

Regardless of the preferred governance definition, in most cases there will be a consensus on the fact that the global level is an important feature in the debate on governance. In fact, some sources consider this level to be of such importance that they prefer to speak of the concept "global governance", and then talk about the same issue. Of course there is a formal difference between the two concepts: governance can refer to any policy level, while "global governance" limits the scope to this one level.

In practice, however, the fact that the two concepts are sometimes used interchangeably, can be illustrated by the following statement by the Commission on Global Governance (CGG, 1995, p. 335), which outlined that:

Global governance, once viewed primarily as concerned with intergovernmental relationships, now involves not only governments and intergovernmental institutions but also NGOs, citizens" movements, trans-national corporations, academia, and the mass media. The emergence of a global civil society, with many movements reinforcing a sense of human security, reflects a large increase in the capacity and will of people to take control of their own lives.

The resemblance of this statement to the general definitions of governance shows that the distinction between "governance" and "global governance" is not always made. However confusing this mix-up may be, there is no question that governance is a "vertical" term, meaning it applies to all possible policy levels. For this, reason, the term "multi-level governance" is often added to the debate.

---

[1] As of the 1st of October 2001, the World Humanity Action Trust (WHAT), the Stakeholder Forum for Our Common Future, and GLOBE Southern Africa have joined forces to co-ordinate a programme on Governance for Sustainable Development ahead of and after the 2002 World Summit on Sustainable Development.

In order to avoid confusion about the meaning of the concepts used in this chapter, Box 1 provides the most important definitions as they are used in this text.

Box 1. The concepts used in this chapter.

- Governance: the process of how governments and non-state actors interact, with the common aim to organise society.
- Good governance: a mode or model of governance that leads to social and economic results sought by citizens (Plumptre and Graham, 1999).
- Sustainable development: satisfying the needs of the present generation without compromising the ability of future generations to meet their own needs (WCED, 1987).
- Sustainability governance: the process of how governments and non-state actors interact, with the common goal to move towards a more sustainable society.
- Civil society: all people, their activities and their relationships that are not part of the process of government (Willetts, 2002).
- NGO: private organizations that pursue activities to relieve suffering, promote the interests of the poor, protect the environment, provide basic social services, or undertake community development. (World Bank, 1996).
- Participation: Participation is a process through which stakeholders influence and share control over development initiatives and the decisions and resources which affect them (World Bank, 1996).

---

3. Good governance

"Governance" is sometimes used interchangeably with "good governance", although there is a clear difference: governance is a neutral term, which refers to a framework of systems, organisations and relationships, while behind good governance is a normative meaning. Plumptre and Graham (1999), define good governance as "*a mode or model of governance that leads to social and economic results sought by citizens*". The term "good governance" is often used by governments to refer to a reform of governance organisation in a particular country. A widely accepted aspect of good governance is the use of best practices that exist in various institutions.

Some issues are regarded as key issues with regard to good governance, like transparency, accountability and human rights. Plumptre and Graham (1999) provide an extensive list of good governance items, collected in Box 2.

Box 2. Good governance items (Plumptre and Graham, 1999).

Constitutional legitimacy	Judicial independence
Democratic elections	Transparency
Respect for human rights	Absence of corruption
Rule of law	Active independent media
Political openness	Freedom of information
Predictability and stability of laws	Administrative competence
Tolerance, equity	Interests on issues of public concern
Public participation	Accountability to public
Public expenditures directed to public purposes	Administrative neutrality: merit-based public service

---

Despite their apparently anodyne character, the practical implementation of these attributes may give rise to controversy, as excessive emphasis on one may lead to undesirable results. For example, public participation is attractive, but excess may result in the taking of decisions by individuals with little knowledge and no accountability.

Nevertheless, some universal norms or values appear to be generally accepted. The United Nations (UN, 1997) has published a list of characteristics of good governance (Box 3).

Box 3. Characteristics of good governance (UN, 1997).

Participation	Equity
Rule of law	Effectiveness and efficiency
Transparency	Accountability
Responsiveness	Strategic vision
Consensus orientation	

The emergence of the concept of governance is very much related to the profound changes that the functions and powers of governing institutions have undergone during the last decades. The state and its instruments of government remain of central importance nowadays, but it is no longer the sole actor in determining the direction of society. Throughout the world, trends like democratisation, decentralisation, globalisation, governance, trans-national activist networks and sustainable development all have taken their places in the modern global society, and they are all related to each other.

Terms like governance and good governance are important for both the developed and the developing countries as a tool for reaching the objectives of sustainability. Of course, these two groups are faced with a totally different situation, with regard to the ecological, the social and the economic aspects of sustainable development. Consequently, developing countries have other expectations of and set other priorities for good governance.

In Africa, terms like democratisation, battle against corruption, respect for human rights, political pluralism and enforcing the rule of law get high priority. At the same time, countries relate the governance issue mainly to development aspects, like the eradication of poverty and sustainable economic growth in developing countries. Among the industrialised countries, the European Union (CEC, 2001), proposes in its "White Paper on Governance" the following five priorities: openness, participation, accountability, effectiveness and coherence. These principles are often referred to in the context of environmental governance, which usually gets the main focus in developed countries.

4. The role of civil society

It is clear that a lot of different actors are involved in governance for sustainable development. The actors are often called stakeholders; because they have an interest in a particular decision, either as individuals of representatives of a group.

Governance is often linked to multi-stakeholder processes (MSP) (Hemmati, 2001). Indeed, government is considered to be just one of the stakeholders; civil society is given an important role in this framework. But what exactly is "civil society"? UNDP (1997) defines a civil society organisation as the part of society that connects individuals with the public realm and the state. Civil society organisations are generally taken to include industry associations, trade unions, commercial associations, employers" organisations, professional associations, advocacy groups, credit unions, co-operatives, academic and research institutions, the media, community-based organisations (CBOs), NGOs, and not-for-profit and religious groups. In the sustainability debate, these groups are usually called the "major groups". The diversity of the different groups shows that "civil society" is a very heterogeneous group. In the realm of sustainability governance, NGOs are the most prominent actors. According to Gemmill and Bamidele-Izu (2002), NGOs are "Groups of individuals organised for the myriad of reasons that engage human imagination and aspiration. They can be set up to advocate a particular cause, such as human rights, or to carry out programmes on the ground, such as disaster relief. They can have memberships ranging from local to global".

The diversity of civil society and its value to official intergovernmental processes on sustainable development are acknowledged in Agenda 21, the comprehensive sustainable development blueprint adopted at the 1992 Rio Earth Summit (UN, 1992). The document does not make use of the term civil society, although it expressly recognises the members of civil society as a force to be reckoned with.

Some authors question whether business and industry should be included in the definition of civil society (Gemmill and Bamidele-Izu, 2002). They contend that, because they already have considerable influence over international governance processes through informal lobbying opportunities and formal influence channels, business and industry should not be included. It is not the main objective of this paper to resolve this question; however, according to the above definitions of governance, it seems unreasonable to exclude any group from the process.

Civil society can have an influence on policy decisions. It can use the mechanisms of governmental accountability, provide checks on government power and monitor social abuses. But there is also an evolution in which civil society can do a whole lot more than this: it can also make valuable contributions to policy formulation, and really be included in the policy formation process. As mentioned before, the role of NGOs as an important stakeholder tends to be emphasised, because they raise awareness on development, environment and other issues and increase pressure on governments through media campaigns and raising public awareness.

This conclusion is not new. Policy influence of certain stakeholders, e.g. religious organisations, the military, businesses or other interest groups through formal or informal channels has existed for centuries. A modern and institutionalised example is the tripartite labour consultation model, which has gradually surfaced in most western European countries since the end of the nineteenth century, and which is still in place. In this model, government, employers" and employees" organisations meet and make joint decisions after consultation and discussion.

The organisation of effective multi-stakeholder processes constitutes a big challenge. On the international level, WHAT (2001) suggests that it would help to gather NGO interests in one or a few international organisations, like Environment Liaison Centre International (ELCI). In this way, an analogy would be created with other major groups that are also represented by umbrella organisations, e.g. ICFTU for unions, ICLEI for local government and WBCSD for businesses.

In the next paragraph we will elaborate further on participation by civil society and by NGOs.

5. Participation

Agenda 21 is the first United Nations document to properly address the role of different stakeholders in the implementation of a global agreement. In each of its Chapters, Agenda 21 refers to the role that stakeholder groups should take up in order to put the blueprint into practice. Stakeholder involvement is described as absolutely crucial for sustainable development. On top of this, the nine Chapters of Section 3 deal with the nine different major groups and with the role they have to play in the effective implementation of the objectives of sustainable development. The nine groups are women, youth, indigenous peoples, non-governmental organisations, business and industry, workers and trade unions, the science and technology industry, farmers and local authorities.

Since UNCED, participation is considered to be a necessary condition for success, or, as Agenda 21 states in Chapter 23: *"One of the fundamental prerequisites for the achievement of sustainable development is broad public participation in decision-making".*

In 1992, the United Nations Commission on Sustainable Development (UNCSD) was created to ensure effective follow-up of UNCED. The CSD is itself one of the most advanced forums for multi-stakeholder discussions. Well-prepared multi-stakeholder dialogues take place there every year.

Other existing international forums have incorporated some of them only reluctantly, multi-stakeholder processes in their operations, e.g. ICLEI, OECD, UNEP, World Bank, IMF and WTO.

But also on the regional, national, and local levels, the Rio-Conference has increased the awareness of actors about the need of creating and institutionalising more participation structures. About 100 countries created a National Council for Sustainable Development (Earth Council, 2002), a participatory advisory body that reports to the UNCSD. At local level, similar councils are being established.

As mentioned in the paragraph on civil society, NGOs play an increasingly important role in the governance processes. However, participation of civil society in global governance is not new (Gemmill and Bamidele-Izu, 2002); it has taken place for over two centuries. What is new is the wide proliferation of non-governmental organisations. In 1948, the UN listed forty-one consultative groups that were formally accredited to participate in consultative processes; in 1998, there were more than 1,500 organisations with varying degrees of participation and access. Several factors explain the rise of the NGOs. The development of information technology, greater awareness of global interdependence, the spread of democracy, the way the UN started to collaborate more openly with NGOs (Weiss, 1999), and finally the call in Agenda 21 for new forms of participation; these are probably only some of the explanatory factors.

After UNCED, NGO activity within UN processes has intensified further. NGOs participated (informally) in the Habitat II-conference (1996), and took part in the negotiations that led to the 1998 Aarhus Convention on Public Access to Information, Participation in Decision-making and Access to Environmental Justice.

The advantages of and the need for NGO participation in the sustainability governance debate are widely accepted: civil society can help build the political will for a new (sustainable) approach; NGOs can serve as alternatives for weak or inadequate democratic institutions. Yet, some doubts can be heard: many feel that the drawbacks of civil society participation may outweigh the benefits. Some fear that NGOs, supported and financed by their governments, might become too powerful, and that intergovernmental decision-making processes would become bogged down by NGOs, which are not necessarily representative of or accountable to their particular constituencies (Gemmill and Bamidele-Izu, 2002). Finally, some decision makers are anxious that NGOs may seek to usurp the sovereign powers of governments.

However, weighing up the pros and the cons of civil society participation, it is clear and generally accepted that NGO involvement is both indispensable and useful for making progress to governance for sustainable development.

5.1. TYPES OF PARTICIPATION

Participation in environmental and sustainability policy and governance can be interpreted and organised in many different ways. For a better understanding of the meaning of participation, it is useful to reflect on the term in greater detail and to consider the different types of participation that one can discern.

The first typology of participation mechanisms relates to the number of participants. *Direct* public participation is a mechanism in which every individual of the public has the opportunity to participate. Evidently, this approach brings along major practical problems, particularly when participation is considered on a global level. The World Summit on Sustainable Development itself is a good example for this: so far, 60,000 people participated at the Summit; but if every citizen who feels affected by the topics discussed in Johannesburg would want to participate, the process would very soon become unworkable and even impossible. This is why participation on most levels is usually organised in an *indirect* way: citizens participate through representation. Citizens can join up to a non-governmental organisation that defends their ideas, workers are represented by trade unions, farmers by agricultural organisations, etc. Moreover, these organisations are usually accountable to their supporters through mechanisms like membership and elections.

Apart from the direct/indirect distinction, we can discern a second typology of participation: *input* participation and *output* participation (Bruyninckx and Bachus, 2001). This breakdown is particularly relevant for indirect participation. Input participation is the extent to which organisations like NGOs are admitted to take part in policy and governance processes and allowed to express their opinion. Output participation on the other hand, is the degree to which the participation process allows stakeholders to actually change the output and outcome of the processes they are participating in. It is not hard to imagine a participatory process in which all the stakeholders are allowed to pass comment (high input participation), while at the end, when everybody has finished talking, it is the government or the political authority that draws the conclusion it would have done anyway, regardless of the spouted opinions (low output participation).

It is important not to focus exclusively on input participation, but to always keep the output participation in mind. In recent years, civil society has been invited to take part in an exponentially growing number of advisory councils, both on the local, national, regional and global level, organised particularly for matters relating to sustainable development. It is not always clear what influence these councils have on the actual policy outcome.

A third way of acknowledging the dynamics of participation is by making a distinction between *formal* and *informal* participation. Informal participation (or lobbying) has been exercised for centuries by a large number of lobby groups who try to influence policy processes without formally taking part. It can be a very effective way of participating, provided that government respects the vision of the stakeholder and regards its approval as important. However, the weakness of informal participation is the lack of a guarantee of being considered: at any time, the government can decide not to take the stakeholder's view into consideration. Although it can still be rather difficult for a government to implement a certain policy without the support of the major stakeholders, this type of participation is likely to be situated hierarchically at a lower level than formal participation. In a democratic regime, a formal participation structure gives to the included stakeholders the guarantee that they can have (at least input) participation on every one of the discussed themes. Besides, history has shown that informal participation seldom shows favour to the whole of civil society. Usually, one or two groups (for instance the military or certain selected economic sectors) are given preference. Consequently, this system leads to a very narrow one-way participation, leaving out several of the civil society groups.

To conclude these thoughts on participation: sometimes it can be more advantageous to the party that grants the participation than to the stakeholder who is allowed to participate. This is the case when the collaboration of the target group of a certain policy rule is expected to contribute greatly to its success, e.g. a law forcing waste companies to report every shipment of waste abroad will most likely be more effective, when the economic sector of waste companies is allowed to have a say in the terms and conditions of the new legal rule. Bruyninckx and Bachus (2001) call this type of participation "planner-centred participation". The counterpart is "people-centred participation", which is participation as it was originally meant to be, i.e. to the advantage of the stakeholders who get the opportunity to participate.

6. Link between governance and participation

The two main subjects of this paper, participation and governance, are closely linked. Governance for sustainable development implies a new way of implementing decisions, of pursuing goals; the main aspect of this process is the involvement of the major groups, the various stakeholders who are gathered throughout the world in numerous organisations, and they all have to be heard in the sustainability debate, whether it is directly or through representation.

The connection between the two concepts can be illustrated with the following statement from the UN Secretary General's Millennium Report (Annan, 2000, p13):

> *"Better governance means greater participation, coupled with accountability. Therefore, the international public domain – including the United Nations – must be opened up further to the participation of the many actors whose contributions are essential to managing the path of globalisation. Depending on the issues at hand, this may include civil society organisations, the private sector, parliamentarians, local authorities, scientific associations, educational institutions and many others."*

7. Governance and democratisation

Democracy, or at least democratisation, is often seen as an important prerequisite for effective governance and participation. This matter is monitored closely by the UN. Between 1988 and 1997 the UN organised three international conferences on new or restored democracies. Another conference in 1997, the International Conference of Governance for Sustainable Growth and Equity, organised by the UNDP, also paid special attention to democratic governance.

The relationship between democracy and participation works in two ways: democracy may be important to achieve effective participation, but the opposite should be studied as well. Rueschemeyer *et al.* (1998) examined this relationship, and they concluded that democracy in a narrow sense of the term may be able to function without broad-based participation. They added, however, that this model is likely to lead to the subordinate classes being neglected unless they have a strong and responsive organised representation.

8. The United Nations Commission on Sustainable Development

The United Nations Commission on Sustainable Development (UNCSD) plays an important role in both the governance and participation debates. The Commission was created in December 1992 to ensure effective follow-up of UNCED, and consequently to monitor and report on implementation of the Earth Summit agreements at local, national, regional and international levels. The CSD is a functional commission of the UN Economic and Social Council (ECOSOC), with 53 members. The Commission consistently generates a high level of public interest. Over 50 ministers attend the CSD each year and more than one thousand non-governmental organisations (NGOs) are accredited to participate in the Commission's work.

One of the roles of the Commission is directly related to participation of civil society: "to promote dialogue and build partnerships for sustainable development with governments, the international community and the major groups identified in Agenda 21". During the period 1998-2002, the Commission has tackled mainly issues on poverty and sustainable consumption and production patterns. One of the features of the yearly CSD-sessions is a multi-stakeholder dialogue on a selected topic.

Agenda 21 also called on countries to establish their own national council for sustainable development (NSCD); their role is to facilitate countries" follow-up of the implementation of Agenda 21 at national level and assist in the preparation of national reports to be presented regularly to the UNCSD. Several studies on national cases have been carried out to evaluate the role of the national councils (Maurer, 1999, Bruyninckx & Bachus, 2001).

9. Treatment of "governance" at the World Summit

9.1. OCCURRENCE IN POLITICAL DECLARATION AND PLAN OF IMPLEMENTATION

Governance is not a typical WSSD-topic. Topics like "energy" or "poverty" refer to the *state* of a certain sustainability subject. They relate to the final goals of the sustainability debate in the field of interest; for example: "reducing dependency on fossil energy sources" or "eradicating poverty in the world".

Governance can be situated on a different discussion level: like IUCN (2002) stated: *"Governance is the means to an end, not an end in itself"*. It is a *tool* that can play an important role in fulfilling the objectives in every one of the sustainability fields, but it is not an objective in itself.

One consequence of this characteristic of the governance debate, is that it didn't get as clear a place on the summit as the other "state-related" fields, like "biodiversity", "sanitation", "poverty", or "consumption and production patterns". It is a subject that is relevant to any of the above-mentioned issues. Consequently, it is not always easy to get an overview of the achievement at the summit on this subject. The survey in this section is based on a collection of results of different debates.

Despite the scattered attention for the concept of governance at the summit, one of the thirty-seven points of action of "The Johannesburg [political] Declaration on Sustainable Development" (UN, 2002d) is dedicated to it. It states:

> *"We undertake to strengthen and improve governance at all levels, for the effective implementation of Agenda 21, the Millennium Development Goals and the Johannesburg Plan of Implementation".*

This explicit reference to governance underlines its importance for sustainable development. Furthermore, the formulation of this statement, with explicit reference to three very important sources of sustainability objectives (Agenda 21, the Millennium Development Goals and the Plan of Implementation), directly confirms that governance should be seen as a tool to tackle challenges, not as a goal in itself.

TABLE I. Occurrences of "governance" in the WSSD Plan of Implementation

Paragraph	Formulation	Context
4	Good governance within each country and at the international level is essential for sustainable development.	General introduction, national level, globalisation, democratisation
43f	Create [...] forest law enforcement and governance at all levels	Forests and trees
56a	Achieving sustainable development includes actions at all levels to: [...]support African efforts for peace [...]good governance [...]	Sustainable development for Africa
65	[...]the development of efficient and effective governance systems in cities and other human settlements [...]	Consequences of Habitat and Istanbul Declaration for Africa
120bis	Good governance is essential for sustainable development.	Institutional framework for sustainable development
122	[...]Fully implement the outcomes of decision I on international environmental governance	the institutional framework for sustainable development at the international level
123	Good governance at the international level is fundamental for achieving sustainable development	the institutional framework for sustainable development at the international level

9.2. A CHAPTER ON "GOVERNANCE" OR ON "INSTITUTIONAL FRAMEWORK"?

The list of table 1 illustrates that a lot of attention was given to governance, both as a general concept and as a tool for achieving sustainability in several policy domains. However, a close observer would notice that the concept is mentioned occasionally in the fifty-four-page document, but it is never thoroughly dealt with. The reason for this fragmentary appearance of the term can be found in the preparatory process for the WSSD. The place of the governance debate has been marked by some changes during the process of the four Summit Preparatory Committees (Prepcoms). At the second Prepcom in New York, governance appeared on the agenda as the tenth topic to be covered at the Johannesburg Summit. The subject was allocated the title: "Strengthening governance for sustainable development at the national, regional and international levels". The theme was elaborated during the third (New York) and fourth (Bali) Prepcoms. During this process, many parts of the proposed text were modified. A drastic modification, which occurred from the fourth Prepcom onwards, was the abolition of the term "governance" in the title of the text. The term "sustainable development governance" was replaced by "institutional arrangements for sustainable development". Later on, it was changed to "institutional framework for sustainable development", which eventually became the title of the tenth Chapter of the Johannesburg Plan of Implementation. Accordingly, the subtitles on "sustainable development governance at the international, regional and local level" were transformed into "the institutional framework for sustainable development at the international, regional and local level". These subtitles were likewise included in the Plan of Implementation. It is clear that, originally, the term governance was given a more central place in the debate than it eventually ended up with.

Whatever the title of the Governance Chapter of the Plan may be, in the introduction to the tenth Chapter (par. 120bis), good governance is mentioned as "essential". With this, a list of what should be understood as "good governance" is added: sound economic policies, solid democratic institutions and improved infrastructure; further freedom, peace and security, domestic stability, respect for human rights and the rule of law, gender equality, market-oriented policies, and an overall commitment to just and democratic societies are also essential and mutually reinforcing.

Paragraph 121 names the objectives of governance (institutional arrangements) for sustainable development. Several references are made to Agenda 21.

In paragraph 122e, the international community commits itself to the timely completion of the negotiations on a comprehensive United Nations convention against corruption.

Paragraph 122f is dedicated to the promotion of corporate responsibility. Paragraph 122g calls on actions to implement the Monterrey Consensus (financing of development).

The rest of the chapter on the institutional framework at the international level is dedicated to the role of institutions: the General Assembly, the Economic and Social Council, the Commission on Sustainable Development, the World Trade Organisation and the Global Environmental Facility.

Paragraphs 145 to 149 are dedicated to strengthening the institutional frameworks for sustainable development at national level. All states are called on to strengthen governmental institutions, e.g. by providing the necessary infrastructure and by promoting transparency, accountability and fair administrative and judicial institutions.

The plan urges national governments to "take steps to make progress in the formulation and elaboration of national strategies for sustainable development and begin their implementation by 2005" (par. 145b). This formulation means a remarkable weakening of the formulation of the Prepcom IV-paper on governance, which stated that "all countries should ensure that they have national sustainable development strategies in place by 2005".

This abolition aside, several other modifications were made to the Prepcom IV-paper; abolition of certain formulations is compensated by addition of other issues. Sometimes, modifications are subtle and inconspicuous, for example: "the international community should commit to **_designate_** GEF as the permanent financial mechanism of UNCCD" (Prepcom IV paper on governance for sustainable development) was changed to "**_consider_** making GEF a financial mechanism of the Convention [to Combat Desertification]" (Plan of Implementation, par. 39f). In some cases, it is clear that certain texts are compromises following negotiations between representatives of national governments.

A last interesting modification concerns the concept of "sustainable development governance": we mentioned earlier that this term was abandoned as a title for Chapter 10 of the Plan of Implementation. It appears that the term was not acceptable at all, because it does not appear a single time in the Plan, while it was mentioned no less than eight times in the Prepcom IV-paper, although even then it was already banned from the titles.

It is my judgement that the process of the negotiations at the World Summit (including the preparatory talks) have eroded the good intentions that were clearly there at the beginning of the process. Negotiations with such a large group of countries inevitably lead to an agreement with such a high degree of compromise and vagueness that the call for action can easily be ignored or be given a minimalistic implementation by the partners that insisted on the weakening of the formulation. Obviously, this is a backward step for the countries that would be willing to make progress, but that aren't urged to do so because of the weak formulation. Therefore, the abandoning of the principle of making a plan of implementation supported by all states could be advantageous for the potential outcomes of large international conferences.

9.3. THE VITAL ROLE OF PARTNERSHIPS

An important issue related to governance that has been given an important place at the summit, is the term "partnership". As "governance" expresses a way in which governments and civil society should collaborate with the objective of making steps towards sustainable development, many types of partnerships are put forward in the Johannesburg Plan of Implementation: public-private partnerships, partnerships that give priority to the needs of the poor, partnerships between international financial institutions, partnerships for the marine environment, partnerships for the conservation and sustainable use of biodiversity, partnerships among interested governments and stakeholders, partnerships to enhance health education, etc.

After the fourth Preparatory Committee, it was already clear that these partnerships would get a strong emphasis in the rest of the process. In particular, new partnerships between government and major groups are encouraged through the so-called "type 2-outcomes". These are partnerships, developed within the frame of the WSSD-process, which intend to contribute to and reinforce the implementation of the Political declaration and the Plan of Implementation of the World Summit. They also aim to facilitate the further implementation of Agenda 21 and the Millennium Development Goals.

During the process of the summit, some 220 partnerships were launched; however, the partnerships were also criticised from the beginning, especially by a number of NGOs, in the fear that the type 2-outcomes of the summit will make it easier for national governments to refuse to come to international agreements (the so-called type 1-outcomes). Officially, the partnerships are not at all meant as a substitute for governmental action. It is recognised that commitments by governments are the cornerstone of national, regional and global efforts to pursue sustainable development. The Partnerships are meant to supplement and not to supplant actions and commitments by governments.

Two of the proposed partnerships deal specifically with governance issues (public participation, good governance); many others don't focus on governance, but have an implicit reference to items like access to information.

9.4. PRINCIPLES OF GOOD GOVERNANCE

The outcome of the UNCED, the Rio Principles, outlined some of the core elements of good governance for sustainable development. Ten years later, it was more difficult to reach an agreement even on a status quo vis-à-vis the Rio Declaration. In particular, no agreement could be reached on the inclusion of the Precautionary Principle (Principle 15), the Principle of Common but Differentiated Responsibility and the Ecosystems Approach (Principle 7). The problems in negotiations were so far-reaching that a conflict resolution (contact group) was set up to deal exclusively with the Rio-principles. In the end, the disagreement was resolved by a general reiteration of commitment to the Rio-principles as a whole. However, it was fairly clear that not all representatives acknowledged the need for these principles to be widely applied.

Two principles of good governance mentioned earlier , namely accountability and transparency, appear several times in the text, particularly in relation to key sectors, e.g. water, energy, finance and trade. Paragraph 4 of the Plan of Implementation explicitly states that good governance is essential and that it should be based on "sound environmental, social and economic policies, democratic institutions responsive to the needs of the people, the rule of law, anti-corruption measures, gender equality and an enabling environment for investment". The Plan also commits governments to ensure the finalisation of the UN Convention against Corruption (par. 122e and 124).

References to human rights are scarcer. They deal with the general issue (par. 5), the rights of workers (par. 9), standard of living (par. 38a), good governance (par. 120bis) and others. In paragraph 120bis the need for better infrastructure is mentioned as the basis for sustained economic growth, poverty eradication, and employment creation. Freedom, peace and security, domestic stability, respect for human rights and the rule of law, gender equality, market-oriented policies, and an overall commitment to just and democratic societies are also regarded as essential and mutually reinforcing.

With respect to the issue of corporate governance, voluntary initiatives remained the preferred option. On the other hand, a new version of the Global Reporting Initiative (GRI) corporate sustainability reporting guidelines was launched at the Summit.

9.5. GOOD GOVERNANCE FOR AFRICA

In theory, most of the African countries embrace the concept of good governance. However, in spite of the new wind that blew across most of sub-Saharan Africa in the 1990s which brought along with it the notion of democratic and pluralistic systems of government, several of these countries are still faced with a legacy of corruption and decadence, a syndrome that bedevils most of Africa's Development initiatives (Banda, 2002). Africa is clearly in need of an evolution towards good governance.

Yet, Africa is making efforts to set an agenda for good governance. In 2001, the "New African Initiative" (NAI) was created. Some of the ideals to which the African leaders pledged themselves were associated with democracy, human rights, accountability, transparency, participatory governance, good fiscal and monetary policies, transparent frameworks for financial markets, auditing of private companies and the public sector, setting and enforcing a legal framework, and maintaining law and order.

The NAI was the inspiration for five heads of state (from Algeria, Egypt, Nigeria, South Africa and Senegal) to initiate the New Partnership for Africa's Development (NEPAD, 2002). Principles of participation and good governance are central for the partnership, particularly as *conditions* for development. These are some of the references to the two concepts that can be found in the statements of NEPAD:

- Development is impossible in the absence of true democracy, respect for human rights, peace and good governance.
- Facilitating countries" follow-up of the implementation of Agenda 21 at the national level, and assisting in preparation of national reports to be presented regularly to the UNCSD.
- Promoting participatory decision-making;
- NEPAD has six priorities:
  1. Peace, Security, Democracy and Political Governance
  2. Economic and Corporate Governance
  3. Infrastructure development
  4. Central Banks, African Development Bank and Financial structure
  5. Market Access and Agriculture
  6. Debt reduction and Foreign Direct Investment

NEPAD is generally considered as a serious attempt to come to better governance for Africa, in order to achieve the final goal of development. The partnership also gained an important place at the WSSD. It is expressly welcomed by the international community in paragraph 56 of the Plan of Implementation, and mentioned on a couple of other occasions.

10. Treatment of "participation" at the World Summit

Like "governance", "participation of major groups" gets an explicit mention in the political declaration of the summit. Action point 26 states that:

"We recognise sustainable development requires a long-term perspective and broad-based participation in policy formulation, decision-making and implementation at all levels. As

social partners we will continue to work for stable partnerships with all major groups respecting the independent, important roles of each of these. "

Moreover, the important role of major groups is repeated in paragraph 34:

"We are in agreement that this must be an inclusive process, involving all the major groups and governments that participated in the historic Johannesburg Summit. "

In the fifty-four paged WSSD Plan of Implementation, the notion of participation is mentioned on about twenty-five occasions. About one third of these refer to the participation of developing countries in international processes and mechanisms. However important this issue may be, these references are not included in this analysis, since they are not relevant to the issue that is being dealt with in this paper.

This leaves us with 17 relevant occurrences; the table below gives an overview.

TABLE II. Occurrences of "participation" in the Plan of Implementation.

Paragraph	Formulation	Context
6d	Promote women's equal access to and full participation, on the basis of equality with men,	Poverty eradication
19g	Promote rural community participation, including local Agenda 21 groups	Consumption and production
21	with the participation of government authorities and all stakeholders	Consumption, production and waste
24b	Facilitate access to public information and participation, including by women, at all levels,	Safe drinking water
38f	Enhance the participation of women in all aspects and at all levels	sustainable agriculture and food security
40e	Promote full participation and involvement of mountain communities in decisions that affect them	Mountain ecosystems
42l	Promote the effective participation of indigenous and local communities	Biodiversity
43h	Recognise and support indigenous and community-based forest management systems to ensure their full and effective participation	Sustainable forest management
44b	Enhance the participation of stakeholders	Mining, minerals and metals
67	Participation of civil society	Sustainable development in Latin America and the Caribbean
119ter	Ensure [...]public participation in decision-making, so as to further principle 10 of the Rio Declaration on Environment and Development, taking into full account principles 5, 7 and 11 of the Declaration	Means of implementation
121g	Enhancing participation and effective involvement of civil society and other relevant stakeholders in the implementation of Agenda 21, as well as promoting transparency and broad public participation	Institutional framework for sustainable development
126c	The Council should encourage the active participation of major groups in its high-level segment	Role of the Economic and Social Council
143d	Continue to promote multi-stakeholder participation and encourage partnerships to support the implementation of Agenda 21	Governance at the regional level
146bis	All countries should also promote public participation	Governance at the national level
147	Further promote the establishment or enhancement of sustainable development councils and/or co-ordination structures at the national level, including at the local level, in order to provide a high-level focus on sustainable development policies. In that context, multi-stakeholder participation should be promoted	Governance at the national level (NCSDs)
153	Promote and support youth participation	Governance at the national level / participation of major groups

In the majority of the references, the plan refers to participation of all stakeholders, or of civil society. Women and indigenous or local stakeholders get four references each, while youth is mentioned once expressly. The other major groups come under "all stakeholders".

Participation is mentioned in the context of a wide variety of subjects: poverty, consumption and production, water, agriculture, mountain ecosystems, biodiversity, forests, economic development, and means of implementation. On top of this, the term participation is used six times in relation to governance. Here we find a clear illustration of the connection between the two issues.

Paragraph 119ter of the Plan of Implementation includes a notable reference to the Rio-Declaration on Environment and Development. It refers to Principle 10, which states that *"environmental issues are best handled with the participation of all concerned citizens, at the relevant level"*.

In the Governance –Chapter of the Plan (Chapter X), one of the objectives of governance refers to participation (121g): *"Enhancing participation and effective involvement of civil society and other relevant stakeholders in the implementation of Agenda 21, as well as promoting transparency and broad public participation"*. This formulation leaves no doubt that participation of civil society has to remain a crucial point, both in the governance debate and in the whole of the sustainability process.

The Economic and Social Council is explicitly called on to encourage the participation of major groups in its high-level segment and the work of its relevant functional commissions (par. 126c).

The regional and sub-regional bodies are equally called on to promote multi-stakeholder participation and encourage partnerships to support the implementation of Agenda 21 at the regional and sub-regional levels (par.143d). Besides, participation is linked with the implementation of Agenda 21 on several occasions in the Plan of Implementation.

Paragraph 147, which is part of the chapter on governance for sustainable development at national level, expressly refers to the need for continued promotion of the establishment of sustainable development councils, both at national and at local level. The plan calls for promotion of multi-stakeholder participation for these councils. The Prepcom-IV governance paper stipulated that the national councils should be involved in the process of the formulation of the national strategies for sustainable development (par. 145b), but the Plan of Implementation makes no mention of this link.

The tenth and last Chapter of the Plan of Implementation contains a subtitle called "Participation of Major Groups". This last part of the Plan contains only three articles:

- 150: [States should:] Enhance partnerships between governmental and non-governmental actors, including all major groups, as well as volunteer groups, on programmes and activities for the achievement of sustainable development at all levels;
- 152: is about participation of member states of the UN;

- 153: is an explicit reference to youth participation.

The tight link between governance and participation clearly reappears in the Plan of Implementation: participation gets a very prominent place in the Governance Chapter. We could say that participation is one of the tools to attain good governance, while good governance is in itself a tool to achieve the final objective of sustainable development.

A couple of phrases on participation of the Prepcom IV-paper on governance for sustainable development did not make it to the final Plan of Implementation:

- Par. 3c of the paper: the international community should facilitate participation of civil society in the work of the WTO;
- Par. 27d: Negotiation and implementation capacity, for enhancing effective participation in international environmental and economic agreements and instruments.

Although one article of the Paper from the paragraph entitled "participation of major groups" was omitted from the Plan, this paragraph as a whole was retained; a clear sign that the issue was given high priority.

11. Post WSSD

As for most subjects discussed at the 2002 World Summit on Sustainable Development, it was not easy for the official delegates to come to an agreement on governance and participation. This was in particular the case for governance for sustainable development. As we stated before, it is a subject on which it was a nearly insurmountable problem to reach only a status quo compared to the Rio-Declaration from 1992. The precautionary principle was omitted, as were a couple of other governance-related references.

According to Speth (2002), the scenario needed to meet the global environmental and sustainability governance challenges is called "Jazz". Jazz is a model in which business conduct is enforced by public opinion and consumer behaviour. The role of governments in this model is to facilitate; while NGOs also play a very active role in forging corporate initiatives. Other actors able to contribute are local governments, universities and other identities.

Speth further stresses the need to address more directly the underlying drivers of environmental degradation, like population growth, poverty and underdevelopment, technology (opportunities particularly lie in the energy sector), market signals (elimination of improper subsidies and internalisation of external costs are required).

11.1. GLOBAL ENVIRONMENTAL GOVERNANCE: NEED FOR A WORLD ENVIRONMENT ORGANISATION?

There is a general consensus that, in recent years, the world has been suffering from a "global environmental governance crisis". Since 1992, UNEP is said to have become weaker (Dodds, 2002). Moreover, excessive fragmentation undermines the effectiveness of the existing global environmental regime. Today, over 500 international treaties and agreements relating to the environment are in existence, more than 300 of which have been adopted since Stockholm 1972, and 41 of which are considered core conventions (UNEP, 2001).

Feeling unable to respond accurately to the challenges of globalisation, UNEP unsuccessfully started pushing for the creation of a stronger World Environmental Organisation (WEO). To date, there still is no consensus on whether the world would need a WEO or not. Yet, nowadays NGOs, think tanks and governments are addressing the idea more seriously. Whalley and Zissimos (2000) already propose a possible form of such an organisation, complete with structure, mandate, authorities and activities. They claim that such a body is needed to resolve conflicts between trade agreements and the environment. It would be able to address the relative lack of internalisation of cross-border and global externalities, which is a feature of the present global environmental regime. The central activity of a WEO would consist of generating internalisation deals between countries on global environmental issues. Deals would involve verifiable environmental commitments exchanged across countries in return for various forms of compensation, including cash. For example, a developing country could commit itself to preserve a certain forest area for a number of years, and in return make some cash, or ask for improved trade access. It would make the system advantageous to developing countries.

Not everybody has as clear a vision of the WEO as Whalley and Zissimos do. Many questions remain unresolved, concerning membership, the multilateral character, the relationship with UN, WTO, CSD, etc.

Others are simply opposed to the creation of a World Environmental Organisation. Najam (2002), for instance, claims there is not even an agreement on whether global environmental governance would be a good thing. He claims that replacing UNEP with something that might look more like the World Trade Organisation (WTO) would be an attack on legitimacy. Moreover, Najam fears that the installation of a WEO could lead to the exclusion of civil society concerns, largely from the South. After all, the existing UNEP structure has always been rather open for civil society participation.

Since the debate on the WEO has been going on for a couple of years now, and since it is being discussed in increasingly more forums, the WSSD may be a good moment for a breakthrough, or at least a reference to the topic. Nevertheless, the Plan of Implementation does not record any discussion on the topic, although paragraph 122b deals with the collaboration within the United Nations. A summit on sustainable development may not be regarded as the ideal opportunity to raise this environmental issue. In paragraph 120, the strengthening of the international bodies and organisations is even supplemented with the reflection "while respecting their existing mandates".

11.2. THE ROLE OF CIVIL SOCIETY IN THE FUTURE SUSTAINABILITY DEBATE

In the last decades, civil society organisations have played an increasingly important role in the international debate on sustainable development. Gemmill and Bamidele-Izu (2002) identify five major roles civil society organisations should play:

The first is the collection, dissemination and analysis of information. NGOs often produce a wealth of research and policy documents at conferences, which are sometimes very valuable and inspiring for the conference delegates. Yet, there is often little feedback on these documents and limited opportunities for constructive dialogue. The contributions from civil society participation at the international level need to be enhanced through a strengthened, more formalised structure for engagement. This would equally strengthen civil society in its second role, which is to give input into the policy development process. Apart from this, NGOs and other civil society organisations can be called in for assessment and monitoring. Particularly in developing countries, which sometimes have limited monitoring capacities, NGOs could make a positive contribution. Other areas where NGOs can be called in are operational functions and advocacy for environmental and sustainability justice.

11.3. ENVIRONMENTAL GOVERNANCE AND TRADE

The current global governance system is characterised by a number of shortcomings, which prevent the system from being really sustainable. One could make recommendations to remove the distortions in all areas related to sustainable development. Rather than give a full overview, this paragraph focuses on environmental governance and trade.

The reforms necessary to reach a sustainable global governance system, are also called for by NGOs. They are, of course, disputed by organisations defending other interests. The discussion of whether these interventions are justifiable, taking into account all the interests at stake, will not be dealt with here. Simply selected measures which would contribute to a governance system more concentrated on the sustainability issues than the current one, are listed.

In order to attain environmental governance, countless measures could be listed. These are the suggestions offered by the Heinrich Böll Foundation (2002):

- Recognise communities" rights to the natural habitat and resources;
- Establish a (participatory) World Commission on Mining, Gas and Oil Extraction;
- Promote citizen's democratic rights;
- Globalise the Aarhus Convention (access to environmental information);

- Reinforce the Rio principles of environmental management (prevention, polluter pays...).
- Remove subsidies to resource extraction, transport and chemical agriculture;
- Shift the tax base from labour to resources pollution and waste; ensure the right pricing of goods;
- Introduce user fees for global commons.

It is clear that these principles are not supported by a number of (powerful) societal groups, and will not be implemented in the short run. Civil society tries to change the culture of these and other groups, hoping to generate some changes in the long run.

The measures for trade are also well known: fair trade instead of free trade, reform of WTO, encouragement of Multilateral Environmental Agreements (MEAs), elimination of environmentally harmful subsidies and promotion of socially accountable production.

12. Conclusion

The issues of governance, participation and civil society are very closely connected. The term "governance" refers to a way of organising and managing a society's affairs. This way of organisation consists of the involvement of a large number of actors who are not part of government nor are they a state actor. These new partners are usually called civil society actors. Governance is about the participation of these civil society organisations in governing the world, a country, a community or another level of society.

Good governance refers to a number of basic principles that are accepted by most actors as crucial in governance for sustainability, like rule of law, and the absence of corruption.

Participation has gained a lot of attention at the international sustainability scene since the United Nations Conference of Environment and Development (1992). In Agenda 21, the action plan adopted at the conference, participation of all "major groups" was given high degree of priority.

At the World Summit on Sustainable Development (2002), the importance of governance, participation and civil society was re-emphasised. A whole Chapter of the Johannesburg Plan of Implementation was dedicated to governance, although the word disappeared from the title. Many good intentions concerning good governance were determined. The same can be said about the issue of participation of major groups and civil society. However, it has to be said that most of the commitments made in the plan don't go beyond the ones that were made ten years earlier. This is related to the conclusion that the principles set out in Agenda 21 failed to generate that many changes in practice. The decreased willingness of certain countries or groups of countries to make strong commitments has resulted in a plan that can be considered as a standstill vis-à-vis Agenda 21 instead of progress. Serious doubts can be raised over the question whether the Johannesburg Conference and its Plan of Implementation will be implemented with the same amount of enthusiasm by national governments as when it was drawn up.

With regard to participation and civil society, the type 2-partnerships are put forward as an important issue. On the one hand, NGOs consider these partnerships as opportunities, but on the other hand, the criticism is raised that the type II-outcomes are used to compensate for the lack of type I-outcomes (international agreements between governments).

It is clear that the World Summit on Sustainable Development was not a big success in terms of new commitments and a greater willingness to tackle the world's sustainability problems. However, civil society is getting more organised, and it is also increasingly involved in consultations with governments. If it succeeds to influence the policy process somewhat further than they do today, this may be the proper way to make progress towards a more sustainable society.

References

Annan, K.: 2000, *UN Secretary-General's Millennium Report*, United Nations, New York.

Banda, F.: 2002, "Governance and corruption", *The Future is now*, IIED, pp. 92–98.

Blackburn, J. and Holland, J. (Eds): 1998, *Who Changes? Institutionalizing participation in development*, Intermediat Technology Publications, Exeter.

Bruyninckx, H. and Bachus, K.: 2001, *Naar een sociaal pact over duurzame ontwikkeling*, HIVA – Hoger Instituut voor de Arbeid, Leuven.

CEC – Commission of the European Communities.: 2001, *European Governance. A White Paper*, Brussels.

CGG – Commission on Global Governance.: 1995, *Our Global Neighbourhood*, University Press, Oxford.

Dodds, F. (Ed): 2000, *Earth Summit 2002. A New Deal*, Earthscan, London.

Earth Council.: 2002, *NCSD Report 2001*, Earth Council, San José.

Esty, D. and Ivanova, M.: 2002, "Vitalizing global environmental governance: a function-driven approach", in D. Esty and M. Ivanova (Eds), *Global Environmental Governance. Options & Opportunities*, Yale School of Forestry & Environmental Studies, New Haven, pp. 181–205.

Gemmill, B. and Bamidele-Izu, A.: 2002, "The role of NGO's and civil society in global environmental governance", in D. Esty and M. Ivanova (Eds), *Global Environmental Governance. Options & Opportunities*, Yale School of Forestry & Environmental Studies, pp. 77–101.

Hales, D. and Prescott-Allen, R.: 2002, "Flying blind: assessing progress toward sustainability", in D. Esty and M. Ivanova (Eds), *Global Environmental Governance. Options & Opportunities*, Yale School of Forestry & Environmental Studies, New Haven, pp. 31–53.

Heinrich Böll Foundation.: 2002, *The Jo'burg Memo. Fairness in a Fragile World*, Heinrich Böll Foundation, Berlin.

Hemmati, M.: 2001, *Multi-Stakeholder Processes for Governance and Sustainability – Beyond Deadlock and Conflict*, Earthscan, London.

IUCN – International Union for Conservation of Nature and Natural Resources: 2002, *IUCN Statement on Governance for Sustainable Development*, IUCN, Gland.

LIFE.: 1997, *Participatory Local Governance. LIFE's Method and Experience 1992–1997*, United Nations Development Program, New York.

Maurer, C.: 1999, *Rio+8, An Assessment of National Councils for Sustainable Development*, Washington.

Najam, A.: 2002, *Global Environmental Institutions. Perspectives on Reform*, Royal Institute of International Affairs, London.

Nayyar, D. and Court, J.: 2002, *Governing Globalization: Issues and Institutions*, United Nations University, World Institute for Development Economics Research, Helsinki.

NEPAD – New Partnership for Africa's Development.: 2002, *The New Partnership for Africa's Development (NEPAD). Declaration on Democracy, Political, Economic and Corporate Governance*, NEPAD.

OECD – Organisation for Economic Co-operation and Development.: 2002, *Governance for Sustainable Development. Five OECD case studies*, OECD, Paris.

Plumptre, T. and Graham, J.: 1999, *Governance and Good Governance: International and Aboriginal Perspectives*, Institute on Governance, Ottawa.

Rueschemeyer, D., Rueschemeyer, M. and Wittrock, B. (Eds): 1998, *Participation and Democracy. East and West*, M.E. Sharpe, London.

Speth, J.: 2002, "The global environmental agenda: origins and prospects", in D. Esty and M. Ivanova (Eds), *Global Environmental Governance. Options & Opportunities*, Yale School of Forestry & Environmental Studies, New Haven, pp. 11–31.

Streck, C.: 2002, "Global public policy networks as coalitions for change", in D. Esty and M. Ivanova (Eds), *Global Environmental Governance. Options & Opportunities*, Yale School of Forestry & Environmental Studies, New Haven, pp. 121–141.

UN – United Nations.: 1992, *Agenda 21: Programme of Action for Sustainable Development*, United Nations, New York.

UN – United Nations: 1997, *Report of the Secretary-General*, General Assembly October 21, 1997, United Nations, New York.

UN – United Nations: 2002a, *Compilation Text on Sustainable Development Governance/Institutional Arrangements for Sustainable Development*, 15 May 2002, United Nations, New York.

UN – United Nations: 2002b, *Draft Text of the Monterrey Consensus, Prepared by the Co-Chairpersons, with the Assistance of the Facilitator*, United Nations, New York.

UN – United Nations: 2002c, Consideration of the chairman's paper transmitted from the second session of the Commission acting as the preparatory committee, together with other relevant inputs to the preparatory process, United Nations, New York.

UN – United Nations: 2002d, *From our Origins to the Future. The Johannesburg Declaration on Sustainable Development*, United Nations, New York.

UN – United Nations: 2002e, *Global Challenge, Global Opportunity; Trends in Sustainable Development*, United Nations, New York.

UN – United Nations: 2002f, *Highlights of Commitments and Implementation Initiatives*, United Nations, New York.

UN – United Nations: 2002g, *Institutional Framework for Sustainable Development. Paper Prepared by the Vice-Chairs Mr. Ositadinma Anaedu and Mr. Lars-Goran Engfeldt for Negotiation at the Fourth Session of the Preparatory Committee for WSSD*, United Nations, New York.

UN – United Nations: 2002h, *Key Outcomes of the Summit*, United Nations, New York.

UN – United Nations: 2002i, *Sustainable Development Governance. Paper Prepared by the Vice-Chairs Mr. Ositadinma Anaedu and Mr. Lars-Goran Engfeldt for Consideration in the Second Week of the Third Session of the Preparatory Committee for WSSD*, United Nations, New York.

UN – United Nations: 2002j, *World Summit on Sustainable Development. Plan of Implementation*, United Nations, New York.

UNDP – United Nations Development Programme: 1997, *Governance and Sustainable Human Development*, New York.

UNEP – United Nations Environment Programme: 2001, *International Environmental Governance: Multilateral Environmental Agreements (MEAs)*, United Nations Environment Programme, Nairobi.

Van Heffen, O., Kickert, W. and Thomassen, J. (Eds): 2000, *Governance in Modern Society. Effects, Change and Formation of Government Institutions*, Kluwer Academic Publishers, Dordrecht.

WCED – World Commission on Environment and Development.: 1987, *Our Common Future*, Oxford University Press.

Whalley, J. and Zissimos, B.: 2000, "Trade and Environment Linkage and a Possible World Environmental Organisation", *Environment and Development Economics*, **5**(4), 510–516.

WHAT – World Humanity Action Trust.: 2000, *Governance for a Sustainable Future*, Russell Press Ltd., Nottingham.

Willetts, P.: 2002, "What is a Non-Governmental Organization?", in UNESCO, *Encyclopaedia of Life Support Systems*, London.

Young, O.: 1997, *Global Governance. Drawing Insights from the Environmental Experience*, the MIT Press, London.

CHAPTER 15

PARTNERSHIPS

ROBERT WHITFIELD
Envirostrat
66, Weltje Road, London, W6 9LT, UK
(E-mail: robert@envirostrat.org.uk)

Abstract. Partnerships for sustainable development were seen by some countries as a key means of making progress towards sustainable development. As a result, partnerships were proposed as one of the WSSD outcomes and were the subject of much debate during the run-up to, and at, the Summit. This paper first seeks to understand why partnerships are perceived as having so much to offer sustainable development, and then goes on to assess the impact that the WSSD has had and can be expected to have on the use of partnerships for sustainable development. The conclusion reached is that whilst the initial step taken by the WSSD Bureau in proposing partnerships as an outcome was a bold move, nevertheless much more needs to be done to promote and provide support for partnership development. Some further actions are proposed, at the level of promotion and support and including the need to clarify the UN's role regarding partnerships for sustainable development. In addition, some suggestions regarding the politics of partnerships are put forward. It is further argued that the contribution the business sector could be expected to make to sustainable development, not least through the medium of partnerships, would be greatly enhanced by the realignment of the framework within which businesses operate, so as to make it more conducive to sustainable behaviour.

Key words: Development, equity, framework, partnership, multi-stakeholder, process, stakeholder, sustainable, Type II Partnership, WSSD.

Abbreviations: CSD – Commission on Sustainable Development; JPoI – Johannesburg Plan of Implementation; MSP – Multi-stakeholder Process; NGO – Non-Governmental Organisation; PFSD/pfsd – Partnership for Sustainable Development; PrepCom – Preparatory Committee (of the WSSD); TNC – Trans National Corporation; UN – United Nations; US – United States of America; WBCSD – World Business Council for Sustainable Development; WSSD – World Summit on Sustainable Development.

1. Introduction

Partnerships have been around in one form or another since time immemorial, that is two or more parties working in close co-operation together and bound, whether legally or not, by joint rights and responsibilities. Partnerships for sustainable development have equally been in existence for centuries in any society which was not truly totalitarian, though only described as such in recent years. Here partnerships for sustainable development describe partnerships where two or more people or organisations join together on a project with a common aim of benefiting society. In the last decade or so, however, a whole new focus has been placed upon this area of social activity. This reflects an increasing awareness of the concept of sustainability and the extent of the challenge that

L. Hens and B. Nath (eds.), The World Summit on Sustainable Development, 347–372.

mankind faces. Agenda 21 reflected the realisation that Governments could not hope to meet the challenge of sustainable development alone. Governments understood that the efforts of all the other elements of society needed to be harnessed successfully if sustainable development was to be achieved. There was awareness that partnerships were one of the few tools that could make that happen. Partnerships for sustainable development are now defined as voluntary multi-stakeholder partnerships which contribute to the implementation of inter-governmental commitments in Agenda 21, the Programme for the Further Implementation of Agenda 21 and the Johannesburg Plan of Implementation.

This paper seeks to explore what it is about partnerships for sustainable development that is significant and then to evaluate the impact that the World Summit for Sustainable Development (WSSD) has made on the role that such partnerships will play in the struggle to achieve sustainable development.

The paper starts by reviewing in Section 2 the state of partnerships for sustainable development before the summit. The different approaches towards partnerships will be identified together with the alternative ways in which they can be established. The key features of multi-stakeholder partnerships are then summarised and the case for partnerships for sustainable development is set out.

The paper goes on to evaluate the role that the WSSD has had in promoting partnerships for sustainable development. Partnerships were made a point of major focus at the WSSD and the means by which this was done will be explored in Section 3. The question remains as to whether what was done will prove sufficient. This question will be addressed not only by an assessment of what occurred during WSSD but also by evaluating the role of partnerships in the year following the Summit. First of all, a number of features of partnerships within the context of the WSSD will be explored in Section 4. A number of shortcomings will be identified. Section 5 then seeks to address these shortcomings, identifying possible solutions that should be implemented.

The WSSD did much to promote partnerships for sustainable development, but as yet not enough for their potential to deliver sustainable development to be fully realised.

2. Partnerships

Back in 1994, it was suggested (PoWBLF, 1994: 9) that the term partnership had become one of the most widely used words in the debate on sustainable development. If it was widely used then, it is even more widely used today. There is a broad concept of partnership meaning working together. This broad concept is advocated widely, whether in the business, political or social sphere. It underpins the concept of "partnerships for sustainable development", and yet this latter term has a distinct meaning, and it is that distinct meaning that will be addressed in this paper.

2.1. PARTNERSHIPS FOR SUSTAINABLE DEVELOPMENT

Agenda 21 reflected a global consensus on environment and development co-operation and stated clearly that its successful implementation is first and foremost the responsibility of governments (UN, 1992: Ch1). It was made equally clear in subsequent Chapters however that formal and informal organisations and grass root movements representing the 9 major groups identified in Agenda 21 should also be recognised as partners in the implementation of Agenda 21 (UN, 1992: Ch23-32). This section addresses different aspects of partnerships for sustainable development, quite independent of the WSSD. The different types of social and economic interaction sometimes called partnerships are identified in Sub-section 2.2 below, whilst subsequent sub-sections explore how such partnerships are formed, their key features and why they are worth the effort.

2.2. PARTNERSHIP CONCEPTS

There is a spectrum of types of multi-stakeholder process or interaction ranging from dialogues at one end through to action and implementation at the other. It is this end of the spectrum that is the focus of the partnerships for sustainable development. There are several different ways of describing such partnerships, including public private partnerships, the tripartite model, multi-stakeholder partnerships and public policy networks that vary in their appropriateness. Each of these models is described briefly below.

2.2.1. Public-private partnerships
Public-private partnership is a concept that has evolved comparatively recently and the 1990s has seen the rapid development of such partnerships across the world (Osborne, 2002: 1). Historically they have been used particularly for social purposes such as urban regeneration and social inclusion. Necessarily, they involve both government and civil society. As such they are a sub-set of the broader concept of partnerships for sustainable development, which is not prescriptive as to whether any particular sector of society is represented. Riley (2002) advocates the concept of "critical collaboration" between Government and NGOs where the NGOs retain their right and ability to be critical of the Government in areas outside the specific subject of the collaboration.

2.2.2. Tripartite model
The Tripartite model (not to be confused with the consultation model with government, business and labour unions with the same name) is a framework often put forward for analysing Governance issues by organisations such as the Centre for the Study of Global Governance, which sees society as having three components, namely Government, Business and Civil Society, where civil society is defined in this case as all sectors of society excluding Government and business. Others have used this framework to describe partnerships for sustainable development. The Prince of Wales Business Leaders Forum

(PoWBLF, 1994) have used this framework, but using the term Non-Governmental Organisation (NGO). In this case they have used in the same sense as civil society above, in contrast to the sense used in Agenda 21 (UN, 1992) that uses a narrower definition, excluding Major Groups such as Trade Unions, Women, Youth, Scientists and Local Authorities. The model has the attraction of simplicity, but as a result can be misleading. The PoWBLF (1994) booklet was seeking to describe business' role in partnerships for sustainable development. As such the model suited their purposes as it gives clear prominence to business whilst not making such partnerships appear too daunting. Such a framework, as shown in Figure 1, has less appeal however to members of some of the other stakeholder groups such as the trade unions, indigenous peoples groups or the churches – groups that would not immediately see themselves as NGOs. More importantly, the model implies that a partnership for sustainable development necessarily includes a representative from each of the three components of society. But this is not the case. There is a significant role for partnerships that do not include Government or do not include business. There is a significant role for partnerships between different stakeholder groups within the NGO grouping. The tripartite model does not encourage this view.

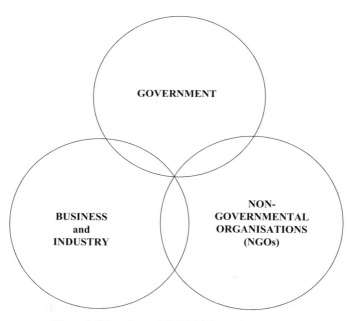

Figure 1. Tripartite model (Aloisi de Larderel et al, 1994).

2.2.3. Multi-stakeholder partnerships

A broader concept is that of multi-stakeholder partnerships. A definition of stakeholder that is widely used is that

"Stakeholders are those who have an interest in a particular decision, either as individuals or representatives of a group. This includes people who influence a decision, or can influence it, as well as those affected by it."

(Hemmati, 2002: 2)

A multi-stakeholder partnership is one involving the different stakeholders in a decision, or area of activity. Preferably, the partnership should include representatives of all the main stakeholders though that is often not the case. It is not possible to be prescriptive about precisely which stakeholders should be involved. A way of visualising such partnerships draws more on the major group concept set out in Agenda 21 (see Figure 2). Here each of the stakeholder groups is represented together with other stakeholder groups not specifically identified. A partnership is a particular sub-set of representatives of these different stakeholders.

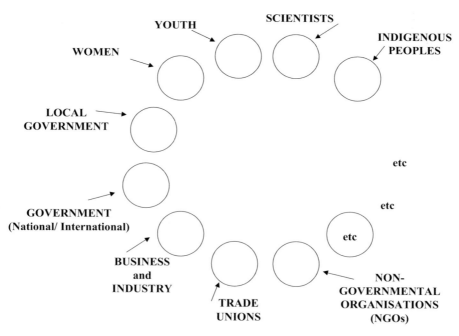

Figure 2. Multi-stakeholder approach.

Figure 2 portrays the potential diversity within a partnership for sustainable development, reflecting the need within a multi-stakeholder partnership, not to have a representative from a particular box per se, but to have a representative of each group of stakeholders that is significantly involved in the issue that is the basis of the proposed partnership, whatever category they might fall into.

Box 1. Global Reporting Initiative. An example of a multi-stakeholder approach to dialogue and partnership.

The concept of a global reporting system which becomes the global standard for reporting started to move towards reality with the establishment of the Global Reporting Initiative by the Coalition for Environmentally Responsible Economies (CERES) in 1997. Under the leadership of Robert Kinloch Massie, funding was secured from a number of donors, particularly the Foundations and UNEP was soon secured as a key partner. In due course a Steering Committee was formed with 17 organisations from 7 nations and this committee managed the development of the GRI for the next five years. The focus of the initiative was the development of reporting guidelines that embrace the environmental and social aspects of performance as well as the economic. The process employed has been described as intensive, multi-stakeholder and international. Agreement was sought through working groups, using consensus rather then majority decision-making.

In 2002 the GRI was formally established as a legal entity. Amsterdam was the chosen location and the GRI was established as a Foundation in Holland. The Board of Directors is appointed by a 60 strong Stakeholder Council made up of representatives of the 6 regions of the world. Each representative is required to maintain effective multi-stakeholder dialogue with the stakeholder groups and networks in the regions they are representing. The current council includes representatives from business, unions, fund managers, NGOs, rating agencies, science and academia.

By March 2004 there were 418 organisations from 43 countries engaged in GRI reporting, including a significant proportion of the world's largest companies.

Hemmati, 2002 (140-144); Global Reporting Initiative, 2004

2.2.4. Networks

Global public policy (GPP) networks is a term used by Reinicke, Deng et al (2000) and they can be seen as one extreme form of partnership. They are described as multi-sectoral collaborative alliances, often involving governments, international organisations, companies and NGOs. Witte *et al.* (2003: 64) see partnerships and networks as distinct. They give three examples however, one of which describes itself as a partnership (Global Water Partnership). Similarly Eco-agriculture Partners or the International Rainwater Harvesting Alliance, both registered Type II partnerships have already some of the attributes of a network due to the number of partners who have joined.

2.3. PROCESS OF FORMING A PARTNERSHIP FOR SUSTAINABLE DEVELOPMENT

Partnerships may be formed in a number of different ways. Whilst there is a traditional approach this has its weaknesses and there are a number of alternatives approaches to partnership formation, which are now emerging.

2.3.1. Traditional approach

The formation of a partnership typically has resulted from one party conceiving the idea and discussing it with one or more other parties. The partnership may be formed in one single step, that is by assembling the potential partners at one time and discussing the proposal with them all together, or alternatively, the partnership may be developed step by step, with people/ organisations being introduced to the idea individually or in groups. A problem with such an approach is that it tends to reflect the power asymmetry within the group. It is often the most powerful organisation that takes the initiative with the risk that the other potential partners are suspicious and may be reluctant to become engaged. If the partnership is successfully formed, there remains the danger that the power asymmetry is institutionalised within the partnership, that is to say that the partnership continues to be driven by the agenda of the powerful party. This danger can potentially be addressed however and this form of partnership has the advantage of there being a clear champion, who will provide the necessary leadership.

2.3.2. Facilitation, mediation, brokering

In order to address the type of the problems outlined above, there is an increasing trend towards the use of an independent third party to act as facilitator, mediator or broker. Whilst these terms are largely synonymous, there can be a range of approaches as to how to play this role and the nature of the skills to be deployed. Neutrality is perhaps the most significant characteristic but the individual will require significant interpersonal/facilitation skills. There are different views as to the degree to which the facilitator should have specific knowledge of the problems being addressed. Whilst such knowledge would appear a clear advantage, there is a risk that the facilitator/ broker loses his/her objectivity and does not reflect the wishes of the group. Knowledge of inter group dynamics is crucial as is the fact that people will tend to only take ownership of a particular plan of action if they have been a party to its design. Hemmati (2002: 222 and 237) addresses the issues relating to the facilitator in some detail and Calder (2002: 11) briefly describes how an effective facilitator can help establish a multi-stakeholder network and lead the network through to commitment to concrete action.

2.3.3. Facilitation framework

The above approaches somewhat beg the question as to who appoints the facilitator. Frequently it will be the more powerful organisation that is seeking to address some of the problems outlined above. The appointment of an "independent" facilitator is likely to be constructive, but the fact that the facilitator has been appointed by the powerful party remains, with associated risks of mistrust.

In an alternative approach being explored by Stakeholder Forum for Our Common Future for example (Stakeholder Forum, 2002b), the partnership development and facilitation role in relation to a specific action area is carried out by a third party organisation not directly involved in the partnership. This role is performed under the guidance of a multi-stakeholder advisory group convened to address a particular issue and/or geographic area. The multi-stakeholder group identifies action areas where it considers there is scope for collaborative action and then provides support for the stakeholder analysis and stakeholder engagement prior to the facilitation. The key point with respect to facilitation is that it is not sponsored by any one of the partners. Here the challenge is to identify, or allow to emerge, an individual or organisation who will take on the role of champion and provide leadership to the group.

2.3.4. Summary

These three approaches are shown graphically in Figure 3. The two circles reflect the multi-stakeholder model shown in Figure 2 where each capital letter represents a particular stakeholder grouping. In each case the partnership being brokered involves four stakeholders, the difference being how their interaction is facilitated. The lines represent personal interaction. Approach 1 is the traditional approach. Approach 2 is the independent facilitation framework approach whilst brokering lies somewhere in between. Each needless to say has its strengths and weaknesses. The weakness of the traditional approach has been described above; it has the advantage however that there is a clear champion for the partnership, prepared to drive it forward and to put resources into it. The framework approach depends upon the identification of one or more organisations willing to play this role.

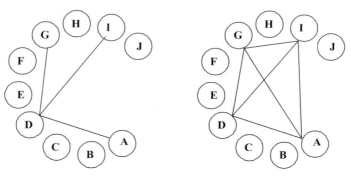

APPROACH 1   APPROACH 2

Figure 3. Partnership initiation and facilitation.

2.4. KEY FEATURES OF SUCCESSFUL MULTI-STAKEHOLDER PARTNERSHIPS

In some societies, whether at a sports day at a primary school or at a village party, there is a so-called three-legged race, where the contestants run in pairs. The two participants in a pair stand side by side with their adjoining legs tied together. The pairs then have to race a hundred metres or so. There are two main strategies for winning this race. One is for one partner to be much bigger and stronger than the other, and in this case the bigger partner runs the race almost as if on his own, effectively carrying the smaller partner. The other quite distinct strategy is for the two participants to be fairly evenly matched and to run in a totally synchronised manner with one "leg" being the combined middle legs and "the other" being the two outer legs. In most cases it is the latter strategy that proves most successful, and both prove more successful than any intermediate strategy (though some claim that the former strategy hardly counts as the outer leg of the smaller participant sometimes rarely touches the ground).

There is a clear analogy here with partnerships. It is possible to organise a partnership along the lines of the first strategy, but multi-stakeholder partnerships reflect the second strategy, that is the principle of equity.

The key principles and strategies of multi-stakeholder partnerships have been articulated by a number of authors. The following list draws heavily upon the work of Hemmati (2002: 248-251) and PoWBLF (1994: 14/15). It has been prepared before there has been a proper assessment of the most effective features of a successful multi-stakeholder partnership.

Agreed management structure and systems – decision-making guideline, processes for conflict resolution and systems to review and evaluate progress

Effectiveness – creating commitment through the participants identifying with the outcome and thus increasing the likelihood of a successful implementation

Equity – operating on the basis of equally valued contributions from all

Flexibility – remaining flexible over time

Good Communication – aiming at overcoming mistrust and suspicion and generating a shared vision.

Good Governance – further developing the role of stakeholder participation and collaboration in (inter) governmental systems as supplementary and complementary vis a vis the roles and responsibilities of governments, based on clear norms and standards; providing space for stakeholders to act independently where appropriate

Inclusiveness – providing for all views to be represented

Learning – taking a learning approach throughout the partnerships design and operation

Legitimacy – requiring democratic, transparent, accountable, equitable processes in the partnership design; requiring partners to adhere to those principles

Ownership – people-centred processes of meaningful participation, allowing ownership for decisions and thus increasing the chances of successful implementation

Participation and engagement – bringing together the principal actors

Partnership/co-operative management – clear focus and purpose, based on creating mutual benefit: "win-win" situations rather than "win-lose"

Societal gains – creating trust through honouring each participant as contributing a necessary component of the bigger picture; helping participants overcome stereotypical perceptions and prejudice

Strengthening of (inter) governmental institutions – developing advanced mechanisms of transparent, equitable and legitimate stakeholder participation strengthens institutions in terms of democratic governance and increased ability to address global challenges

Strong leadership and vision – finding a way of providing vision and leadership when needed whilst building at the same time a sense of shared ownership

Transparency – publishing activities in an understandable manner to non-participating stakeholders and the general public

2.5. THE CASE FOR PARTNERSHIPS FOR SUSTAINABLE DEVELOPMENT

The whole thrust of partnerships for sustainable development is based upon the notion that they genuinely have a major role to play in progressing towards sustainable development. The hypothesis is that this is through the manifestation of the characteristics outlined above but this remains to be proved. Partnerships are clearly not appropriate in every situation. Where their features are relevant to the problem to be addressed, several benefits have been identified as accruing. These benefits include the ability to mobilise greater amounts of skills and resources, problem analysis in a more integrated, multidisciplinary and comprehensive manner, elimination of unnecessary duplication of cost and effort, helping to reconcile the views of traditional adversaries, the facilitation of dialogue, creativity and trust and the facilitation of information flow and technology transfer (PoWBLF, 1994: 8). In addition one can identify the ability of partnerships to enable leaders in sustainable development to proceed without being held back by laggards (Calder, 2002: 8). This issue is at the heart of the discourse discussed later reflecting the advantages put by the US and others, and the concerns expressed by many developing countries and those who might fear being left behind.

Nelson and Zadek (1999: 25, 30) cite some studies of partnership costs and benefits and go on to analyse the benefits of social partnerships. They see the case for a partnership approach being when the additional benefits of the approach exceed the additional costs (Nelson and Zadek, 1999: 24) and distinguish two distinct types of benefit, namely societal benefits and participant benefits whilst emphasising also the importance of process gains. These benefits they see as distinct, though overlapping.

Potential participant benefits that they identify (Nelson and Zadek, 1999: 28) for social partnerships include the development of human capital, improved operational efficiency, organisational innovation, increased access to resources, better access to information, more effective products and services, enhanced reputation and credibility and the creation of a stable society. Societal benefits vary with the precise purpose of the partnership in question and range from job creation to overall improvements in quality of life. The authors concede that "in practice there is relatively little empirical research into the systematic realisation of these wider societal benefits" (Nelson and Zadek, 1999: 29) and go on to note that "the same anecdotal evidence can be used to argue both increased societal benefits and increased costs, depending on the perspective of different participants and observers" (Nelson and Zadek, 1999: 29). They acknowledge however that "societal benefits are more likely to be attained if the partnership process is one that enables a combination of greater efficiency with improved effectiveness and equity" (Nelson and Zadek, 1999: 29).

3. Sustainable development partnerships at WSSD

Partnerships (for sustainable development) became a major component of the WSSD both in its preparation and its outcomes. They were however controversial when they were originally proposed and have remained so since the WSSD. This section summarises briefly what actually happened regarding partnerships at the WSSD and in the following 12 months.

3.1. TYPE II PARTNERSHIPS

During 2001 preparations were underway within different sectors to promote partnerships within WSSD. Business started to plan for its Virtual Partnerships initiative and for its main partnership celebration event whilst other stakeholder groups and organisations prepared for their own events. In addition, in July, Stakeholder Forum (2002b) announced its Implementation Conference process (designed to launch a series of new partnerships at WSSD).

In December there was the announcement by the WSSD Secretariat (2001) on behalf of the WSSD Bureau, proposing a "possible framework for strengthening linkages between the expected outcomes of WSSD". They suggested that two types of outcome were expected from the Johannesburg Summit and its preparatory process, namely a "first type" of outcome, that is a negotiated text and political declaration similar in nature to all previous summits, and a "second type of outcome" which would consist of

"a series of commitments, targets and partnerships made by individual governments or groups of governments, at the regional and/or inter-regional level, as well as with involvement of or among major groups".

(WSSD Secretariat, 2001: 1)

This new initiative was a radical and brave proposal, breaking new ground for the UN.In particular it proposed that

> "In order to ensure tangible and credible nature of the overall "package" of WSSD outcomes, all of the outcomes described above need to be developed through transparent and participatory processes. The "inter-governmentally agreed" outcome of the Summit should also include concrete and participatory mechanisms for monitoring progress, which should be seen as an integral part of [a] future strengthened system of sustainable development governance (at local, national, regional and global levels) and which would be based on principles of shared responsibility and accountability."
>
> (WSSD Secretariat, 2001: 2)

The Secretariat did not advocate "the second type of outcomes" but sought to show how those outcomes that they expected from the summit should be handled. It is significant to note that this original articulation of "the second type" of outcomes included three categories, including commitments and targets, which could be made by individual Governments/ stakeholders or groups. It is regrettable that in the subsequent vociferous debate during the PrepComs concerning "the second type of outcome", these two additional outcomes were lost and the debate focussed entirely on partnerships for implementation. That is not to say that no governments or other organisations made any unilateral commitments or established any targets, but rather that there was limited encouragement of such action and less spotlight on it than would otherwise have been the case.

The objective of the WSSD Bureau was to ensure that there would be significant tangible outcomes to announce at the Summit. Over the following nine months, the discourse centred on the criteria that needed to be applied for such Type II (or 2) partnerships as they were variously called, and how they should be monitored.

3.2. CRITERIA/GUIDING PRINCIPLES DEBATE

The response to the Type II initiative was very mixed. Whilst many supported the radical step, others saw it as evidence that the UN was being dominated by business and that the US and other countries were seeking to escape the rigours of multilateral agreement by establishing an alternative approach where they could do what they liked. The proposal was shrouded in suspicion. Some NGOs rejected it outright whilst others sought to allay their fears by insisting on the establishment of stringent criteria and reporting requirements. It tended to be the policy and campaigning NGOs such as Third World Network (Utting, 2002) who were most concerned in contrast to the development NGOs that were involved in implementation who have been actively participating in such collaborative action for many years.

Stakeholder Forum (2002a) produced a paper that sought to build upon the original WSSD Outcomes document and suggest how a "package" of Type I and II outcome documents could be put together, introduced by a preamble outlining the roles and responsibilities of governments, intergovernmental bodies and stakeholders. Recommendation regarding Type II partnerships per se fell into three categories, embracing content criteria, process criteria and follow up procedures. It was proposed that the content criteria address the priorities and content, the beneficiaries and approach and should not exclude smaller partnerships. The recommended process criteria embraced principles for developing partnership initiatives, a common framework for the plans, the securing of resources and implementation and monitoring criteria. Finally a review process performed by the CSD was recommended together with opportunities for learning and replication.

The debate raged through PrepComs 2 and 3 and there was a clear danger that the criteria debate would continue to the point that there would be no time to determine whether the partnerships proposed met those criteria, let alone create new partnerships in the light of the criteria. Finally, following further discussion at the fourth Prep-Com in Bali, the debate was brought to an official close with the circulation of "the Bali Guiding Principles" (Kara and Quarless, 2002). In this document, Jan Kara and Diane Quarless, the Vice Chairs of the WSSD Bureau responsible for partnerships, set out the principles that they proposed should be adopted for Type II partnerships. This provided the basis for review of Type II partnerships submitted to the UN though not all parties were satisfied.

3.3. WSSD OUTCOMES

Partnerships are mentioned 46 times in the Johannesburg Plan of Implementation (UN, 2002a) and the Johannesburg Political Declaration (UN, 2002b). As such this could be seen as a victory for the supporters of partnerships. By the time of the summit, 254 partnerships had been accepted by the Secretariat as meeting the criteria set out in the Bali Guiding Principles. Significantly however, no wording was agreed that formally bound the Type II partnerships into the formal summit outcome. Nor did the inter-governmentally agreed (first type of) outcome include "concrete and participatory mechanisms for monitoring progress" (WSSD Secretariat, 2001), delegating any decision in this controversial area to the CSD. These developments could be seen as a "victory for partnerships" detractors. The fact was that many developing countries were uncomfortable with partnerships and the risk of them allowing the developed countries to escape from the need for stretching Multi-lateral commitments. Politically, at the summit itself, it was seen as too sensitive to open up at that late stage.

3.4. CSD-11

In the twelve months since the summit, there is clear evidence that the difference of views regarding partnerships for sustainable development remains. For example, Witte *et al.* (2003) rehearse similar views from stakeholders to those heard during the run-up to the summit, with business seeking to avoid restrictions and NGOs being more focussed on criteria and monitoring. Nevertheless, some significant steps have been taken including those associated with CSD-11 and steps taken by the UN itself relating to partnerships.

By CSD-11 (April/May 2003) the number of Type II partnerships registered was up to 264. A "partnership fair" was introduced at CSD-11, running in parallel with the main debate, designed to facilitate progress reports on specific Type II partnerships. One element of the formal negotiation related to partnerships and agreement was finally reached on the criteria and guidelines for partnerships in the WSSD process and its follow up, superseding all previous documents such as the Bali Guiding Principles UNDESA 2003). The criteria and guidelines are set out in four paragraphs of the final text. By and large, the new criteria and guidelines (paragraph 22) reflect the spirit of the Bali Guiding Principles. A multi-stakeholder approach is no longer defined but it is made clear in paragraph 21 that the partnerships being discussed are multi-stakeholder initiatives. New paragraphs include (g), (i) - (l), stressing greater national links (respecting national laws, national reporting) and some greater constraints on international institutions. Paragraph 23 elaborates in greater detail on the reporting requirements to the CSD, whilst, significantly, paragraph 24 calls for activities aimed at strengthening partnership in the context of the WSSD.

3.5. REGIONAL IMPLEMENTATION MEETINGS

The Secretary General produced a report before the CSD (UN, 2003), proposing not only a 10+ year programme for the CSD based on a two year cycle, but also a regional process with regional implementation fora (subsequently changed to "meetings"). The main thrust of these proposals was agreed at CSD-11 including a role for partnerships in these new regional processes.

3.6. OTHER UN PARTNERSHIP ACTIVITIES

In 1999, UN Secretary General Kofi Annan set up the Global Compact, a partnership with business, providing the opportunity for dialogue and encouraging business to showcase their partnership for sustainable development activities. As with the WSSD partnership activities, the Global Compact has been attacked from the beginning as a vehicle for business "green wash" and "blue wash". Following CSD-11, the UN Global Compact held a policy dialogue on partnerships for sustainable development. A new partnership web-site was described and some different tools for establishing partnerships put forward, but the same conflict of views remained with a number of the NGOs present expressing continued concern.

An announcement by the Secretary General of the formation of a new UN Partnership in the autumn of 2002 sounded significant though it is suggested that it will mainly be an amalgamation of the UN Global Compact Secretariat and the UN Fund for International Partnerships, set up to collaborate with the UN Foundation. The significance of this mechanism remains therefore to be seen.

4. Analysis

In light of the involvement of partnerships in the WSSD and the evidence of the 12 months following the Summit, it is possible to make some observations regarding the impact that the WSSD will have on sustainable development in terms of partnerships.

4.1. BOLD NEW MOVE IN THE DARK

As suggested above, the WSSD Bureau are to be applauded for the bold move that they made in giving partnerships for sustainable development such a high profile within WSSD even though there was a great deal of uncertainty and risk.

One inevitable disadvantage of such a move is that there is very little data on which to make forecasts and judgements, and certainly no targets within the Johannesburg Plan of Implementation in relation to partnerships for sustainable development. The CSD Secretariat have made an assessment that the value of the funds committed to the current Type II partnerships is $1,250 million. Though a substantial sum, it is dwarfed by the magnitude of the challenge and as such therefore tells us very little.

Another feature of the uncertainty is that there are many more questions than answers. Writers on partnerships for sustainable development tend to conclude with a series of questions (Hemmati, in press).

4.2. CONFUSION

There is a remarkable degree of confusion regarding Type II/2 partnerships and "partnerships for sustainable development in the context of the WSSD". There are two quite distinct notions, and yet it is frequently unclear as to which one is intended. These two notions are:

a. Partnerships for sustainable development, that are linked to the UN and for which globally established selection criteria and monitoring procedures are relevant (which I shall call PFSD).
b. Partnerships for sustainable development that do not seek any such links with the UN (which I shall call pfsd).

I should emphasise that I have not introduced PFSD and pfsd as terms that are intended to last, quite the contrary. They facilitate a clear distinction that needs to be made, and at the same time underline the need for suitable terms.

There is good evidence to suggest that small pfsds at the community level can be highly effective. It would be wholly unrealistic to suppose that all pfsd could become PFSDs since the bureaucracy required to monitor such an exercise would be out of all proportion to the benefit. It is realistic to conclude therefore that the PFSDs will be a small, albeit important sub-set of the wider set of pfsds.

Having made that distinction, one can start to see PFSDs in a clearer light. The object is not to maximise the number or value of PFSDs per se, but rather to maximise the real contribution that pfsds can make to global sustainable development (having netted off any reduction in progress towards sustainable development resulting from a shift of resources or reduction in commitment to the multi-lateral negotiated approach). In this light, it is clear that PFSDs need to be examples of good practice and something that people aspire to. Whether the partnership is large or small, quality and integrity need to be the key features of PFSDs rather than size and number.

With this distinction, the question of promotion also becomes clear. In many ways, PFSD becomes a brand, the UN brand in relation to pfsds. Whilst a brand is normally associated with a business the concept is equally relevant to institutions. The UN brand is unique, but vulnerable. The concerns being expressed by some in the NGO community regarding the Global Compact reflect in part concern for the UN brand and the danger that it will be damaged. As any marketeer will know, once a brand has become tarnished, it takes a great deal of time and resources to recover. Type II (2?) was a brand for a while, even though the brand name emerged by default and was never too distinct (II/2). There is little doubt that the CSD Secretariat were right to abandon Type II/2 as such but "partnerships for sustainable development in the context of the WSSD" is no credible substitute. The UN needs to call on marketing skills from within or outside the organisation to develop and implement a credible marketing plan.

With this distinction, the question of criteria becomes clear. The brand needs to be protected from bad experiences, that is examples of where a pfsd has been abused by one or more partners, at the expense of other partners, or where a pfsd has been established for the sake of being seen to have been created, without any real intent to implement. Such dangers can be reduced by the application of effective selection criteria though such criteria, if they are realistic, will not guarantee success.

Finally, there is the issue discussed in Sub-section 4.3 below, namely the question of evidence of the merits of the partnership approach. If it is deemed that a set of test case partnerships need to be launched, then this group may represent a distinct category in themselves in terms of selection criteria and monitoring and evaluation.

This analysis is summarised in Table I.

Table I. Categorisation of partnerships for sustainable development.

Partnership type	Selection criteria	Monitoring and evaluation	Promotion
Tests to evaluate the benefit of partnerships for sustainable development	Carefully controlled procedure. Wider sample on a less rigorous basis.	Post hoc and propter hoc evaluation, agreed on a case by case basis	Those responsible for the tests to communicate results. In addition, onus on all to come up with convincing evidence.

PFSD	Implementation of CSD-11 paragraph 23	Implementation of CSD-11 paragraph 24. The process should preferably be further refined through a global Multi-Stakeholder Dialogue.	Good brand management. Currently Type II (2) replaced by "partnerships for sustainable development within the context of WSSD".
Pfsd	CSD-11 para. 23 can be used for reference but up to partners to decide.	None other than that decided by the partners.	General encouragement/ incentives/ aspiring to PFSD

4.3. HAS THE CASE FOR PARTNERSHIPS BEEN ESTABLISHED?

The question is, whether the case has been successfully made that there is a significant portion of the sustainable development challenges that the world faces which are best addressed through partnership. The case was set out in Section 2.5 above, but the evidence has tended to be more anecdotal than scientific. It is a difficult case to prove. It is not a scientific proof, which once seen is conclusive. The question is whether the bulk of the delegates at the WSSD believed that partnerships do have a major role to play in achieving sustainable development. The wide spread advocacy of partnerships for sustainable development within the JPoI is an indication that the answer may be yes, but far from conclusive. The subject is highly politicised. Developing countries have difficulty in answering the question, given that it is shrouded in suspicion that partnerships are a US ploy to escape multi-lateral negotiation.

This is something that does need to be clarified. If the case has been made, attention can be focussed on implementation, whilst paying due heed to the concerns expressed. However if the case has not been adequately made, then the priority is to do so. It is probable that the world is sufficiently convinced by now of the merits of partnership to continue with implementation. Nevertheless, there is sufficient uncertainty on this point for there to need to be a significant continued demonstration of examples of where the partnership approach has proved successful (particularly in cases where other approaches have proved unsuccessful). The PFSD community and database will help in this, in addition to more scientific studies.

4.4. THE INTEREST OF THE PARTIES

Actors will only act in partnership if it is in their interest to do so. The question here is whether this statement poses any problems for pfsds. Is it a significant hindrance to the formation and effective implementation of otherwise worthwhile pfsds? Or is it not such a hindrance? If a stakeholder's clear objective is sustainable development (whether individually or enshrined in some charter, corporate objective or whatever) then if he/she is invited to participate in a relevant partnership, a priori there is no reason why he/she would not decide to participate. If however the stakeholder's objectives were not to progress towards sustainable development, then it is likely that there would be many constructive partnership opportunities that he/she would choose to decline.

4.4.1. The interest of business

Take the case of business. A business' prime motivation, within the typical legal and fiscal framework across the world, is economic. There is a very active Corporate Social Responsibility movement, which is promoting a triple bottom line approach (Elkington, 1998). Such activity is clearly supportive of sustainable development, making companies, managers, employees and shareholders alike aware of the impact that the company has socially and on the environment as well as economically. Knowledge can change behaviour, measuring is often the first step towards actively managing and the movement is to be applauded. But one should not be deluded into believing that CSR alone can transform corporate behaviour into that of maximising progress towards sustainable development.

Acknowledgement of this situation is widespread within the business literature. For instance Charles Holliday, Stephan Schmidheiny and Phil Watts, three leaders of the World Business Council for Sustainable Development (WBCSD), state that

> "If basic framework conditions push us all in the wrong directions, then that is the way society will go."
>
> Holliday, C., Schmidheiny, S. and Watts, P. (2002: 58)

Furthermore, they state in their book published in the run-up to the WSSD that in its report to the Rio Earth Summit years earlier, the Business Council for Sustainable Development, as it then was, called for

> "a steady, predictable, negotiated move towards full-cost pricing of goods and services; the dismantling of perverse subsidies; greater use of market instruments and less of command and control regulations; more tax on things to be discouraged, like waste and pollution and less on things to be encouraged like jobs (in a fiscally neutral setting); and more reflection of environmental resource use in Standard National Accounts."
>
> Holliday, C., Schmidheiny, S. and Watts, P. (2002: 58)

The authors argue that "there has been very little political support for such moves from governments, civil society organisations, or frankly business" (Holliday et al., 2002: 58). The advocacy for modifying the framework is well articulated in the literature (e.g. Ayres, 1998; Hawkin et al., 1999). Authors on partnerships (PoWBLF, 1994: 10; Witte et al., 2003) acknowledge the inappropriate framework within which businesses currently operate. The movement that is advocating greater accountability addresses several aspects of the framework outlined by the WBCSD, calling for binding rules for corporations which establish, amongst other things, "accountability to the highest social, labour and environmental standards,…mechanisms to identify and eliminate perverse subsidies to corporations" (Global Policy Forum, 2002).

Attempts were made within the context of the WSSD to establish a Convention for Corporate Accountability but there was insufficient support. A major opportunity to start to better shape the framework may have passed, but the need for action has not gone away.

In the Global Compact Policy Dialogue on Partnerships in June 2003, the opportunity of creating partnership such that Sustainable Development becomes a core business activity was emphasised. Where such opportunities exist, clearly they should be seized. Without an appropriate transformation of the business framework however, such opportunities are sadly likely to prove ephemeral.

4.4.2. Other stakeholder groups

The significance of the framework for business is clear. The concept applies also to other stakeholders/ stakeholder groups to the extent that the actions of those stakeholders are significantly affected by the regulatory framework within which they operate. The key issue for business is that it is constrained to operate in the interests of its shareholders, which are currently interpreted as economic. This can preclude it from say entering partnerships that would not directly provide such an economic benefit. The same limitation can occur for an organisation that is constrained by its objectives to act in the interests of its members in a prescribed manner. An organisation might be constrained to promote environmental improvement and therefore might not be in a position to pursue some initiative of a more social nature, even though it was in the ideal position to do so. Such constraints can act as blockers to sustainable development and would be best removed, that is objectives of organisations should wherever possible include the promotion of sustainable development. The decision would still be in the hands of the relevant individuals concerned, but they would not be adversely constrained.

4.4.3. Action within WSSD to transform the framework for business and others

The sad reality is that very little was agreed within the JPOI that will lead to an amelioration of the frameworks within which business and other stakeholders operate. There were widespread calls by NGOs for greater accountability and transparency, particularly of Trans-national Corporations (TNCs) (Friends of the Earth, 2002). The issue is however wider than TNCs, even though they may be the most visible and easiest target. Changing the business framework is typically the responsibility of the Trade and Industry Ministries, with a strong involvement of the Finance Ministries. An inescapable problem of sustainable development is that its ramifications are so widespread, and yet it is not realistic for entire governments to be represented. The common response is to say that that is why the heads of state participate as they do indeed represent their entire administrations. Whilst that is clearly the case, it is not realistic to expect a Prime Minister in a few days at a conference to launch a major new commitment in one sphere of activity if the entire conference has been prepared by staff from a quite different ministry. Perhaps more fundamentally, the key issue is whether there is the political will.

4.5. POLITICAL ASSESSMENT

Lempert (2002) argues that the US, as the lead supporter of pfsds, had a major opportunity to gain acceptance to its agenda by agreeing to commit to some relevant targets. The fact that the US resisted virtually every single target proposed at the WSSD and did not come up with any of its own, meant that many of the other participants were not prepared to listen to what the US was proposing. He suggests that if only President Bush had agreed to some targets in the field of sustainable development, just as his administration has targets in most of the other spheres of government, then the US would have earned the right to have other WSSD participants listen seriously to their proposals. This did not happen – but it remains an option for the US to adopt. The US would be likely to ensure that the timing of the new target commitments was such as to gain maximum leverage, but such action should not be ruled out as impossible, especially not for the administration after George W. Bush.

5. What more should be done?

As suggested above, in many ways there are more questions than answers regarding partnerships for sustainable development. Nevertheless, certain steps can be identified which should increase the likelihood of their making a significant contribution towards attaining the goal of sustainable development. Several authors have addressed this subject such as Witte *et al.* (2003) and Hemmati (in press).

5.1. RESOLVE PFSD CONFUSION

The confusion between PFSD and pfsd needs to be addressed as a matter of urgency. This is best achieved by using some form of branding for those pfsds which are going to be given some formal UN recognition. It would similarly be preferable if there was convergence on one term for pfsds in general.

5.2. THE CSD AND "PARTNERSHIPS FOR SUSTAINABLE DEVELOPMENT"

The CSD has been tasked by the WSSD with the responsibility of giving "more emphasis on actions that enable implementation at all levels, including promoting and facilitating partnerships involving Governments, international organisations and relevant stakeholders for the implementation of Agenda 21 (UN, 2002a: 51). To discharge this responsibility the CSD can do a number of things.

5.2.1. PFSDS
The main ground rules for PFSDs have been established at CSD-11 (see Box 1). Now these ground rules need to be applied and critically, an appropriate brand for such partnerships developed and implemented. The brand needs to be protected,

but at the same time, care needs to be taken to ensure that the barriers to protect the brand are not so great that appropriate pfsds are deterred from applying to the UN at all. Once the brand has been established, it needs to be effectively managed and promoted, bearing in mind that the prime objective of PFSDs is to maximise the positive net contribution to sustainable development made by all pfsds.

5.2.2. Partnership learning network

The PFSDs and the organisations and people that make them up represents a significant community in terms of size, influence, diversity and commitment. Engagement with this community provides many opportunities for learning and the effective management of that learning. There is significant scope for learning from the PFSDs in terms of the potential for partnerships to make a major contribution to sustainable development, the ways in which to launch partnerships, how they should be supported and how they should be managed. This learning could also draw upon the experience of the development community, and seek to form a Community of Practice. This learning could then be disseminated beyond the inner learning network to the wider body of stakeholders who are engaged in pfsds or could be.

5.2.3. Broaden the CSD group managing the partnership responsibility

Over the last decade the CSD has shown itself to be open to new ideas and receptive to ways in which stakeholders can be more effectively engaged in sustainable development in general and the CSD in particular. With their new partnership responsibility, they could decide to engage stakeholder representatives in the exercise of this responsibility. One way would be through the introduction of a multi-stakeholder council, an idea raised during the WSSD PrepCom process. The idea is that the council, appointed by the stakeholder groups through a transparent process, could seek to build the partnership knowledge base, maintaining close links with the stakeholder groups and promoting best practice (Hemmati, in press). Alternatively a joint 50:50 group could be formed between the UN and the stakeholders, with staff seconded from the UN, governments and stakeholder groups (Hemmati, in press). Such a group could be given a significant role in the development of pfsds across the world.

5.2.4. Different use of UN time and space

The CSD has started to innovate with its Partnership Fair during CSD-11. This provided the opportunity for progress reports on PFSDs to be presented, though there was some concern that it was in parallel with the negotiations meaning that it detracted from the Multilateral process to the extent that it lured Government delegation members away. The CSD needs to be encouraged to go further, to experiment, to take risks in its meeting format and processes, to make full use of facilitation skills and more advanced meeting techniques.

5.3. PROMOTION OF "PARTNERSHIPS FOR SUSTAINABLE DEVELOPMENT" (PFSD)

Partnerships for sustainable development will be of every different shape and size and there should not be any sense of a straight jacket, requiring any heavy bureaucracy. Nevertheless, it is important that people embarking upon such partnerships understand the nature of the journey that they are setting out on and should appreciate what constitutes best practice in terms of partnership formation and operation. In particular, would-be partners need the right multi stakeholder concept, ensuring that those people who need to be involved to successfully address the objective are involved. An indication of the appropriate notion is set out in Section 2 on Multi-stakeholder Partnerships.

Reflecting the newness of multi-stakeholder partnerships for sustainable development, there is a need for capacity building, both at the level of those who will be assisting pfsd formation and also for the participants in newly forming partnerships. For instance, training and learning networks for partnerships are needed (Benner, 2003).

5.3.1. Facilitation/partnership broking training courses
Courses are required that can provide training in the skills necessary to facilitate the establishment of successful pfsds. Courses such as the Postgraduate Certificate in Cross-Sector Partnership (PCCP) at Cambridge University is an example of what is needed.

5.3.2. Capacity building for pfsd participants
At the same time there is the need to ensure that within the proposed partnership, there is sufficient capacity to proceed through to a successful partnership. Such capacity is best built when partnership is being developed, as proposed in a partnership development programme in Africa (Hemmati and Whitfield, 2003).

5.3.3. Pfsd support services
Partnerships are not easy to establish successfully nor operate effectively. They benefit greatly from the availability of support services covering such areas as business plan preparation, links to local professional support, designing effective processes, knowledge building and management, funding sources and fund raising, governance structures, power gaps within the partnership and means of collectively feeding experience into the policy making process (Hemmati and Whitfield, 2003: 6). As an example, the SEED Award Scheme (Supporting Enterprise in Environment and Development, a joint scheme being developed by Stakeholder Forum, UNEP DTIE and IUCN together with the German Government) is planning to offer a tailor made version of such support services to award winners (Hemmati, in press).

5.4. POLITICAL

In order to address the scepticism and mistrust that surrounds PFSDs, and to a lesser extent pfsds, several steps need to be taken.

5.4.1. More effective framework for business, including incentives

Far greater goal congruence needs to be achieved between corporations and society such that corporations have an incentive to act in the interests of sustainable development. A concerted move by the sustainable development community, both Government Ministers and officials and stakeholders, needs to be made to bring about this long overdue structural change in the corporate framework.

Such a transformation would benefit from greater transparency and accountability, though such changes would need to be brought about within NGOs and other stakeholder organisations and not only within business. Such a transformation would not affect partnerships alone but would have very wide ramifications for corporate behaviour overall. Nevertheless, without such a transformation, partnerships cannot be expected to realise their potential.

5.4.2. Additional evidence needed of US commitment to sustainable development

As much the most prominent proponent of pfsds, the US is directly associated with the whole policy initiative. The developing world is looking for evidence that the US is committed to engaging with the rest of the world in achieving sustainable development. Significant progress along the lines proposed in 5.4.1 above would no doubt achieve this, though such a major restructuring of the business framework will no doubt take time. It is significant that a major thrust of the US and other countries at the WSSD was on improving governance, on ensuring that effective institutions were in place that would underpin trade and efforts to bring about sustainable development. Restructuring the business framework along the lines suggested above can be seen as part of that broader move towards appropriate institutions and governance. If these can be in place, then the full energies of society can be harnessed for the benefit of society. Whilst the rule of law is weak, whilst there is corruption and significantly, whilst business is not encouraged by the framework within which it operates to engage in delivering sustainable development, we cannot hope to reach the goal of sustainability. Real progress needs to be made in each of these areas. Radically amending the business framework must not be avoided any longer.

Independent of any change in the business framework, any significant step by the US in the direction of sustainable development and multilateral engagement would greatly assist in the sceptics starting to hear the partnership ideas.

5.4.3. Additional cash

Finally there is the question of cash. The sceptics regarding the motives of the US and others in promoting pfsds fear that cash will be diverted from supporting the achievement of multilaterally negotiated commitments in favour of more unilateral partnerships. There have been calls from across the spectrum of world opinion, declaring that pfsds must be financed with additional cash or else the overall initiative will fail. It is important that the politicians heed this advice.

6. Conclusions

There is good reason to believe that partnerships are a key means of achieving sustainable development. They have the ability to succeed in many instances where other solutions have failed and it is important that they be promoted, encouraged and supported.

The WSSD has significantly raised public awareness of the benefits of partnerships for sustainable development. There remains concern however, amongst many developing countries and stakeholders, that these partnerships will tend to follow a business agenda and that they will provide a screen behind which the US and others will seek to distance themselves from the need for multi-lateral negotiation and commitment. This has led to some confusion as to whether comments being made by such critics relate to all partnerships for sustainable development or only those linked to the UN.

The confusion can be remedied quickly, with appropriate branding for the UN linked partnerships. Nevertheless, partnerships will need support. The diversity inherent in multi-stakeholder partnerships brings, in addition to the benefits, additional complexity. The CSD has the responsibility to promote and facilitate partnerships and it requires international support to enable it to discharge this responsibility. It has the opportunity to continue to be innovative in the manner in which it engages people and organisations and facilitates their interaction. Not only should the CSD be promoting its own brand of accredited partnerships but it should also be focussing on steps that it can take to stimulate partnerships for sustainable development across the world. The facilitation role goes beyond promotion. It embraces supporting the evolution of existing partnerships and the conception of new partnerships. Partnerships will need support both in their creation and their operation, through the appropriate training of brokers, innovation in partnership development processes and the establishment of accessible on-going support systems. Whilst the CSD have been given an overall responsibility in this regard, they will need the support of the international community if they are to succeed.

The partnership concept is based upon the implicit view that by bringing different interests together the problem will be more effectively addressed. Partnerships are fundamentally voluntary in nature however. If the framework or ground rules by which one major sector of society is constrained to operate deters that sector from certain courses of action, then that sector will remain deterred by and large unless and until the ground rules are modified. This represents a key challenge for Governments across the world, which will have a profound affect upon the approach taken by business towards sustainable development, transforming it from the realm of a fringe activity for most of business today to that of mainstream corporate activity. One key element of business's relationship with its environment that will be transformed and enhanced as a result will be that of partnerships, both the nature and the significance of the partnerships that business will choose to enter into.

Finally, the politics surrounding partnerships will need to be carefully addressed. The concerns of the NGOs and many developing countries need a response. Clearly a major move in relation to the business framework could constitute such a response, but the time scales of a framework change will be relatively long. There remains a need, left over from the WSSD for the supporters of partnerships to convince the doubters that partnerships are not a ploy to escape the constraints of the multi-lateral system.

To conclude, partnerships have a major contribution to make to sustainable development. The WSSD made a significant impact in promoting this potential, though less progress was made and less agreement reached on partnerships than was needed. The situation is not irrecoverable, though a number of significant steps need to be taken before partnerships for sustainable development can make the level of contribution of which they are capable.

References

Aloisi de Larderel, J., Bennet, N., Rappaport, A., Flaherty , M., Nelson, J. (eds) 1994 *Partnerships for sustainable development: the role of business and society,* UNEP, Paris; Prince of Wales Business Leaders Forum, London

Ayres, R.: 1998, *Turning Point – the End of the Growth Paradigm*, Earthscan, London.

Benner, T., Ivanova, M., Streck, C. and Witte, J.: 2003, "Moving the partnership agenda to the next stage: key challenges", in J. Witte, C. Streck, and T. Benner, *Progress of Peril? Partnerships and Networks in Global Environmental Governance, The Post-Johannesburg Agenda*, GPPi Washington/Berlin.

Calder, F.: 2002, *What Will it Take to Make "Type 2" Partnerships for Implementation Work?* Royal Institute of International Affairs http://www.worldsummit2002.org/texts/RIIATypeIIOutcomeDiscussion.pdf.

Elkington, J.: 1998, *Cannibals with Forks: the Triple Bottom Line of the 21ˢᵗ Century*, Capstone, Oxford.

Friends of the Earth.: 2002, *Towards Binding Corporate Accountability*, FoEI Position Paper for the WSSD, available at http://www.foe.co.uk/resource/briefings/corporate_accountability.pdf.

Global Policy Forum.: 2002, *Resolution: People's Action for Corporate Responsibility*, http://www.globalpolicy.org/reform/business/2002/2002action.htm.

Global Reporting Initiative.: 2004, available at http://www.globalreporting.org/index.asp.

Hawkin, P., Lovins, A. and Lovins, H.: 1999, *Natural Capitalism, the Next Industrial Revolution*, Earthscan, London.

Hemmati, M.: 2002, *Multi-stakeholder Processes for Governance and Sustainability; Beyond Deadlock and Conflict*, Earthscan, London.

Hemmati, M.: in press, "Social engagement: participation and collaboration in sustainable development", in R. Gardiner, *et al.* (Eds), *Governance for Sustainable Development: After Johannesburg* (draft title), Earthscan, London.

Hemmati, M. and Whitfield, R.: 2003, *Sustainable Development Partnerships in the Follow-up to Johannesburg Stakeholder Forum for Our Common Future*, available at http://www.earthsummit2002.org/es/preparations/global/partnerships.pdf.

Holliday, C., Schmidheiny, S. and Watts, P.: 2002, *Walking the Talk: The Business Case for Sustainable Development*, Greenleaf, Sheffield.

Kara, J. and Quarless, D.: 2002, *Explanatory Note by the Vice-Chairs: Guiding Principles for Partnerships for Sustainable Development*, PrepCom IV Bali, 7 June 2002, available at http://www.johannesburgsummit.org/html/documents/prepcom4docs/bali_documents/annex_par tnership.pdf.

Lempert, R.: 2002, *Missed Opportunities in Johannesburg*, 22 October 2002, UPI Press, accessible at http://www.upi.com/view.cfm?StoryID=20021021-095122-6514r.

Nelson, J. and Zadek, S.: 1999, *Partnership Alchemy: New Social Partnerships in Europe*, Copenhagen Centre, Copenhagen.

Osborne, S. (Ed): 2002, *Public-Private Partnerships: Theory and Practice in International Perspective*, Routledge, London & New York.

Reinicke, W., Deng, F., Witte, J.M., Benner, T., Whitaker, E. and Gershman, J.: 2000, *Critical Choices: The United Nations, Networks and the Future of Global Governance*, IDRC, Ottawa, accessible at http://www.idrc.ca/acb/showdetl.cfm?&DS_ID=2&Product_ID=534&DID=6.

"Resolution: People's action for corporate responsibility", signed by more than 100 NGOs during the "Corporate Accountability Week" Held in Sandton, South Africa, in August 2002 on the eve of the World Summit on Sustainable Development, available at http://www.globalpolicy.org/reform/business/2002/2002action.html.

Riley, J.: 2002, *Stakeholders in Rural Development: Critical Collaboration in State-NGO Partnerships*, Sage Publications, New Delhi.

Stakeholder Forum.: 2002a, *Comments on the Proposed Framework of Outcome Documents for Earth Summit 2002*, available at http://www.earthsummit2002.org/ic/process/summit.htm.

Stakeholder Forum.: 2002b, *Implementation Conference: Stakeholder Action for Our Common Future*, available at http://www.earthsummit2002.org/ic.

Tennyson, R.: 1998, *Managing Partnerships: Tools for Mobilising the Public Sector, Business and Civil Society as Partners in Development*, The Prince of Wales Business Leaders Forum, London.

UN – United Nations.: 1992, *Agenda 21*, available at http://www.un.org/esa/sustdev/documents/agenda21/english/agenda21toc.htm.

UN – United Nations.: 2002a, *The Johannesburg Plan of Implementation*, available at http://www.un.org/esa/sustdev/documents/WSSD_POI_PD/English/POIToc.htm.

UN – United Nations.: 2002b, *The Johannesburg Declaration on Sustainable Development*, A/CONF.199/L.6/Rev.1, available at http://www.un.org/esa/sustdev/documents/WSSD_POI_PD/English/POI_PD.htm.

UN – United Nations.: 2003, *Report of the Secretary General: Follow-up to Johannesburg and the Future Role of the CSD – The Implementation Track*, E/CN.17/2003/2, available at http://www.un.org/esa/sustdev/csd/csd11/sgreport.pdf.

United Nations Department of Economic and Social Affairs.: 2003, *CSD-11 Decision on Partnerships*, http://www.un.org/esa/sustdev/partnerships/csd11_partnerships_decision.html.

Utting, P.: 2002, *UN-Business Partnerships – Whose Agenda Counts?* Third World Network, available at http://www.globalpolicy.org/socecon/tncs/2001/0727twn.htm.

Witte, J., Streck, C. and Benner, T.: 2003, *Progress of Peril? Partnerships and Networks in Global Environmental Governance, The Post-Johannesburg Agenda*, GPPi, Washington/Berlin.

WSSD – World Summit for Sustainable Development Secretariat.: 2001, Possible Framework for Strengthening Linkages between the Expected Outcomes of WSSD, available at http://www.johannesburgsummit.org/html/documents/strengtheninglinkages.doc.

Note: all documents on the web were accessed during September 2003 except for Calder (2002) where the actual access was March 22nd 2004.

CHAPTER 16

IS MULTILATERALISM THE FUTURE? SUSTAINABLE DEVELOPMENT OR GLOBALISATION AS 'A COMPREHENSIVE VISION OF THE FUTURE OF HUMANITY'

MARC PALLEMAERTS

Faculty of Law & Institute of European Studies, Vrije Universiteit Brussel; Faculté de Droit & Institut de Gestion de l'Environnement et d'Aménagement du Territoire, Université Libre de Bruxelles
(e-mail: mpallema@vub.ac.be)

Abstract. This paper provides an overall evaluation of the outcomes of the World Summit on Sustainable Development (WSSD), which took place in Johannesburg from 26 August to 4 September 2002, in a historical perspective, against the background of earlier major United Nations conferences and General Assembly resolutions on environment and development. It focuses on the political and institutional context of the WSSD and its preparatory process and explores its policy implications for future international cooperation on sustainable development in a globalizing world. Both the results of the formal intergovernmental negotiations and the new phenomenon of 'partnerships for sustainable development' between governments, international organizations, the private sector and other major groups are analysed. The Johannesburg Declaration and the WSSD Plan of Implementation are shown to contain little in the way of political vision, credible new commitments and innovative approaches, likely to reinvigorate the implementation of the objectives of sustainable development as formulated in Rio. Though ostensibly designed to give a new political impetus to multilateralism, the WSSD rather revealed the inadequacy of intergovernmental political governance structures to address the social and environmental consequences of economic globalization.

Key words: Agenda 21, biodiversity, chemicals, Doha Development Agenda, governance, Johannesburg Declaration, major groups, marine resources, Millennium Declaration, multilateralism, multilateral environmental agreements, partnerships, poverty eradication, production and consumption patterns, renewable energy, United Nations, water resources, WSSD Plan of Implementation.

Abbreviations: CBD – Convention on Biological Diversity; CCD – United Nations Convention to Combat Desertification; CSD – United Nations Commission on Sustainable Development; EU – European Union; FAO – Food and Agriculture Organization of the United Nations; ILO – International Labour Organization; IMO – International Maritime Organization; MEAs – multilateral environmental agreements; OPEC – Organization of Petroleum Exporting Countries; PoI – Plan of Implementation of the World Summit on Sustainable Development; UNCED – United Nations Conference on Environment and Development; UNCHE – United Nations Conference on the Human Environment; UNCTAD – United Nations Conference on Trade and Development; UNECE – United Nations Economic Commission for Europe; UNEP – United Nations Environment Programme; UNFCCC – United Nations Framework Convention on Climate Change; UNGA – United Nations General Assembly; UNGASS – Special Session of the United Nations General Assembly; WSSD – World Summit on Sustainable Development; WTO – World Trade Organization.

Readers should send their comments on this paper to: BhaskarNath@aol.com within 3 months of publication of this issue.

L. Hens and B. Nath (eds.), The World Summit on Sustainable Development, 373–393.
© 2005 *Springer. Printed in the Netherlands.*

1. Introduction

Since the United Nations Conference on the Human Environment (UNCHE), held in Stockholm in June 1972, the United Nations General Assembly (UNGA) has made it a practice to convene high-level intergovernmental meetings at regular intervals to commemorate that seminal event, evaluate progress made in international cooperation in the field of environment and development and provide renewed political impetus for such cooperation. The latest in this series of conferences, the World Summit on Sustainable Development (WSSD), which met in Johannesburg from 26 August to 4 September 2002, once more provides an opportunity for taking stock. This paper is a first and necessarily incomplete attempt to evaluate the outcomes of the WSSD in a historical perspective and to explore its policy implications for future international cooperation aimed at achieving the objectives of sustainable development. In a separate publication, I provide a more detailed analysis of the relevance of the WSSD for the implementation and development of international law in the field of sustainable development (Pallemaerts: 2003).

2. The vague mandate and unfocused agenda of the WSSD

The decision to convene a new World Summit ten years after Rio and thirty years after Stockholm was taken by the UNGA in its Resolution 55/199, adopted on 20 December 2000. The principle of a ten-year review of progress achieved since the United Nations Conference on Environment and Development (UNCED) had already been decided by the Special Session of the General Assembly (UNGASS) held in June 1997 to mark the fifth anniversary of the Rio Conference and complete the first 'comprehensive review' of the implementation of its recommendations, initiated within the framework the Commission on Sustainable Development (CSD). In the Programme for the Further Implementation of Agenda 21 adopted by the 1997 Special Session, heads of State and Government had indeed committed themselves 'to ensuring that the next comprehensive review of Agenda 21 in the year 2002 *demonstrates greater measurable progress in achieving sustainable development*' (United Nations: 1998, 2). The modalities of this second 'comprehensive review' remained to be determined at a later stage.

So, in Resolution 55/199, the General Assembly 'reaffirm[ed] the political importance of the forthcoming ten-year review' and decided that this review should be organized 'at the summit level *to reinvigorate the global commitment to sustainable development*' (UNGA Resolution 55/199, 20 December 2000, para. 1 – emphasis added). It decided to call the event the 'WSSD' and entrusted the CSD with the task of preparing it. Though stressing 'the importance of early and effective preparations for the Summit' (*ibid.*, para. 5), the General Assembly resolution was much less specific as to its agenda and purpose than the corresponding resolution that convened the Rio Conference and laid down the broad outlines of its agenda as early as December 1989 (UNGA Resolution 44/228, 22 December

1989). The General Assembly decided to leave all the substantive work to the CSD, which was effectively given less than one year to complete its task.

The CSD itself was to 'consider a process for setting the agenda and determining possible main themes for the Summit in a timely manner' (UNGA Resolution 55/199, 20 December 2000, para. 16(d)). It was given no more specific instructions by the General Assembly than to identify 'major accomplishments and lessons learned in the implementation of Agenda 21' as well as 'major constraints hindering the implementation of Agenda 21' and to 'propose *specific time-bound measures* to be taken and institutional and financial requirements' (*ibid.*, para. 15). In doing so, the Commission was also invited to 'address *new challenges and opportunities* that have emerged since the [Rio] Conference' and to 'address ways of *strengthening the institutional framework for sustainable development*' (*ibid.*).

Based on the preparatory work of the CSD, the WSSD was to result in '*action-oriented decisions*' for the further implementation of Agenda 21, including also the further implementation of the 'Rio + 5' programme adopted by UNGASS in 1997, and 'renewed political commitment and support for sustainable development' (*ibid.*, para. 3). In addition to the 'conclusions and recommendations for further action' resulting from the preparatory process, the Summit was to adopt 'a *concise and focused document* that should emphasize the need for a global partnership to achieve the objectives of sustainable development' and 'reinvigorate (...) the global commitment to a North/South partnership and a higher level of international solidarity and to the accelerated implementation of Agenda 21 and the promotion of sustainable development' (*ibid.*, para. 17(b)). From the outset, the General Assembly had also made it clear that '*Agenda 21 and the Rio Declaration on Environment and Development should not be renegotiated*' (*ibid.*, 13th preambular para.) and that the Summit 'should ensure a balance between economic development, social development and environmental protection' (*ibid.*, para. 4).

With such an unfocused and ambitious-sounding agenda, it is not surprising that the preparatory process did not at all unfold as planned. After a round of regional preparatory meetings in 2001, the CSD started its substantive work in January 2002, but did not succeed in completing even the first stage of the preparations – the actual review of implementation and formulation of 'conclusions and recommendations for further action' (*ibid.*, para. 17(a)) – by the end of its fourth preparatory session, which concluded in Bali at the ministerial level on 7 June 2002. Less than three months before the Summit was due to convene in Johannesburg, the CSD had not produced a single agreed document and had not yet effectively started negotiations on a draft political declaration for consideration by heads of State and Government. In fact it had not even drawn up a proper agenda for the Summit that was any more focused than the initial vague guidance from the General Assembly itself. This political free-for-all and drawn-out agenda-setting process worked to the benefit of those governments which had low ambitions for the WSSD and viewed it essentially as an 'implementation summit' rather than as a forum for the formulation of new political commitments. The political climate of the WSSD was also strongly influenced and constrained by developments in other international fora, especially by the Monterrey Conference on Financing for Development, which took place just prior to the third preparatory meeting of the CSD, as well as the World Trade Organization's (WTO) Ministerial Conference in Doha in November 2001, whose 'Doha Development Agenda' effectively prejudged the possible outcome of the WSSD's deliberations concerning the relationship between trade, environment and sustainable development.

3. The international political ritual of review and re-commitment: a never-ending story

In global environmental politics, the practice of periodical high-level political gatherings to take stock of past achievements and unresolved problems and solemnly pledge stronger and more determined action in the future has developed into a self-perpetuating process.

In fact, the Stockholm Conference itself had already recommended 'that the General Assembly of the United Nations decide to convene a second UNCHE', but the General Assembly had refrained from taking immediate action on this recommendation. In 1981, however, the General Assembly set in motion the first post-Stockholm review exercise through its Resolution 36/189 of 17 December 1981. Noting 'important changes in the perception of the environment and of environmental problems' since 1972 and expressing its concern 'that there is *need to revive the sense of urgency and commitment by Governments for national and international co-operative action to protect and enhance the environment*, which found expression at the UNCHE' (UNGA Resolution 36/189, 17 December 1981, 2nd & 3rd preambular paras – emphasis added), the General Assembly decided to convene a 'session of a special character' (SSC) of the Governing Council of the United Nations Environment Programme (UNEP) in Nairobi in May 1982, which would provide 'a unique opportunity for Governments to *re-emphasize their continued commitment and support* to the cause of the environment' (*ibid.*, 4th preambular para. – emphasis added). Member states of the United Nations were specifically urged 'to participate in the session of a special character *at the highest political level*' (*ibid.*, para. 4).

Though the agenda of the SSC was still focused primarily on 'action and international co-operation *in the field of the environment*, and major *environmental* trends to be addressed by the UNEP over the next ten years' (*ibid.*, annex, section I, para. 7 – emphasis added), Resolution 36/189 also explicitly recognized 'the importance of the interrelationships between people, resources, environment *and development*' (*ibid.*, 3rd preambular para. – emphasis added). Thus, the conference held in Nairobi to mark the tenth anniversary of the Stockholm Conference addressed environmental issues in the context of social and economic development even more directly than had already been the case in Stockholm itself. The Nairobi Declaration and the resolutions of the SSC duly stressed the interrelationship between environment and development and already contained references to the emerging concept of sustainable development.

At the same time as governments gathered in Nairobi reaffirmed their 'commitment to the implementation of the Action Plan for the Human Environment adopted by the Stockholm Conference' and their conviction 'that the principles of the Declaration of the UNCHE are as valid today as they were in 1972', they also concluded that 'environmental protection consists not only of pollution abatement, but also of the rational use of natural resources *for sustainable development*' (UNEP Governing Council, Session of a Special Character, Resolution I, 18 May 1982) and proposed the establishment of a 'special commission' with a mandate 'to propose *long-term environmental strategies for achieving sustainable development* to the year 2000 and beyond' (UNEP Governing Council, Session of a Special Character, Resolution II, 18 May 1982). This recommendation of the UNEP Governing Council led to the launch, in 1984, of the World Commission on Environment and Development (WCED), chaired by Gro Harlem Brundtland, whose activities further conceptualized and popularized the notion of sustainable development.

Though its establishment was endorsed by the UNGA in Resolution 38/161, and its chairman and vice-chairman were appointed by the United Nations Secretary-General in accordance with political guidance given in that resolution, the WCED was not an intergovernmental body entrusted with any negotiating mandate. The General Assembly, while calling on the Commission 'to make available a report on environment and the global problematique to the year 2000 and beyond, *including proposed strategies for sustainable development*' (UNGA Resolution 38/161, 19 December 1983, para. 10 – emphasis added), stressed that this report would only serve as input for future intergovernmental deliberations, 'it being understood that the report of the Special Commission will not be binding on Governments' (*ibid.*, para. 13). In the end, it turned out that the Brundtland Commission's report *Our Common Future*(WCED: 1987), despite its lack of any official status, received far more public and political attention and had considerably more influence on global policy-making than the formal outcome of its consideration by an intergovernmental process within the framework of the UNEP Governing Council and the UNGA, the 'Environmental Perspective to the Year 2000 and Beyond' adopted by General Assembly Resolution 42/186 of 11 December 1987 as 'a broad framework to guide national action and international co-operation on policies and programmes aimed at achieving environmentally sound development' (UNGA Resolution 42/186, 11 December 1987, para. 2). The Brundtland Report and the resulting political debate effectively set the stage for the next round of high-level policy-making within the United Nations, which culminated in the 1992 Rio Conference on Environment and Development (UNCED).

In December 1989, 'deeply concerned by the continuing deterioration of the state of the environment and the serious degradation of the global life-support systems (. . .) and recognizing that decisive, urgent and global action is vital to protecting the ecological balance of the Earth' (UNGA Resolution 44/228, 22 December 1989, 9th preambular para.), the UNGA decided to organize a UNCED in Rio de Janeiro, twenty years after Stockholm, with 'the highest possible level of participation' (*ibid.*, para. I.1). In Resolution 44/228, UNCED was instructed to 'address environmental issues in the developmental context' (*ibid.*, para. I.15) and given the task 'to elaborate strategies and measures to halt and reverse the effects of environmental degradation in the context of increased national and international efforts to *promote sustainable and environmentally sound development in all countries*' (*ibid.*, para. I.3 – emphasis added). The resolution also listed a series of more specific objectives, including in particular the important mandates to 'promote the further development of international environmental law, taking into account the Declaration of the UNCHE, as well as the special needs and concerns of the developing countries' (*ibid.*, para. I.15(d)) and to 'examine the relationship between environmental degradation and the international economic environment, with a view to ensuring a more integrated approach to problems of environment and development in relevant international forums' (*ibid.*, para. I.15(h)).

As we know, the Rio Conference produced important results both in political and legal terms. It adopted Agenda 21, a voluminous programme of action for the international community covering all aspects of sustainable development, and two political declarations: the Rio Declaration on Environment and Development enunciating 27 'principles' purporting to lay down, in a 'soft law' context, the 'general rights and obligations of States (. . .) in the field of the environment', building upon the earlier Stockholm Declaration on the Human Environment, and the highly controversial 'Non-legally Binding Authoritative Statement of Principles for a Global Consensus on the Management, Conservation and Sustainable Development of All Types of Forest' (United Nations: 1993). In addition, the UNCED process resulted in the adoption and signing of three major legally binding international agreements, the United Nations Framework Convention on Climate Change (UNFCCC), the Convention on Biological Diversity (CBD) and, two years later, the United Nations Convention to Combat Desertification (CCD), all of which have meanwhile entered into force and been ratified by the vast majority of States. The Rio Conference firmly established the objective of sustainable development as a central tenet of national and international political discourse. According to the preamble of Agenda 21, the Rio process 'marks the beginning of a *new global partnership for sustainable development*' (United Nations: 1993, 13), but the objective of this partnership, sustainable development, remained a vague concept without any agreed, unambiguous definition being provided in any of the UNCED documents. As I have noted elsewhere (Pallemaerts: 1995), it is the very ambiguity of the concept which made it possible to reach such universal, but shallow consensus.

In its Resolution 47/190 of 22 December 1992, the General Assembly formally 'endorsed' the texts adopted by UNCED, while 'reaffirming the need for a balanced and integrated approach to environment and development issues', as well as the 'new global partnership for sustainable development', and calling upon 'all concerned' to 'implement all commitments, agreements and recommendations' made in Rio (UNGA Resolution 47/190, 22 December 1992, 4th & 5th preambular paras & para. 5). In the same resolution, it also already decided that it would hold 'a special session for the purpose of an overall review and appraisal of Agenda 21' five years later (*ibid.*, para. 8). When it later confirmed this intention, determined the dates of this Special Session and decided to convene it 'at the highest political level of participation', the General Assembly solemnly 'stresse[d] that *there should be no attempt to renegotiate*' Agenda 21, the Rio Declaration and the Statement of Principles on Forests and that the Special Session 'should *focus on the fulfilment of commitments and the further implementation of Agenda 21* and related post-Conference outcomes' (UNGA Resolution 51/181, 16 December 1996, paras 1 & 5 – emphasis added). This language sounds strikingly similar to that of the later General Assembly resolution convening the Johannesburg Summit.

The 1997 UNGASS was generally decried as a failure because this gathering of heads of State and Government in New York did not succeed in producing a high-sounding political declaration and revealed deep disagreements between governments on the interpretation of the UNCED outcomes and their respective responsibilities for the implementation of the Rio commitments. Despite the acrimony, 'Rio+5' nevertheless adopted a rather elaborate 'Programme for the Further Implementation of Agenda 21' and thus succeeded in keeping alive the process initiated in Rio and the illusion of progress. In the aftermath of Johannesburg, this 1997 statement of intergovernmental consensus and commitment makes rather interesting reading. It provides a useful benchmark against which to measure what progress, if any, has been achieved in the last five years. As was quite obvious in Johannesburg, UNGASS had manifestly failed in its stated intent 'to re-energize (. . .) commitment to further action on goals and objectives set out by the Earth Summit' (United Nations: 1998, 1). The much-hailed 'new global partnership for sustainable development' launched in Rio and reaffirmed in New York appears more as a recurrent exercise in political discourse than as an effective vehicle for reform and concrete action on the ground.

But this did not deter the international community from reiterating the exercise and indulging once more in the art of environmental summitry in Johannesburg. As described above, the scene was set by General Assembly resolution 55/199. Let us now consider the results.

4. The multi-faceted 'outcomes' of the WSSD: the new face of multilateralism?

4.1. PARTNERSHIPS

The Johannesburg Summit concluded its proceedings on 4 September 2002 with the formal adoption by consensus of two political documents: a four-page political declaration, entitled 'Johannesburg Declaration on Sustainable Development' (United Nations: 2002a, 1–5, hereafter referred to as 'Johannesburg Declaration'), and a 70-page action plan, entitled 'Plan of Implementation of the WSSD' (United Nations: 2002a, 7–77, hereafter referred to as 'PoI').

But the United Nations itself does not consider these classical products of multilateral consensus diplomacy to be the only results of the WSSD. A public information document from the United Nations Department of Economic and Social Affairs (DESA) identifying the *'Key Outcomes of the Summit'* also lists the more than 200 'partnerships' between governments and other actors announced within the framework of the WSSD as one of its notable results:

> The concept of partnerships between governments, business and civil society was given a large boost by the Summit and the PoI. Over 220 partnerships (with $235 million in resources) were identified in advance of the Summit and around 60 partnerships were announced during the Summit by a variety of countries (United Nations: 2002b, 1).

The Johannesburg Declaration itself refers to the 'decisions on targets, timetables *and partnerships*' (Johannesburg Declaration, para. 18) taken at the Summit. In the declaration, Governments, 'as social partners', pledge to 'continue to work for *stable partnerships with all major groups* respecting the independent, important roles of each of these' (Johannesburg Declaration, para. 26 – emphasis added). In the draft political declaration tabled by South African President Thabo Mbeki in his capacity as President of the WSSD, it had actually been proposed to give formal standing to the partnerships as part of a 'coherent and integrated *Johannesburg Commitment on Sustainable Development*', a concept coined to denote collectively all the outcomes of the Summit as 'the product of distinct and comprehensive processes that comprised intergovernmental negotiations, multi-stakeholder dialogues *and partnership announcements*' (United Nations Doc. A/CONF.199/L.6, 2 September 2002, para. 22 – emphasis added. See also *ibid.*, para. 47). But the apprehensions of many developing countries about giving too much prominence to these 'non-intergovernmental' outcomes of the Summit eventually prevailed and the novel notion of the all-encompassing 'Johannesburg Commitment' eventually disappeared from the final version of the political declaration.

In view of the importance given to partnerships in the WSSD process, the genesis and background of this special form of international cooperation deserve special attention. While the concept of 'global partnership' in the Rio texts referred to a partnership of States within the framework of the United Nations, the multiple 'partnerships' launched in Johannesburg clearly and deliberately involve a broader range of 'partners' and a different nature of reciprocal commitments distinct from intergovernmental political commitments. Of course, there is nothing new about various forms of cooperation between States and non-State actors, but what is new is the special prominence given to them in the context of an intergovernmental political process. This particular legitimacy granted to 'outcomes of a second type' was actively promoted by the Bureau of the preparatory committee and various actors within the United Nations system. Thus, from the outset, the United Nations itself was, somewhat paradoxically, downplaying the significance of intergovernmentally negotiated results, and lowering the level of public expectations with respect to such traditional 'first type outcomes' of United Nations conferences, as if in anticipation of disappointing negotiating results from the intergovernmental process.

The origin of the United Nations' political interest in partnerships can be traced back to the Millennium Declaration of September 2000, in which heads of State and Government resolved 'to develop strong partnerships with the private sector and with civil society organizations in pursuit of development and poverty eradication' (UNGA Resolution 55/2, 8 September 2000, para. 20). A few months later the General Assembly adopted a resolution entitled '*Towards global partnerships*' in which it called for further consideration of the concept, 'stressing that efforts to meet the challenges of globalization could benefit from enhanced cooperation between the United Nations and all relevant partners, in particular the private sector, in order to ensure that globalization becomes a positive force for all', while at the same time 'underlining the intergovernmental nature of the United Nations' (UNGA Resolution 55/215, 21 December 2000, 2nd & 4th preambular paras). Based on a report from the Secretary-General and further discussions, a second resolution on the subject was passed in December 2001, which lends further legitimacy to partnerships as a means of contributing 'to the realization of the goals and programmes' of the United Nations, such as those contained in the Millennium Declaration and in the outcomes of major United Nations conferences, and 'invites the United Nations system to adhere to a common approach to partnership' based on a number of general principles (UNGA Resolution 56/76, 11 December 2001, paras 1 & 2). The WSSD became the first large-scale testing ground for the new partnership approach.

Though the report of the first meeting of the CSD acting as preparatory committee for the WSSD contains no mention at all of any discussion on the respective role and importance of 'first type' versus 'second type' outcomes, the Secretariat, at the request of the Bureau, produced a note in December 2001 in which it explained that 'in accordance with decisions' of the CSD meeting, 'two types of outcomes are expected from the Johannesburg Summit and its preparatory process'. Whereas the first type 'would be in the form of documents to be negotiated by all States in the global Preparatory Committee', 'the second type of outcomes would consist of a series of *commitments, targets and partnerships made by individual governments or groups of governments*, at the regional and/or interregional level, *as well as with the involvement of or among major groups*,' which, though not negotiated with the involvement of all States, would nevertheless 'be released as part of the Summit's outcomes' (United Nations: 2001d – emphasis added). It seems that the idea of emphasizing voluntary partnerships as a major part of the expected results of the WSSD emerged during the regional roundtables of 'eminent persons' organized by the WSSD Secretariat in the course of 2001, and was later endorsed by the Bureau, by the General Assembly and by the preparatory committee, which at its first meeting had only in very general terms 'encourage[d] further preparatory initiatives by major groups, in particular those which result in new partnerships and commitments to sustainable development' (CSD Resolution 2001/PC/1, para. 11, United Nations Doc. A/56/19, p. 27).

The first roundtable, involving participants from Europe and North America, took place in the United States in June 2001. Its report refers to partnerships in the following terms:

> In an increasingly globalised world, new partnerships are critical between governments, NGOs, trade unions and the private sector. *The role of states and their institutions will become less and less relevant. Adequate engagement of stakeholders will lead to the most constructive results* (United Nations: 2001a, para. 39 – emphasis added).

In its summary of the results of all the regional roundtables, the Secretariat referred to the *'development of more sustainable development partnerships between governments, businesses, finance institutions, academics, NGOs, indigenous groups and trades unions'* as as one of the key recommendations arising from these multi-stakeholder meetings (United Nations: 2001b).

At its meeting in November 2001, the Bureau then agreed on the importance of partnerships, describing them as 'innovative mechanisms for strengthening the linkage between intergovernmental decisions and commitments by major groups, civil society and the private sector' (United Nations: 2001c, para. 23). When it reviewed the progress of the WSSD preparatory process in December 2001, the General Assembly jumped on the bandwagon and, subtly re-arranging some language from Resolution 55/199, decided to 'encourag[e] *new initiatives* that would contribute to the full implementation' of Agenda 21, the Rio Declaration and other outcomes of UNCED. The concept of a 'global partnership to achieve the objectives of sustainable development', as referred to in Resolution 55/199, suddenly acquired a new meaning, when the General Assembly, in Resolution 56/226, called for the Summit to 'reinvigorate, at the highest political level', not 'the global commitment to a North/South partnership and a higher level of international solidarity', as in its earlier resolution, but 'global commitment *and partnerships*, especially between Governments of the North and the South, on the one hand, and *between Governments and major groups* on the other' (UNGA Resolution 56/226, 24 December 2001, 6th preambular para. – emphasis added).

At an informal 'brainstorming session' held in advance of the second meeting of the CSD acting as preparatory committee, the United States delegation made a strong plea for the voluntary partnership approach, which it described as 'forging coalitions of the willing'. This approach was clearly presented as an alternative to the traditional mode of multilateral intergovernmental negotiations, in the following terms: 'We're used to negotiating text, but maybe there is another model. Maybe we can *conceptualize the role of government in another way, as a galvanizer* to forge coalitions' (United Nations Feature Story, 28 January 2002 – emphasis added).

The importance of various kinds of partnerships and voluntary initiatives by governments and major groups was quite extensively debated during the second meeting of the preparatory committee, especially in the framework of its multi-stakeholder dialogue segment. According to the chairman's summary of that segment, 'all participants stressed partnership initiatives as essential to implementation' (United Nations Doc. A/CONF.199/PC/2, Annex II, para. 39). The committee decided to include in its report a note elaborated by the Secretariat on 'proposals for partnerships/initiatives to strengthen the implementation of Agenda 21', including a rather vague description of the general concept and an 'indicative list' of proposals announced by various participants, in order to 'encourage interested parties to initiate action' (*ibid.*, Annex III). Shortly after the meeting, an additional explanatory note by the chairman of the preparatory committee was posted on the Summit website, appealing for the development of further proposals for partnerships and initiatives which, it was announced, 'are expected to become *one of the major outcomes*' of the WSSD.

While governmental negotiators were still struggling to establish the basic structure and content of the negotiated policy documents explicitly called for by the General Assembly in Resolution 55/199, the framework for the production of 'type 2 outcomes' was more or less formalized and institutionalized through 'informal' consultations led by two vice-chairs of the Bureau during the third meeting of the preparatory committee. These consultations resulted in 'further guidance' in the form of another 'explanatory note' annexed to the report of the meeting, supplementing the chairman's initial one, and intended to clarify questions concerning the 'scope and modalities' of 'type 2' partnerships and their 'relationship with the globally agreed, negotiated outcomes' of the WSSD. This note raised more questions than it answered and triggerred an animated multi-stakeholder discussion and a second round of consultations at the fourth and final meeting of the preparatory committee in Bali. These were concluded by informal consensus on yet another 'explanatory note', this time entitled 'Guiding principles for partnerships for sustainable development', which was appended to the official report of the meeting without ever being formally approved by the committee (United Nations Doc. A/CONF.199/4, Annex III, Appendix). Interestingly, these 'guiding principles' contain no reference at all to the general principles for partnerships laid down a few months earlier in General Assembly Resolution 56/76.

Despite its final ministerial segment, the fourth session of the CSD acting as preparatory committee failed to complete work on the actual mandate initially given to it by the General Assembly, merely transmitting to the Summit a bracket-ridden draft 'PoI' for further consideration and authorizing its chairman 'to prepare elements for' a political declaration, on the basis of very preliminary and non-committal discussions.

In this overview article, there is no space for a detailed substantive analysis of the debate on partnerships and the resulting guiding principles. But as this debate mobilized considerable attention during the preparatory process, which did not at all unfold as planned in the initial General Assembly resolution, an analysis of this process and of the future implications of the WSSD would have been incomplete without an account of the parallel process leading to the emergence and legitimation of the novel concept of 'second type outcomes'. In Johannesburg too, significant time was devoted to a series of 'partnership events' designed to provide 'recognition' to voluntary partnerships and initiatives and generating further ones. In this forum, the multi-stakeholder debate focusing on 'coalitions of the willing' bore virtually no relationship to the real-world intergovernmental negotiating arena in which coalitions of the *unwilling* were effectively preventing meaningful multilateral agreement on concrete and time-bound political commitments to further the objectives agreed in Rio ten years earlier. Did Johannesburg inaugurate a new form of multilateralism: multilateralism *à la carte* in a global 'multi-stakeholder bazaar'?

At any rate, the Johannesburg PoI recognizes that the implementation of the outcomes of the Summit 'should involve all relevant actors through partnerships', stressing that 'such partnerships are key to pursuing sustainable development in a globalizing world' (PoI, para. 3). The further promotion and follow-up of such partnerships was put on the agenda of the CSD for the coming years, as the WSSD expressly mandated the Commission, *inter alia*, to 'serve as a focal point for the discussion of partnerships that promote sustainable development, including sharing lessons learned, progress made and best practices' (PoI, para. 148(b)). The 'Guiding principles' annexed to the report of the Bali preparatory meeting refer to a 'follow-up process' and stipulate that partnerships should keep the CSD informed 'about their activities and progress in achieving their targets', but this 'self-reporting' requirement was not formalized in the texts adopted by the WSSD. The PoI, in a provision which does not directly address the role of the CSD, merely provides that 'further development of partnerships and partnership follow-up should *take note* of the preparatory work for the Summit' (PoI, para. 156(b) – emphasis added). It remains to be seen how this will be done and what kind of arrangements will be established by the CSD and/or other United Nations bodies to ensure the effectiveness, transparency and accountability of the 'partnerships for sustainable development' which the United Nations helped to promote and legitimize. At any rate, the General Assembly, in its most recent resolution welcoming the outcomes of the WSSD, explicitly called for 'further discussion' of the matter of partnerships within the CSD (UNGA Resolution 57/253, 20 December 2002, para. 5).

4.2. THE JOHANNESBURG DECLARATION

In order to examine the significance of the Johannesburg Declaration in a historical perspective of incremental policy development and reform, its provisions must be analysed against the background of those of earlier declaratory instruments of a universal nature elaborated within the institutional framework of the United Nations, not only the Rio Declaration on Environment and Development, adopted by UNCED in June 1992, but also such instruments as the Stockholm Declaration, adopted by the UNCHE in June 1972, and the Millennium Declaration, adopted by the UNGA in September 2000. In doing so, one must bear in mind that the WSSD, contrary to UNCED or UNCHE, had no specific mandate to contribute to the development of international environmental law, nor even to the further elaboration of general principles of a non-binding nature to guide the conduct of States with respect to sustainable development. WSSD was to be a 'summit of implementation', not a normative exercise.

The substantive content of the Johannesburg Declaration must therefore primarily be evaluated in political terms, in terms of its impact on the international policy discourse on environmental protection and sustainable development. According to the terms of Resolution 55/199, the political declaration of the WSSD was to be 'a concise and focused document that should emphasize the need for a *global partnership* to achieve the *objectives of sustainable development*, reconfirm the need for an integrated and strategically focused approach to the implementation of Agenda 21, and *address the main challenges and opportunities faced by the international community* in this regard', as well as 'reinvigorate, at the highest political level, the global commitment to (. . .) *a higher level of international solidarity*' (UNGA Resolution 55/199, 20 December 2000, para. 17(b) – emphasis added). Does the Johannesburg Declaration live up to these expectations?

The introductory part of the Johannesburg Declaration positions it in a historical policy continuum:

> Thirty years ago, in Stockholm, we agreed on the urgent need to respond to the problem of environmental deterioration. Ten years ago, at the UNCED, held in Rio de Janeiro, we agreed that the protection of the environment, and social and economic development are fundamental to sustainable development, based on the Rio Principles. To achieve such development, we adopted the global programme, Agenda 21, and the Rio Declaration, to which we reaffirm our commitment. The Rio Summit was a significant milestone that set a new agenda for sustainable development.
>
> Between Rio and Johannesburg the world's nations met in several major conferences under the guidance of the United Nations, including the Monterrey Conference on Finance for Development, as well as the Doha Ministerial Conference. These conferences defined for the world a comprehensive vision for the future of humanity. (Johannesburg Declaration, paras 8–9).

According to this retrospective *exposé*, international policy underwent a gradual progress towards full maturity. Stockholm addressed environmental issues but disregarded development. Rio achieved the synthesis of environmental and developmental objectives by setting 'a new agenda for sustainable development'. But it is only after Doha and Monterrey that 'a comprehensive vision for the future of humanity' emerged. The way the text is formulated, one has the impression that sustainable development as articulated in Rio is not actually part of this 'comprehensive vision', but that true comprehensiveness, integration and vision was only achieved through the post-Rio conferences, of which only those related to finance and trade are specifically mentioned, but which presumably also include other major United Nations conferences on social and economic issues held between 1992 and 2002. This revisionist interpretation of the international political history of the last three decades seems rather heavily biased towards the trade and development agenda which has dominated global political discourse since the second half of the 1990s. While the General Assembly called for stronger political commitment to achieving the objectives of *sustainable development*, as defined in Agenda 21 and the Rio Declaration, the Johannesburg Declaration, though formally reaffirming this commitment, at the same time reduces Rio to a mere 'significant milestone'.

According to the United Nations Secretariat's summary of the WSSD outcomes, *'the understanding of sustainable development was broadened and strengthened* as a result of the Summit, particularly the important linkages between poverty, the environment and the use of natural resources' (United Nations: 2002b, 1). In fact, the recognition of the interrelationship between human development, environmental quality and natural resources dates back to the Stockholm Conference and its preparatory activities, such as the 1971 Founex seminar on environment and development (UNEP: 1981). The Stockholm Declaration already clearly recognized that 'both aspects of man's environment, the natural and the man-made, are essential to his well-being and to the enjoyment of basic human rights – even the right to life itself' and that 'economic and social development is essential for ensuring a favorable living and working environment for man and for creating conditions on earth that are necessary for the improvement of the quality of life' (UNEP: 1981, 41, 45). And the Rio Declaration unambiguously proclaimed that the 'essential task of eradicating poverty' was 'an indispensable requirement for sustainable development'. Does the Johannesburg Declaration really break new ground and deepen our understanding by 'recogniz[ing] that poverty eradication, changing consumption and production patterns, and protecting and managing the natural resource base for economic and social development are overarching objectives of, and essential requirements for sustainable development' (Johannesburg Declaration, para. 11)?

As to the goal of reinforcing the political commitment to 'a higher level of international solidarity', let us note that the declaration rhetorically 'recogniz[es] the importance of building *human* solidarity' (Johannesburg Declaration, para. 17 - emphasis added) – not *international* solidarity – and more specifically states the commitment of world leaders 'to build a humane, *equitable and caring global society* cognizant of the need for human dignity for all' (*ibid.*, para. 2 – emphasis added) as well as to 'assume a collective responsibility to advance and strengthen the interdependent and mutually reinforcing pillars of sustainable development – economic development, social development and environmental protection – at local, national, regional and global levels' (*ibid.*, para. 5). In the declaration, heads of State and Government even solemnly 'declare' their 'responsibility to one another, to the greater community of life and to our children' (*ibid.*, para. 6). This definitely sounds like solidarity and partnership, but the text lacks precision as to the concrete implications of these buzzwords for the policies of national governments and international institutions. The multiple connotations and diverse conceptions of the notion of 'partnership' in the WSSD outcomes have already been extensively analysed above.

As regards the readiness and ability of the international community to 'address new challenges' that have arisen since the early 1990s, the Johannesburg Declaration is not very reassuring. To be sure, it acknowledges that 'the benefits and costs of globalization are unevenly distributed' (*ibid.*, para. 14) and that this 'has added a new dimension' (*ibid.*) to the well-known challenges of narrowing the 'ever-increasing gap between the developed and developing worlds' (*ibid.*, para. 12) and reverting the ominous trends of global environmental degradation. But how do the world's governments propose to address these challenges? By reiterating all the solemn pledges that they have already made in Rio and at subsequent intergovernmental conferences but failed to live up to. And by taking 'extra steps' – which are not further specified in the declaration beyond the repetition of language already agreed elsewhere – 'to ensure that (. . .) available resources are used to the benefit of humanity' (*ibid.*, para. 21) – implying, in a welcome but rare lapse of candour, that this has not been the case so far. Beyond that, the Johannesburg Declaration refers humanity to the WSSD PoI by way of a commitment 'to expedite the achievement of the time-bound, socio-economic and environmental targets contained therein' (*ibid.*, para. 36). So, to complete our assessment, we must analyse those targets and the added value they bring to earlier international action plans like Agenda 21.

4.3. THE POI OF THE WSSD: ACTION-ORIENTED DECISIONS AND TIME-BOUND MEASURES?

What are the time-bound, action-oriented targets in the socio-economic and environmental field laid down in the PoI adopted by the Johannesburg Summit?

As far as the social and economic 'pillars' of sustainable development are concerned, the targets contained in the plan are merely a reaffirmation of the so-called 'Millennium Development Goals', a set of targets endorsed by heads of State and Government at the summit-level meeting of the UNGA held in September 2000. The Millennium Declaration, formally General Assembly Resolution 55/2, apart from setting concrete time-bound targets in the field of development and poverty eradication, also enunciates a number of universal values and principles 'essential to international relations in the twenty-first century' (UNGA Resolution 55/2, 8 September 2000, para. 6), as well as more specific political commitments, termed 'key objectives' (*ibid.*, para. 7), in all areas of activity of the United Nations, such as peace and security, human rights and democracy, humanitarian affairs and disaster relief, strengthening of the United Nations system and environmental protection. Sustainable development, remarkably, is not one of the chapter headings, but is referred to in one of the basic principles and a number of operational provisions.

The development targets from the Millennium Declaration (*ibid.*, para. 19) reaffirmed in Johannesburg include:

- Halving, by 2015, the proportion of the world's people whose income is less than $1 a day and the proportion of people who suffer from hunger.
- Achieving, by 2020, a significant improvement in the lives of at least 100 million slum dwellers.
- Halving, by 2015, the proportion of people without access to safe drinking water.
- Reducing, by 2015, mortality rates for infants and children under 5 by two thirds, and maternal mortality rates by three quarters, of the prevailing rate in 2000.
- Ensuring that, by 2015, all children will be able to complete a full course of primary schooling and that girls and boys will have equal access to all levels of education relevant to national needs.

Directly related to the achievement of the Millennium Development Goals, but not specifically endorsed by the United Nations before WSSD, are the commitments to halve, by the year 2015, the proportion of people who do not have access to basic sanitation (PoI, para. 8) and to improve access to energy services and resources, sufficient to achieve the Millenium Development Goals, especially that relating to poverty reduction (PoI, para. 9), and to improve sustainable agricultural productivity and food security in Africa with a view halving, by 2015, the proportion of people who suffer from hunger (PoI, para. 67). The sanitation target is based on the recommendations of an international conference on freshwater convened in Bonn in December 2001 at the initiative of the German government. One must welcome that this target has now been formally endorsed by the United Nations, but at the same time should not forget that the Programme for the Further Implementation of Agenda 21 adopted by UNGASS in 1997 already called for the full implementation of the Programme of Action of the World Summit on Social Development held in Copenhagen in 1995, including by 'providing *universal* access to basic social services, including (. . .) clean water *and sanitation*' (United Nations: 1998, 17 – emphasis added). The same 'Rio + 5' programme also recognized the need for 'concrete measures to strengthen international cooperation in order to assist developing countries in their domestic efforts *to provide adequate modern energy services, especially electricity, to all sections of their population, particularly in rural areas, in an environmentally sound manner*' (United Nations: 1998, 41 – emphasis added).

The concrete environmental measures in the PoI are mostly to be found in the chapters on 'changing unsustainable patterns of consumption and production', 'protecting and managing the natural resource base of economic and social development' and 'health and sustainable development'.

In the Millennium Declaration, heads of State and Government had solemnly proclaimed that 'the current unsustainable patterns of production and consumption must be changed in the interest of our future welfare and that of our descendants' (UNGA Resolution 55/2, 8 September 2000, para. 6). The European Union (EU), for its part, had declared concerted international action to 'change unsustainable patterns of consumption and production so as to decouple economic growth from environmental degradation and natural resources use' to be one of the 'overarching goals' of the WSSD, alongside poverty eradication (EU Council Conclusions, 30 May 2002, EU Council Doc. 8958/02, para. 10). In the end, it only managed to convince the Summit to agree to 'encourage and promote the development of a 10-year *framework of programmes in support of national and regional initiatives* to accelerate the shift towards sustainable consumption and production' (PoI, para. 15 – emphasis added). The PoI thus stops short of giving a mandate to any international institution for the development of a fully coordinated global programme of action to this effect. It remains to be seen what action the CSD, UNEP and other interested organizations will be able to agree on with a view to establishing some overall framework for national and regional policies and measures.

The promotion of renewable energy worldwide was also a much-debated issue at the Summit. Several countries and regional groups, such as the EU, Switzerland, the Latin American and Carribean states and the small island states had formulated proposals for specific time-bound measures, which were strongly resisted by other countries and groups, especially the United States, Australia and the OPEC countries. In his official statement to the Summit, the Secretary General of OPEC had set the tone by stressing the 'compatibility' of fossil fuels with sustainable development:

> Oil and gas, with their abundant resource base, will be crucial in meeting the global energy needs and challenges for achieving sustainable development. (. . .) The successful development of carbon dioxide sequestration technology will ensure that fossil fuels, including oil, are entirely compatible with sustainable growth. While renewables will be an increasing part of the energy mix in the future, the continued development of clean fossil fuels will be, in most cases, more feasible than costly alternatives (Silva Calderon: 2002, 1).

The compromise that was reached after protracted negotiations calls at the same time for a diversification of energy supply 'by developing advanced, cleaner, more efficient, affordable and cost-effective energy technologies, *including fossil fuel technologies and renewable energy technologies*, hydro included, and their transfer to developing countries on concessional terms as mutually agreed' and for a 'substantial' but unquantified increase, 'with a sense of urgency' but no specified target date, of 'the *global share of renewable energy sources with the objective of increasing its contribution to total energy supply*' (PoI, para. 20(e) – emphasis added). Though no quantified objective could be agreed, the PoI recognizes 'the role of national and voluntary regional targets', thereby implicitly excluding the establishment of global targets for renewable energy (*ibid.*).

However, one may wonder what is the added value of the Johannesburg commitments on renewables, in view of the objectives already laid down in Agenda 21 'to initiate and encourage a process of environmentally sound energy transition in rural communities, from unsustainable energy sources, to structured and diversified energy sources by making available alternative new and renewable sources of energy', an objective which was to be implemented 'not later than the year 2000', *inter alia*, through 'self-reliant rural programmes favouring sustainable development of renewable energy sources and improved energy efficiency' (Agenda 21, para. 14.94, United Nations: 1993, 206–207). In another provision, Agenda 21 also specifically called upon international organizations and bilateral donors to 'support developing countries in implementing national energy programmes in order to achieve widespread use of energy-saving and renewable energy technologies, particularly the use of solar, wind, biomass and hydro sources' (Agenda 21, para. 7.51(b)(i), United Nations: 1993, 85).

With respect to chemicals, the WSSD PoI sets a number of political deadlines, but apart from those relating to the entry into force by 2003 and 2004 of two existing international conventions signed, respectively, in 1998 and 2001 (PoI, para. 23(a)), and to the full operationalization of a new globally harmonized system for the classification and labeling of chemicals by 2008 (PoI, para. 23(c)), these target dates apply to rather vaguely described commitments, such as the development of 'a strategic approach to international chemicals management' by 2005 (PoI, para. 23(b)), and the aim, by 2020, to use and produce chemicals 'in ways that lead to the minimization of significant adverse effects on human health and the environment' (PoI, para. 23).

In the area of natural resource management, it is particularly distressing that the WSSD adopted a time-bound target with respect to biodiversity loss which is considerably less ambitious than the target agreed only a few months earlier at the ministerial-level meeting of the Conference of the Parties to the CBD in The Hague. The Johannesburg PoI refers to the 'achievement by 2010 of a significant reduction in the current rate of loss of biological diversity' (PoI, para. 44), whereas the Hague Ministerial Declaration of 18 April 2002 called on the WSSD to reconfirm the commitment, as reflected in that declaration, 'to have instruments in place to *stop and reverse* the current alarming biodiversity loss at the global, regional, sub-regional and national levels by the year 2010' (United Nations Doc. UNEP/CBD/COP/6/20, Annex II, p. 341, para. 15(d) – emphasis added).

As regards water resources, WSSD's plan is somewhat more positive, as it reinforces the Millennium Declaration commitment 'to stop the unsustainable exploitation of water resources by developing water management strategies at the regional, national and local levels' (UNGA Resolution 55/2, 8 September 2000, para. 23) by a specific target to 'develop integrated water resources management and water efficiency plans by 2005' (PoI, para. 26).

The Johannesburg PoI also contains a number of time-bound commitments with respect to the protection of marine living resources, most notably the rather softly formulated but politically important goal to 'maintain or restore stocks to levels that can produce the maximum sustainable yield with the aim of achieving these goals for depleted stocks on an urgent basis and where possible not later than 2015' (PoI, para. 31(a)) and the undertaking to put into effect agreed Food and Agriculture Organization of the United Nations (FAO) international plans of action for the management of fishing capacity by 2005 and for the prevention, deterrence and elimination of illegal, unreported and unregulated fishing by 2004 (PoI, para. 31(d)). It is also worth mentioning the commitments, in the same chapter, to establish representative networks of marine protected areas by 2012 (PoI, para. 32(c)), as well as a regular process under the United Nations for global reporting and assessment of the state of the marine environment by 2004 (PoI, para. 36(b)).

The final chapter of the PoI deals with the institutional framework for sustainable development at the global, regional and national level. This chapter contains a rather detailed set of recommendations on governance issues, addressed to various institutions of the United Nations system, other intergovernmental organizations, national governments and local authorities, which cannot be analysed in detail within the scope of this article. As the introductory section of the WSSD's plan recognizes: 'Good governance within each country and at the international level is essential for sustainable development' (PoI, para. 4). To what extent these WSSD recommendations will effectively result in a 'strengthening' of the multi-level institutional framework for sustainable development, as envisaged by the General Assembly in Resolution 55/199, depends to a large extent on implementing decisions to be taken by the governing bodies of the international organizations concerned, including the General Assembly itself, in so far as some of the institutional measures called for in the WSSD PoI formally fall within its powers under the United Nations Charter. Other decisions will have to be taken by the Economic and Social Council, of which the CSD, technically speaking, is a subsidiary body, and by the CSD itself. Overall, however, the rather modest package of institutional measures contained in the PoI, even if fully implemented, is unlikely to be sufficient to meet the political expectations raised by the Johannesburg Declaration, where world leaders state: 'To achieve our goals of sustainable development, we need more effective, democratic and accountable international and multilateral institutions' (Johannesburg Declaration, para. 31).

5. Conclusion: too little multilateralism to bridge the gap between economic globalization and sustainable development

In spite of the Johannesburg Declaration's profession of faith in multilateralism as 'the future' and the Summit's commitment to strengthen it, one cannot escape a strong impression of *déjà vu* when analyzing the 'outcomes' of WSSD.

Five years ago, at the 'Rio + 5' Summit, the international community already agreed that 'accelerated globalization' since Rio presented 'new opportunities and challenges' for sustainable development (United Nations: 1998, 3). At their Millennium Summit three years later, world leaders solemnly proclaimed their belief 'that the central challenge we face today is to ensure that globalization becomes a positive force for all the world's people' (UNGA Resolution 55/2, 8 September 2000, para. 5). This analysis of the international community's predicament at the dawn of the new century was reaffirmed in the Johannesburg Declaration, as we saw above. But the political will and ability of governments to truly address this 'new challenge' clearly does not match their rhetoric. And the ambiguity of their commitment to sustainable development has, if anything, become even more apparent since Rio, notwithstanding this rhetoric. For the semantic confusion between economic growth, globalization and sustainable development, which was already obvious in Rio, as I argued in earlier publications (Pallemaerts: 1992, 1995, 1996), is all-pervading and continues to be deliberately maintained.

When it first considered the implications of the report of the WCED, the UNGA, in its Resolution 42/187 of 11 December 1987, long before UNCED, already affirmed 'that sustainable development, which implies meeting the needs of the present without compromising the ability of future generations to meet their own needs, should become *a central guiding principle of the United Nations*, Governments and private institutions, organizations and enterprises'

(UNGA Resolution 42/187, 11 December 1987, 2nd preambular para. – emphasis added). This resolution, contrary to Agenda 21 and the Rio Declaration, actually endorsed the Brundtland Report's classical definition of sustainable development, stressed 'the importance of a reorientation of national and international policies towards sustainable development patterns' (*ibid.*, 4th preambular para.) and even emphasized 'the need *for a new approach to economic growth, as an essential prerequisite* for eradication of poverty and for enhancing the resource base on which present and future generations depend' (*ibid.*, 11th preambular para. – emphasis added).

A few years later, at the 'Earth Summit' in Rio, the concept of sustainable development, though it became the buzzword of the new political discourse, was no longer clearly defined, and the explicit call for 'a new approach to economic growth' had faded away. Economic growth, trade liberalization and trade-led globalization suddenly acquired a sort of newfound environmental legitimacy, as Principle 12 of the Rio Declaration affirmed that 'States should cooperate to promote a supportive and open international economic system that would lead to economic growth and sustainable development in all countries, to better address the problems of environmental degradation' (For a more detailed analysis of this paradigm shift, see Pallemaerts: 1992, 1996).

UNGASS in 1997 first explicitly addressed the 'challenges and opportunities' of globalization and added some caveats and nuances to the rather optimistic discourse of Rio:

> It is important that national and *international environmental and social policies be implemented and strengthened in order to ensure that globalization trends have a positive impact on sustainable development,* especially in developing countries. (. . .) Although economic growth - reinforced by globalization - has allowed some countries to reduce the proportion of people in poverty, for others marginalization has increased. (. . .) *There should be a balanced and integrated approach to trade and sustainable development, based on a combination of trade liberalization, economic development and environmental protection.* (United Nations: 1998, 3, 4, 22).

Intergovernmental consensus on such language would no longer be achievable today. In its chapter on 'sustainable development in a globalizing world' and other relevant provisions on finance and trade, the WSSD PoI essentially restates, in marginally 'greener' terms, the new trade and development agenda of Doha and Monterrey. It recognizes 'the major role that trade can play in achieving sustainable development and in eradicating poverty' (PoI, para. 90), but without simultaneously calling for a strenthening of national and international environmental and social policies, as the UNGA did at its Special Session five years ago. While UNGASS invited the WTO, UNEP and UNCTAD 'to consider ways to *make* trade and environment mutually supportive' (United Nations: 1998, 25–26), their 'mutual supportiveness' is nowadays presumed, as something that merely remains to be 'enhanced' and 'promoted' (PoI, paras. 97–98), without indicating how, except through the 'business-as-usual' implementation of the Doha Development Agenda, which, as is well-recognized, fails to provide an adequate mandate to strengthen the role of environmental and social standards in the regulation of international trade. And the WSSD PoI itself will provide little, if any, impetus for the further development of binding international standards in the environmental and social fields, as it merely calls for the implementation of existing international law, but not for the negotiation of new multilateral agreements, unless one agrees to read into it a mandate for the elaboration of international rules on the sharing of benefits from the use of genetic resources (PoI, para. 44(o)) and corporate responsibility and accountability (PoI, para. 49), an interpretation to which the United States has already formally entered reservations (United Nations: 2002a, 148–149).

Against this background, the Johannesburg Declaration's commitment to the 'strengthening of multilateralism' sounds rather hollow and sustainable development, far from becoming 'a central guiding principle', seems doomed to remain nothing but a 'collective hope', to quote a phrase from the closing paragraph of the declaration.

References

Pallemaerts, M.: 1992, 'International environmental law from Stockholm to Rio: back to the future?', *Rev. Eur. Community Int. Environ. Law* **1**, 254–266.

Pallemaerts, M.: 1995, 'De opkomst van het begrip "duurzame ontwikkeling" in het internationaal juridisch en politiek discours: een conceptuele revolutie?', *Recht en Kritiek* **21**, 380–397.

Pallemaerts, M.: 1996, 'International environmental law in the age of sustainable development: a critical assessment of the UNCED process', *J. Law Commerce* **15**, 623–676.

Pallemaerts, M.: 2003, 'International Law and Sustainable Development: Any Progress in Johannesburg?', *Rev. Eur. Community Int. Environ. Law* **12** (forthcoming).

Silva Calderon, A.: 2002, '*OPEC and the World Summit on Sustainable Development*', Official OPEC Statement for the WSSD, Johannesburg, September 2002.

UNEP: 1981, *In Defence of the Earth*, Nairobi, United Nations Environment Programme.

United Nations: 1993, *Report of the United Nations Conference on Environment and Development, Rio de Janeiro, 3–14 June 1992*, Vol. I, New York, United Nations.

United Nations: 1998, *Earth Summit C 5 – Programme for the Further Implementation of Agenda 21*, New York, United Nations Department of Public Information.

United Nations: 2001a, *Report of Regional Roundtable for Europe and North America*, Vail, Colorado, 6–8 June 2001.

United Nations: 2001b, *Secretariat Overview of the WSSD Regional Eminent Persons Roundtables*, Secretariat of the World Summit on Sustainable Development, New York, October 2001.

United Nations: 2001c, *Note on the Outcome of the Fifth Meeting of the Bureau of CSD-10 Acting as the Preparatory Committee for the World Summit on Sustainable Development*, New York, 1–2 November 2001.

United Nations: 2001d, *Possible Framework for Strengthening Linkages between the Expected Outcomes of WSSD*, Paper prepared by the WSSD Secretariat on request of the Bureau, December 2001.

United Nations: 2002a, *Report of the World Summit on Sustainable Development, Johannesburg, South Africa, 26 August–4 September 2002*, New York, United Nations.

United Nations: 2002b, *Key Outcomes of the Summit*, United Nations, New York, September 2002.

WCED: 1987, *Our Common Future*, Oxford/New York, Oxford University Press.

LIST OF ABBREVIATIONS

3P (Triple P) People, Planet, Profit;
or: People, Planet,
Prosperity

AA1000 Account Ability 1000

ABI Association of British
Insurers

ADB African Development
Bank

AG21 Agenda 21

AIDS Acquired
Immunodeficiency
Syndrome

AMCOW African Ministerial
Conference on Water

AOSIS Alliance of Small
Island States

AsrIA SRI in Asia

AWTF African Water Task
Force

BASD Business Action for
Sustainable
Development

BPOA Barbados Programme
of Action

BS7750 British Standard 7750

Btu British thermal units

CAN Andean Community
of Nations

CBD Convention on
Biological Diversity

CBO Community-Based
Organisations

CCD United Nations
Convention to
Combat
Desertification

CDM Clean Development
Mechanism

CDP Carbon Disclosure
Project

CEC Commission of the
European
Communities

CEE Central and Eastern
Europe

CEPAL UN Economic
Commission for Latin
America

CERCLA Comprehensive
Environmental
Response,
Compensation and
Liability Act

CG/HCCS Coordinating Group
for the Harmonisation
of Chemical
Classification
Systems

CGG Commission on
Global Governance

CGIAR Consultative Group
on International
Agricultural Research

CITES Convention on
International Trade in
Endangered Species

COMECON	Council for Mutual Economic Assistance	ELCI	Environment Liaison Centre International
COMSATS	Commission on Science and Technology for Sustainable Development in the South	EMAS	The EU Eco-Management and Audit Scheme
		ENSO	El Niño Southern Oscillation
COP	Conference of the Parties	EOLSS	Encyclopaedia of Life Support Systems (of UNESCO)
CSD	(United Nations) Commission on Sustainable Development	EPI	Environmental Performance Indicator
		ESI	Ethibel Sustainability Index
CSE	Centre for Science and Environment	EU	European Union
CSR	Corporate Social Responsibility	EVI	Environmental Vulnerability Index
CSRR	Corporate Sustainability and Responsibility Research	FAO	Food and Agriculture Organization of the United Nations
CVI	Common Wealth Vulnerability Index	FCCC	(United Nations) Framework Convention on Climate Change
DESA	Department of Economic & Social Affairs (of the UN)	FDI	Foreign Direct Investment
DGD	Decision Guidance Documen	FFA	Framework For Action
E	Global Energy Use	FOE	Friends of the Earth
ECO	Environmental Citizens Organization	GDP	Gross Domestic Product

GEF	Global Environmental Facility	ICLEI	International Council for Local Environmental Initiatives
GEO	Global Environment Outlook	ICSPAC	International Coalition for Sustainable Production and Consumption
GFT250	Top 250 companies of the Fortune 500		
GHG	Green house gasses	ICSU	International Council for Science
GHS	Globally Harmonised System	IFCS	Intergovernmental Forum on Chemical Safety
GNP	Gross National Product	IFS	International Foundation for Science
GRI	Global Reporting Initiative		
GTA	Amazon Working Group (Brazil)	IGO	intergovernmental organisation
GWI	Global Warming Indicator	IGWA	Intergovernmental Agency for Water in Africa
GWP	Gross World Product		
HDI	Human Development Index	IK	Indigenous Knowledge
HIPC	Highly Indebted Poor Countries	ILO	International Labour Organisation
HIV	Human Immunodeficiency Virus	IMF	International Monetary Fund
ICC	International Chamber of Commerce	IMO	International Maritime Organization
ICFTU	International Confederation of Free Trade Unions	INC	Intergovernmental Negotiating Committee

IOMC	Inter-Organisation Programme for the Sound Management of Chemicals	JpoI	Johannesburg Plan of Implementation
IPAM	Amazon Institute of Environmental research (Brazil)	LDC	Least Developed Countries
		LPG	Liquid Petroleum Gas
IPCC	Intergovernmental Panel on Climate Change	MDG	Millennium Development Goals
		MEA	Millennium Ecosystem Assessment; Multilateral Environment Agreements
IRPTC	International Register of Potentially Toxic Chemicals		
IS92a/e	GHG emission scenarios of IPCC	MHAHE	Ministry of Home Affairs, Housing and Environment, Maldives
ISA	Socio-environment Institute (Brazil)		
ISEA	Institute for Social and Ethical Accountability	MoT	Ministry of Tourism, Maldives
ISO 14000	An environmental management standard specification by ISO	MPND	Ministry of National Planning and Development, Maldives
ISO	International Organisation for Standardisation	Mrf	Maldivian Rufiya
		MSP	Multi-stakeholder Process
ISP	International Science Programme	NAI	New African Initiative
IUCN	International Union for Conservation of Nature and Natural Resources	NCSD	National Council for Sustainable Development
JPI	Johannesburg Plan of Implementation	NEPAD	New Partnership for Africa's Development

NGO	Non-governmental Organisation	PIC	Prior Informed Consent
NUFFIC	Netherlands Organization for International Cooperation in Higher Education	PNUD	Programme des Nations Unies pour le Développement (UNDP)
OAU	Organization of African Unity	PoI	Plan of Implementation of the World Summit on Sustainable Development
ODA	Official Development Aid; Official Development Assistance; Overseas Development AiD	POP	Persistent Organic Pollutant
OEA	Organization of the American States	PPP	Purchase Power Parity
OECD	Organisation for Economic Co-operation and Development	PrepCom	Preparatory Committee (of the WSSD)
		PRS	Poverty Reduction Strategy
OPEC	Organization of Petroleum Exporting Countries	PRSP	Poverty Reduction Strategy Paper
PCB	Polychlorinated Biphenyl	PVC	Polyvinyl Chloride
PCC	Intergovernmental Panel on Climate Change	R&D	Research & Development
		REAP	Regional Environmental Accession Programme
PFSD/pfsd	Partnership for Sustainable Development	REC	Regional Environmental Center (Szentendre, Hungary)

Rio+10	10 years after Rio = Johannesburg 2002; ten-year review of progress on Agenda 21	SOPAC	South Pacific Applied Geoscience Commission
Rio+5	five-year review of progress on Agenda 21	SPAC	Sustainable Production And Consumption
SADC	Southern African Development Community	SRI	Socially Responsible Investing; Sustainable and Responsible Investing
SCRES	The Standing Committee for Responsibility and Ethics in Science	SRISTI	Society for Research and Initiatives for Sustainable Technologies and Institutions
SD	Sustainable Development	SST	Sea Surface Temperature
SEE	Social, Environmental, Ethical	TKDL	Traditional Knowledge Libraries
SIDS	Small Island Developing States	TNC	Trans National Corporation
SIF	Social Investment Forum	TRIPs	Trade Related Aspects of Intellectual Property Rights
SiRi	Sustainable Investment Research International Group	TWAS	Third World Academy of Sciences
SME	Small and Medium-sized Enterprise	TWNSO	Third World Network of Scientific Organizations
SMI	Small and Medium-sized Industry	UN	United Nations
		UNCCD	United Nations Convention to Combat Desertification

UNCED	United Nations Conference on Environment and Development (Rio de Janeiro, 1992)	UNESCO	United Nations Educational, Scientific and Cultural Organisation
UNCHE	United Nations Conference on the Human Environment (1972)	UNFCCC	United Nations Framework Convention on Climate Change
UNCSD	United Nations Commission for Sustainable Development	UNFPA	United Nations Populations Fund
		UNGA	United Nations General Assembly
UNCTAD	UN Commission on Trade and Development	UNGASS	Special Session of the United Nations General Assembly
UNDP	United Nations Development Programme	UNICEF	United Nations International Children and Education Fund
UNEC	United Nations Economic Commission	UNIDO	UN Industrial Development Organisation
UNECE	United Nations Economic Commission for Europe	UNITAR	United Nations Institute for Training and Research
UNEP FI	UNEP Finance Initiatives	UNRISD	United Nations Research Institute on Social Development
UNEP FII	UNEP Financial Institutions Initiative	UNSCEGHS	United Nations Economic and Social Council's Sub-Committee of Experts on the Globally Harmonised System of Classification
UNEP III	UNEP Insurance Industry Initiative		
UNEP	United Nations Environment Programme		

UNSIA	United Nations System wide Initiative for Africa	WEHAB	Water, Energy, Health, Agriculture & Biodiversity
US	United States of America	WEO	World Environmental Organisation
USD	United States Dollar	WGBU	German Advisory Council on Global Change
UWICED	University of the West Indies Centre for Environment and Development		
		WHAT	World Humanity Action Trust
VfU	Verein für Umweltmanagement in Banken	WHO	World Health Organisation
VQS	the Voluntary Quality Standard	WHRC	Woods Hole Research Center (USA)
WASAI	Water and Sanitation African Initiative	WIPO	World Intellectual Property Organization
WASH	Water, Sanitation and Hygiene	WMO	World Meteorological Organisation
WB	World Bank	WSSCC	Water Supply and Sanitation Collaborative Council
WBCSD	World Business Council for Sustainable Development	WSSD	World Summit on Sustainable Development
WCD	World Commission on Dams	WTO	World Trade Organisation
WCED	World Commission on Environment and Development	WWF	World Wild Fund for Nature (previously World Wildlife Fund)
WEDC	Water Engineering and Development Centre		

INDEX